Application of Artificial Intelligence in Hydraulic Engineering

Application of Artificial Intelligence in Hydraulic Engineering

Editors

Jie Yang
Chunhui Ma
Lin Cheng

MDPI

Basel • Beijing • Wuhan • Barcelona • Belgrade • Novi Sad • Cluj • Manchester

Jie Yang
Institute of Water Resources
and Hydro-electric Engineering
Xi'an University of Technology
Xi'an
China

Chunhui Ma
Institute of Water Resources
and Hydro-electric Engineering
Xi'an University of Technology
Xi'an
China

Lin Cheng
Institute of Water Resources
and Hydro-electric Engineering
Xi'an University of Technology
Xi'an
China

Editorial Office
MDPI AG
Grosspeteranlage 5
4052 Basel, Switzerland

This is a reprint of the Special Issue, published open access by the journal *Water* (ISSN 2073-4441), freely accessible at: www.mdpi.com/journal/water/special_issues/1S1E74L2LE.

For citation purposes, cite each article independently as indicated on the article page online and using the guide below:

Lastname, A.A.; Lastname, B.B. Article Title. *Journal Name* **Year**, *Volume Number*, Page Range.

ISBN 978-3-7258-1380-3 (Hbk)
ISBN 978-3-7258-1379-7 (PDF)
https://doi.org/10.3390/books978-3-7258-1379-7

Contents

Preface

The intelligent algorithm has become an important research method to solve critical scientific problems in the engineering field It has been widely used in the optimal design, structural simulation, safety monitoring, and safety evaluation of water conservation projects due to its advantages in regression, classification, clustering, and dimension reduction. Experiments and numerical simulations are faced with constraints of time and cost in traditional research methods. With the advancement of sensors and measurement technology, a large amount of safety-monitoring data has been accumulated in water conservation projects. Intelligent algorithms have become a powerful tool for monitoring data, mining information, and constructing data associations quickly and accurately. Combined with traditional computing techniques such as geotechnical tests, non-destructive testing, and numerical simulation, intelligent algorithms will help us further understand various laws and mechanisms in water conservation projects, which is of great significance to improving the safety of water conservation projects and the development level of human society. Therefore, this special theme will focus on applying intelligent algorithms to water conservation projects.

Jie Yang, Chunhui Ma, and Lin Cheng
Editors

water

Editorial

Application of Artificial Intelligence in Hydraulic Engineering

Chunhui Ma *, Lin Cheng and Jie Yang

Institute of Water Resources and Hydro-Electric Engineering, Xi'an University of Technology, Xi'an 710048, China; chenglin@xaut.edu.cn (L.C.)
* Correspondence: machunhui@xaut.edu.cn

1. Introduction to the Special Issue

Water conservancy projects have always been essential throughout the development of human society, including the development and utilization of water resources, the construction and management of water conservancy facilities and flood prevention and control [1]. With the continuous development and popularization of artificial intelligence technology, its application in water conservancy engineering has been increasingly emphasized. Introducing artificial intelligence technology can improve the efficiency and precision of water conservancy projects, lead to the more effective use and protection of water resources, and enhance the safety and intelligence of water conservancy facilities [2]. The application of intelligent algorithms in water conservancy engineering has become an important research area to solve significant scientific problems in the engineering field [3]. Their advantages in regression, classification, clustering and dimensionality reduction have led to their wide use in optimization design [4], structural simulations [5], safety monitoring [6] and water conservancy project evaluations [7]. In traditional research methods, both experiments and numerical simulations are limited by time and cost. However, with the advancement of sensors and measurement technology, a large amount of data has accumulated for use in water conservancy project safety monitoring. Intelligent algorithms have become a powerful tool for mining information, constructing data associations and monitoring data quickly and accurately. Combining intelligent algorithms with traditional computational techniques such as geotechnical testing, nondestructive testing and numerical simulations has been essential for understanding various laws and mechanisms in water conservancy projects.

Therefore, this Special Issue of "*Water*" aims to deeply understand the application value of intelligent algorithms in water conservancy projects and their important role in improving a project's safety and efficiency and to provide new ideas and methods for the development and progress of these projects. This has great significance for improving the safety of water conservancy projects and the development of human society.

Since the call for papers was announced in January 2023, 15 original papers have been accepted for publication after a rigorous peer-review process (contributions 1–15); these papers can be classified into several areas: data-driven intelligent model studies, intelligent algorithms for optimal structural design, nondestructive testing (NDT) sensor applicability studies and simulations and experiments for structural characterization. To better understand this Special Issue, we summarize the highlights of the published papers below.

check for updates

Citation: Ma, C.; Cheng, L.; Yang, J. Application of Artificial Intelligence in Hydraulic Engineering. *Water* **2024**, *16*, 590. https://doi.org/10.3390/w16040590

Received: 22 January 2024
Revised: 4 February 2024
Accepted: 5 February 2024
Published: 17 February 2024

2. Overview of the Contributions to this Special Issue

Regarding data-driven intelligent model research, Shao et al. comprehensively analyze the effects of water level load transfer, rockfill flow and soil properties on settlement during operation (contribution 1). The original position coordinates are replaced by space parameters which are more consistent with the deformation characteristics, such as the elevation of the upper fill, the thickness of the rockfill materials, the distance between the measuring point and the panel, etc. A multi-monitoring point (MMP) model is combined with the XGBoost model, which has a high prediction accuracy for concrete-faced rockfill dams (CFRDs). Then, the importance of the selected factors is determined. This work has a certain reference value for dam safety monitoring. A new method for inverse analyses of permeability coefficients is proposed by Zhao et al. (contribution 2). They constructed an inversion sample set of a permeability coefficient using the combination of a finite element model and orthogonal test design. The random forest (RF) algorithm proxy model is established, and the Harris Hawk optimization (HHO) algorithm is used to determine the optimal value of the permeability parameters in the project area. This method is used to explore and verify the distribution of natural seepage fields in P hydropower stations. An artificial neural network (ANN) is proposed by Saya et al. (contribution 3) to predict future water consumption as a function of some environmental parameters, and the Copernicus Climate Change Service (C3S) is used to determine the water demand trend in the next ten years. Finally, future consumption is predicted based on the ANN model, the continuity equation of the tank is solved through integral discretization and the time series of tank volume changes and the total number of crisis events are obtained, which are used to estimate the optimal capacity of small city reservoirs. In terms of research on structural safety detection methods, an intelligent detection method for concrete dam surface cracks based on two-stage migration learning is proposed by Li et al. (contribution 4). First, two-stage transfer learning across domains and within domains transfers relates domain knowledge to the target domain so that the model can be adequately trained with small data sets. Second, the segmentation capability is enhanced using the residual network 50 (ResNet50), as a UNet model feature extraction network to enhance crack feature information extraction. Finally, multilayer parallel residual attention (MPR) is integrated into the jump connection path to improve the focus on critical information for clearer fracture edge segmentation. This concrete dam surface crack detection method exhibits a high efficiency and accuracy and a strong robustness. A convolutional neural network (CNN) model is used by Peng et al. (contribution 5) to achieve accurate segmentation of coarse-grained soil CT images, exceeding the accuracy of traditional methods and uncovering a new method for soil particle size analyses. The validity of the CNN model is proven by a three-dimensional (3D) model reconstructed from segmented images, highlighting its accuracy and potential for automation in soil particle analyses. New empirical formulas for the ideal particle size and discount coefficient in coarse-grained soil are also revealed and verified. In terms of risk analysis and comprehensive model agent evaluation methods, a feedback, rolling and adaptive operational decision-making mechanism is proposed by Han et al. (contribution 6). The relationship between the work cycle and multiple time scales is considered according to the change in time scale and the dynamics of the work process. The risk transmission mechanisms of urban river ecological governance engineering projects are examined by Xu et al. (contribution 7). Prediction is combined with operation to adapt to dynamic changes in multiple time scales during operation. Based on existing research, the internal mechanism, the influencing factors of risk transmission and the dynamic evolution process of risk are considered. Based on the theory of infectious disease dynamics and the susceptible exposure to infection recovery susceptibility (SEIRS) risk transmission model, a risk transmission delay model of an urban river ecological governance project under a scale-free network is established. This model provides a basis for the effective supervision and control of project participants and theoretical support for indirect and post-event supervision. In-depth literature reviews and brainstorming are used by Xu et al. (contribution 8) to identify 63 risk elements in urban river ecological management projects. An expert survey method is used to identify the correlation between risk factors, and the risk factors are taken as network nodes. Then, the relationships between these nodes are used as network edges (i.e., paths) to construct a complex network model. The risk network's overall characteristic parameters and local characteristic parameters are analyzed using the network visualization tool. According to the parameter characteristics, the risk transmission characteristics of urban river ecological governance projects are analyzed, and the internal relationship of risk transmission in complex networks is revealed. A corresponding comprehensive evaluation index system based on interval number theory is built by Chen et al. (contribution 9). Expert evaluations and the improved analytic hierarchy process (AHP) are combined to propose a reasonable allocation method for index weights in addition to a comprehensive evaluation model. Comprehensive evaluation criteria are formulated. Taking a roller compacted concrete (RCC) dam in China as an example, the applicability of the comprehensive evaluation method based on interval number theory is illustrated.

Regarding the research on intelligent algorithm optimization structure design, a noise reduction method combining the improved adaptive singular value decomposition algorithm (ASVD) and the improved complete ensemble EMD with adaptive noise (ICEEMDAN) is proposed by Wang et al. (contribution 10). A Hankel matrix is constructed based on collected discrete-time signals. After performing SVD on the Hankel matrix, the ASVD algorithm automatically selects the practical singular values to filter out most of the background white noise and retain helpful frequency components with a similar energy in the signal. Then, the ICEEMDAN is combined with the Spearman correlation coefficient method to further to filter out residual white noise and low-frequency water flows. Based on the practically measured vibration signals of a floodgate at a large hydropower station, the results show that the ASVD-ICEEMDAN method exhibits a good noise reduction performance and feature information extraction abilities for floodgate vibration signals and can provide support for operational mode analysis and damage identification for practical structures under complex interference conditions. The characteristics of low-specific-speed centrifugal pumps are simulated and analyzed using CFD by Ke et al. (contribution 11) by considering that the inlet width of the splitter blade is more significant than the inlet width of the primary blade. The influence of the splitter blade's geometric parameters on the flow field is studied in orthogonal experiments, and the geometric parameters of the separator blade are optimized by using the artificial fish swarm algorithm. Finally, the accuracy of the algorithm is verified by experiments. This study provides a reference for the shape selection of splitter blades.

In terms of the applicability study of nondestructive detection sensors, Yu et al. (contribution 12) conduct model tests and quantitative analyses on the leak monitoring performance of a heated temperature sensing cable, and the temperature change curves in different media are obtained. The ability of the heated temperature sensing cable to identify leaks in soil media with additional moisture content is verified. Finally, through a field simulation test at a manganese slag reservoir, it is verified that the cable can be used for leak monitoring in an actual karst depression reservoir basin. A volumetric meter using flow control valves that are ordinarily already present in buildings' hydraulic installations is proposed by Gonçalves et al. (contribution 13). A system is developed for the electromechanical and thermal characterization of the sensor. Computational simulations of the sensor are performed in Ansys® (2021R2) software, and the electronic circuit is designed in LTSpice® (v. 17.0.30.0) software. An artificial neural network is used to estimate the flow and volume from trapezoidal pulses.

In terms of simulations and tests of structural states, a process model of the impact of construction load on a reservoir cutoff wall in a reconstruction section of an expressway is established using FLAC3D two-dimensional and three-dimensional numerical simulation methods by Sun et al. (contribution 14). The stress–strain state of the expressway reconstruction section and the nearby reservoir cutoff wall is simulated in detail, directly reflecting the overall deformation of the reservoir cutoff wall and the interaction and differential deformation between the wall structures. The influence of the construction load on the safety and stability of the cutoff wall of the reservoir is evaluated, and various advanced mechanical behaviors of the cutoff wall are predicted. A simulation method for the mechanical behavior of tight rock is proposed by Tang et al. (contribution 15) by combining physical experiments with numerical analyses and considering fluid mechanics. The validity of the proposed method is verified by comparing the numerical and physical results of triaxial shear tests on sandstone under seepage conditions. Based on this verified method, the stability of the surrounding tight sandstone underground water sealed oil and gas storage caverns during excavation is analyzed.

3. Conclusions

The guest editors envision that the papers published in this Special Issue will be of interest to researchers, designers and practitioners for the safety monitoring and management of reservoirs and dams and will help identify further lines of research. We also hope that readers will find the material in this Special Issue both interesting and stimulating as they explore the application of AI in hydraulic engineering from the perspective of monitoring models, optimization algorithms, nondestructive sensors, numerical simulations and experiments. The research results and methods introduced in this Special Issue, including data-driven models and mechanisms, structural design optimization, nondestructive testing sensors and structural safety simulations, are of great research significance. These technological contributions can help related scholars and project managers analyze and manage the safety of the major structures in reservoir dams.

Author Contributions: Conceptualization, J.Y.; methodology, C.M.; validation, L.C.; formal analysis, C.M.; resources, J.Y.; data curation, C.M.; writing—original draft preparation, C.M.; writing—review and editing, L.C.; visualization, C.M.; supervision, J.Y.; project administration, J.Y.; funding acquisition, J.Y. All authors have read and agreed to the published version of the manuscript.

Funding: This research was partially supported by the National Natural Science Foundation of China (Grant No. 52279140); the Key Scientific Research Project of Shaanxi Provincial Department of Education (Coordination Centre Project) (Grant No. 22JY044); the Natural Science Basic Research Program of Shaanxi (Grant No. 2023-JC-QN-0562); and the Scientific Research Program Funded by Shaanxi Provincial Education Department (Grant No. 23JY058).

Data Availability Statement: Not applicable.

Conflicts of Interest: The authors declare no conflicts of interest.

List of Contributions:

1. Shao, L.; Wang, T.; Wang, Y.; Wang, Z.; Min, K. A Prediction Model and Factor Importance Analysis of Multiple Measuring Points for Concrete Face Rockfill Dam during the Operation Period. *Water* **2023**, *15*, 1081. https://doi.org/10.3390/w15061081.
2. Zhao, W.; Yin, Q.; Wen, L. Intelligent Inversion Analysis of Hydraulic Engineering Geological Permeability Coefficient Based on an RF–HHO Model. *Water* **2023**, *15*, 1993. https://doi.org/10.3390/w15111993.
3. Saya, B.; Faraci, C. Application of Artificial Neural Networks for Predicting Small Urban-Reservoir Volumes: The Case of Torregrotta Town (Italy). *Water* **2023**, *15*, 1747. https://doi.org/10.3390/w15091747.
4. Li, J.; Lu, X.; Zhang, P.; Li, Q. Intelligent Detection Method for Concrete Dam Surface Cracks Based on Two-Stage Transfer Learning. *Water* **2023**, *15*, 2082. https://doi.org/10.3390/w15112082.
5. Peng, J.; Shen, Z.; Zhang, W.; Song, W. Deep-Learning-Enhanced CT Image Analysis for Predicting Hydraulic Conductivity of Coarse-Grained Soils. *Water* **2023**, *15*, 2623. https://doi.org/10.3390/w15142623.
6. Han, C.; Guo, Z.; Sun, X.; Zhang, Y. Dynamic Forecasting and Operation Mechanism of Reservoir Considering Multi-Time Scales. *Water* **2023**, *15*, 2472. https://doi.org/10.3390/w15132472.
7. Xu, J.; Zhu, J.; Xie, J. Study on the Evolution of Risk Contagion in Urban River Ecological Management Projects Based on SEIRS. *Water* **2023**, *15*, 2622. https://doi.org/10.3390/w15142622.
8. Xu, J.; Zhu, J.; Xie, J. Research on Risk Evolution Mechanism of Urban River Ecological Governance Project Based on Social Network Analysis. *Water* **2023**, *15*, 2012. https://doi.org/10.3390/w15112012.
9. Chen, X.; Xu, P.; Liu, X.; Su, C. Comprehensive Evaluation Method for the Safety State of RCC Dams Based on Interval Number Theory. *Water* **2023**, *15*, 2089. https://doi.org/10.3390/w15112089.
10. Wang, W.; Zhu, H.; Cheng, Y.; Tang, Y.; Liu, B.; Li, H.; Yang, F.; Zhang, W.; Huang, W.; Zheng, F. A Combined Noise Reduction Method for Floodgate Vibration Signals Based on Adaptive Singular Value Decomposition and Improved Complete Ensemble Empirical Mode Decomposition with Adaptive Noise. *Water* **2023**, *15*, 4287. https://doi.org/10.3390/w15244287.

11. Ke, Q.; Tang, L.; Luo, W.; Cao, J. Parameter Optimization of Centrifugal Pump Splitter Blades with Artificial Fish Swarm Algorithm. *Water* **2023**, *15*, 1806. https://doi.org/10.3390/w15101806.
12. Yu, B.; Shen, C.; Peng, H.; Guo, F. Laboratory Model Test and Field In Situ Test of Distributed Optical Fiber Monitoring of Seepage in a Karst Depression Reservoir Basin. *Water* **2023**, *15*, 1477. https://doi.org/10.3390/w15081477.
13. Gonçalves, L.d.S.; Medeiros, K.A.R.; Barbosa, C.R.H. Hydrometer Design Based on Thin-Film Resistive Sensor for Water Measurement in Residential Buildings. *Water* **2023**, *15*, 1045. https://doi.org/10.3390/w15061045.
14. Sun, Y.; Lei, A.; Yang, K.; Wang, G. Numerical Simulation Study on the Influence of Construction Load on the Cutoff Wall in Reservoir Engineering. *Water* **2023**, *15*, 993. https://doi.org/10.3390/w15050993.
15. Tang, P.; Zhang, W.; Wang, H.; Zhou, J.; Dang, Y.; Chao, Z. Analysis of the Hydromechanical Properties of Compact Sandstone and Engineering Application. *Water* **2023**, *15*, 2011. https://doi.org/10.3390/w15112011.

References

1. Zhou, W.; Chang, X.L.; Zhou, C.B.; Liu, X.H. Creep Analysis of High Concrete-Faced Rockfill Dam. *Int. J. Numer. Methods Bio.* **2010**, *26*, 1477–1492. [CrossRef]
2. Hariri-Ardebili, M.A.; Mahdavi, G.; Nuss, L.K.; Lall, U. The Role of Artificial Intelligence and Digital Technologies in Dam Engineering: Narrative Review and Outlook. *Eng. Appl. Artif. Intell.* **2023**, *126*, 106813. [CrossRef]
3. Yu, H.; Taleghani, A.D.; Al Balushi, F.; Wang, H. Machine Learning for Rock Mechanics Problems; an Insight. *Front. Mech. Eng.* **2022**, *8*, 1003170. [CrossRef]
4. Zhang, Z.; Zhang, S.; Zhao, Z.; Yan, L.; Wang, C.; Liu, H. HydroBIM—Digital Design, Intelligent Construction, and Smart Operation. *J. Intell. Constr.* **2023**, *1*, 9180014. [CrossRef]
5. Zhang, P.; Sun, X.; Zhou, X.; Zhang, Y. Experimental Simulation and a Reliable Calibration Method of Rockfill Microscopic Parameters by Considering Flexible Boundary. *Powder Technol.* **2022**, *396*, 279–290. [CrossRef]
6. Wei, B.; Liu, B.; Yuan, D.; Mao, Y.; Yao, S. Spatiotemporal Hybrid Model for Concrete Arch Dam Deformation Monitoring Considering Chaotic Effect of Residual Series. *Eng. Struct.* **2021**, *228*, 111488. [CrossRef]
7. Xiong, M.; Wang, W.; Huang, Y. System Dynamic Reliability Evaluation of Multiple Failure Modes of Earth Dams Subjected to Strong Earthquake Excitation. *Soils Found.* **2023**, *63*, 101298. [CrossRef]

water

MDPI

Article

A Combined Noise Reduction Method for Floodgate Vibration Signals Based on Adaptive Singular Value Decomposition and Improved Complete Ensemble Empirical Mode Decomposition with Adaptive Noise

Wentao Wang [1], Huiqi Zhu [1], Yingxin Cheng [1,*], Yiyuan Tang [1], Bo Liu [1], Huokun Li [1], Fan Yang [2], Wenyuan Zhang [2], Wei Huang [1] and Fang Zheng [3]

[1] School of Infrastructure Engineering, Nanchang University, Nanchang 330031, China; wangwentao2323@email.ncu.edu.cn (W.W.); zhuhuiqi@email.ncu.edu.cn (H.Z.); tangyiyuan@email.ncu.edu.cn (Y.T.); liubo1996@email.ncu.edu.cn (B.L.); lihuokun@ncu.edu.cn (H.L.); huangwei0214@ncu.edu.cn (W.H.)

[2] Department of Hydraulics, China Institute of Water Resources and Hydropower Research, Beijing 100038, China; yangfaniwhr@126.com (F.Y.); zhangweny8@163.com (W.Z.)

[3] China Railway Water Conservancy & Hydropower Planning and Design Group, Nanchang 330029, China; jgcskz20110@126.com

* Correspondence: chengyingxin@ncu.edu.cn

Abstract: To address the issue of the vibration characteristic signals of floodgates being affected by background white noise and low-frequency water flow noise, a noise reduction method combining the improved adaptive singular value decomposition algorithm (ASVD) and the improved complete ensemble EMD with adaptive noise (ICEEMDAN) is proposed. Firstly, a Hankel matrix is constructed based on the collected discrete time signals. After performing SVD on the Hankel matrix, the ASVD algorithm is used to automatically select the effective singular values to filter out most of the background white noise and retain the useful frequency components with similar energy in the signal. Then, ICEEMDAN combined with the Spearman correlation coefficient method is used to further filter out residual white noise and low-frequency water flows. The noise reduction performance of this combined method is verified through simulation experiments. Filtered by the ASVD-ICEEMDAN method, the signal-to-noise ratio of the simulation signal (50% noise level) is increased from 4.417 to 16.237, and the root mean square error is reduced from 2.286 to 0.586. Based on the practically measured vibration signals of a floodgate at a large hydropower station, the result shows that the ASVD-ICEEMDAN method exhibits good noise reduction performance and feature information extraction abilities for floodgate vibration signals, and can provide support for operational mode analysis and damage identification of practical structures under complex interference conditions.

Keywords: floodgates; vibration signal; noise reduction; ASVD algorithm; ICEEMDAN

check for updates

Citation: Wang, W.; Zhu, H.; Cheng, Y.; Tang, Y.; Liu, B.; Li, H.; Yang, F.; Zhang, W.; Huang, W.; Zheng, F. A Combined Noise Reduction Method for Floodgate Vibration Signals Based on Adaptive Singular Value Decomposition and Improved Complete Ensemble Empirical Mode Decomposition with Adaptive Noise. *Water* **2023**, *15*, 4287. https://doi.org/10.3390/w15244287

Academic Editors: Jie Yang, Chunhui Ma and Lin Cheng

Received: 19 November 2023
Revised: 8 December 2023
Accepted: 13 December 2023
Published: 15 December 2023

1. Introduction

Recently, the safety diagnosis of hydraulic structures using vibration signals under flood discharge excitation has become one of the hot spots in the research of hydraulic engineering [1–3]. Due to the long-term effect of huge fluctuating loads of water flow, floodgates are prone to problems such as excessive amplitudes and even fatigue damage. Vibration monitoring of floodgates plays a significant role in the safe operation of structures. However, the collected vibration response signal is usually accompanied by the interference of background white noise and low-frequency water flow noise [4,5], which makes the structural vibration feature information easily submerged in noise and greatly affects the extraction of the vibration information of floodgates. Therefore, effective signal processing methods need to be adopted to filter out the noise in the collected signal and retain the characteristic vibration information of floodgates.

The vibration signal of floodgates is a kind of typical nonlinear and non-stationary signal [6], which commonly can be processed by methods including wavelet threshold [7,8], singular value decomposition (SVD) [9–11], empirical mode decomposition (EMD) [12–14], its improved algorithms [15,16], etc. Among them, EMD and its improved algorithms automatically decompose a signal into multiple intrinsic mode functions (IMFs) as well as a residual (RES) based on its own characteristics. Due to the advantages of strong adaptability and no need for predetermined basis functions [17,18], EMD and its improved algorithms are widely used to process nonlinear and non-stationary signals. Zhang et al. [19] used the EMD method to filter high-frequency and low-frequency noise from blasting vibration signals, achieving principal component extraction of blasting vibration signals. However, the traditional EMD algorithm is prone to the phenomenon of mode mixing [20], which makes the IMFs lose their physical significance. In response to this issue, Huang et al. [21] proposed the ensemble empirical mode decomposition (EEMD) algorithm, which introduces white noise with a mean of zero to mask the noise of the signal itself through a multiple-noise-adding process, thereby obtaining more accurate upper and lower envelopes and suppressing the mode mixing phenomenon to a certain extent. Li et al. [22] adopted the EEMD algorithm to filter out the vibration noise of rotating machinery and successfully extracted the vibration components that reflect the true characteristics of the system. In addition, due to the problems of residual noise, as well as the low computational efficiency, in EEMD, Torres et al. [23] proposed the complete ensemble EMD with adaptive noise (CEEMDAN) method, which adds white noise to each EMD decomposition process, making the decomposition process complete and greatly reducing signal reconstruction errors as well as improving the decomposition efficiency. Peng et al. [24] used CEEMDAN to establish a noise reduction model, retaining key information while successfully filtering out high-frequency noise in the collected vibration signals. Colombia et al. [25] highlighted the problem of the weak residual noise and false modes in CEEMDAN; thus, they proposed an improved complete ensemble EMD with adaptive noise (ICEEMDAN) algorithm that adds IMF components of white noise during each EMD process and introduces the concept of the average of local means to further improve the CEEMDAN algorithm. Yuan et al. [26] utilized the ICEEMDAN algorithm to effectively remove random noise and wave noise of the original signal, obtaining high-precision seabed geomagnetic field data. For vibration signals of hydraulic structures, EMD and its improved methods are effective at filtering out low-frequency water flow noise, but are unsatisfactory in filtering white noise. Therefore, some scholars [27–29] proposed combined noise reduction methods. Among them, Li et al. [29] combined CEEMDAN and wavelet threshold methods to filter out noise from the guide wall of a hydropower station. However, the wavelet thresholding method has drawbacks, such as the difficulty in selecting threshold functions and its inability to automatically decompose signals. Zhang et al. [27] combined the CEEMDAN and SVD methods to successfully filter out white noise in the dynamic displacement signals of the Laxiwa arch dam and to extract key vibration information. However, ordinary SVD [30–33] denoising methods rely on certain experience to choose effective singular values (ESVs) rather than automatic selection.

Aiming at the noise characteristics of the collected floodgate vibration signals, a combined adaptive singular value decomposition (ASVD) and ICEEMDAN noise reduction method is proposed. Firstly, the discrete time signal is converted into a Hankel matrix. After performing SVD on the Hankel matrix, the improved hyperparameter ACC method is used to select ESVs. Via signal reconstruction, a signal with removed background white noise can be obtained. Secondly, the ICEEMDAN method is used to decompose the signal into several IMF components, whose center frequencies are arranged from high to low. Effective IMF components are selected through the correlation coefficient method for reconstruction to remove low-frequency water flow noise and residual white noise.

The main structure of this paper is as follows: Section 2 introduces the basic principles of ASVD and ICEEMDAN as well as the process of the combined ASVD–ICEEMDAN noise reduction method. Section 3 validates the noise reduction performance of the ASVD–ICEEMDAN method through simulation experiments. In Section 4, this method is applied to a real-life floodgate structure. Section 5 contains the conclusions of this paper.

2. Basic Principles

2.1. ASVD Noise Reduction

SVD has a strong filtering ability for background white noise, and its principles can be described as follows: convert a one-dimensional signal $\boldsymbol{X} = (x_1, x_2, \cdots, x_L)$ into a Hankel matrix $\boldsymbol{H}$:

$$\boldsymbol{H} = \begin{bmatrix} x_1 & x_2 & \cdots & x_c \\ x_2 & x_3 & \cdots & x_{c+1} \\ \cdots & \cdots & \cdots & \cdots \\ x_r & x_{r+1} & \cdots & x_L \end{bmatrix} \quad (1)$$

The above equation satisfies $1 < c < L$ and $L = r + c - 1$. Based on engineering experience [27,33], in order to achieve good noise reduction effects, the matrix $\boldsymbol{H}$ is usually constructed as a square matrix or a matrix close to a square matrix. Therefore, the variable r needs to satisfy Equation (2).

$$r = \begin{cases} L/2, & \text{When } L \text{ is odd;} \\ (L+1)/2, & \text{When } L \text{ is even.} \end{cases} \quad (2)$$

Then, the matrix $\boldsymbol{H}$ is decomposed via SVD:

$$\boldsymbol{H} = \boldsymbol{UDV^T} \quad (3)$$

where $\boldsymbol{U} \in \boldsymbol{R}^{r\times r}$, $\boldsymbol{V} \in \boldsymbol{R}^{r\times c}$, and $\boldsymbol{U}$ and $\boldsymbol{V}$ are both unit orthogonal matrices. $\boldsymbol{D} \in \boldsymbol{R}^{r\times c}$ can be stated as Equation (4).

$$\boldsymbol{D} = \begin{cases} [diag(\sigma_1, \sigma_2, \cdots, \sigma_a), \boldsymbol{0}] & , r < c \\ diag(\sigma_1, \sigma_2, \cdots, \sigma_a) & , r = c \\ [diag(\sigma_1, \sigma_2, \cdots, \sigma_a), \boldsymbol{0}]^T & , r > c \end{cases} \quad (4)$$

where a is the minimum value of m and n; $\sigma_1, \sigma_2, \cdots, \sigma_a$ are a series of singular values arranged from large to small, that is, $\sigma_1 \geq \sigma_2 \geq \cdots \geq \sigma_a \geq 0$; and $\boldsymbol{0}$ represents a zero matrix. Then, by setting the singular values representing noise to 0 and reconstructing a new Hankel matrix, the denoised signal can be obtained by the diagonal averaging method.

In the process of SVD noise reduction, it is a crucial step to select the ESVs. Li et al. [33] proposed a hyperparameter ACC method after noting the link between the singular values and the frequency components of the signal, which can be summarized into the following three points:

(1) Each frequency component of the signal relates to a singular value group consisting of two adjacent singular values that have small differences between them.
(2) Due to significant energy differences between different frequency components of the signal, a jump phenomenon is observed between adjacent groups of ESVs.
(3) The energy of white noise is uniformly distributed in the broadband frequency domain, which makes the jump phenomenon of singular value inconspicuous.

To sum up, by observing the jump phenomenon between singular values groups, the singular values group can be judged to represent a useful signal or noise. Due to the first few singular values relating to the effective signal components and the latter relating to noise components, at the boundary between the signal and noise, the jumping phenomenon changes from conspicuous to inconspicuous. In the study of Li et al. [33], singular values are processed by differencing and normalization (SVDN), so that the change in singular values is amplified and the singular value boundary can be found more easily. In addition, to enhance the adaptability of the algorithm, the hyperparameter ACC (0.1 suggested) is proposed to automatically select the minimum sequence number that satisfies "acc1 > ACC and acc2 < ACC".

However, the frequency components of the vibration signals of floodgates are usually complex (as shown in Section 4), such as the presence of many frequency components with similar energy levels, which is not consistent with the second law. In order to explain this problem and propose an improvement, a simulation signal with white noise is constructed as follows:

$$\begin{cases} S_1(t) = 4\sin(5\pi t) + 8\cos(6\pi t) + 10\sin(7\pi t) + 4\sin(8\pi t) + 4\cos(9\pi t) \\ Noise_1(t) = std(S_1) \times randn[size(S_1)] \\ X_1(t) = S_1(t) + Noise_1(t) \end{cases} \tag{5}$$

The noisy signal $X_1(t)$ consists of a signal with five frequency components and white noise. In addition, the first (2.5 Hz), fourth (4 Hz) and fifth (4.5 Hz) frequency components possess the same energy. After performing SVD on signal $X_1(t)$, the top 20 singular values are shown in Figure 1. Two adjacent singular values of a singular group are connected by the blue line and the singular groups are connected by red straight lines or dotted lines in Figure 1. It is obvious that the singular groups from one to five represent the five most useful frequency components and other singular groups represent the noise components. After the calculation of SVDN values shown in Figure 2, the hyperparameter ACC is adopted to select ESVs. According to the ACC threshold (0.1), the number of ESVs is 4, which is different to the presumed condition of 10 (twice the number of effective frequencies).

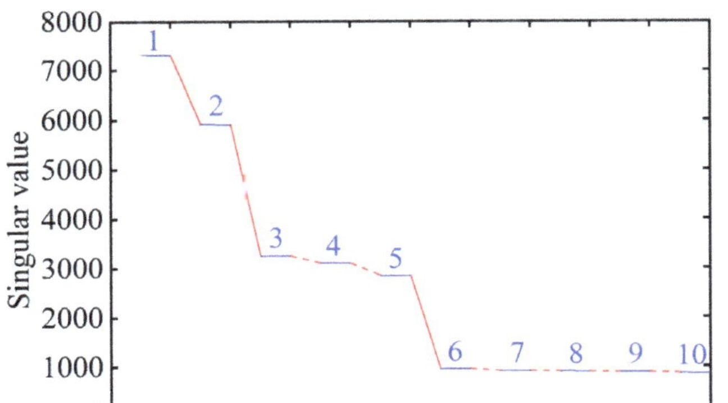

Figure 1. Change in singular values. (Notes: blue lines and numbers represent singular groups, red straight lines represent conspicuous jumping phenomenon, dotted lines represents unconspicuous jumping phenomenon).

By observing Figure 1, we can see that the jump phenomenon can be observed clearly between singular groups 1 and 2, 2 and 3, and 5 and 6, but it is not obvious between groups 3 and 4 and 4 and 5. Therefore, when the sequence numbers are six and eight, the SVDN values are lower than the ACC threshold, which results in the wrong selection. However, the jump phenomenon is still obvious between groups 5 and 6. Thus, an improved hyper-parameter ACC method is developed according to the upper analysis: select the largest sequence number that satisfies "acc1 > ACC and acc2 < ACC".

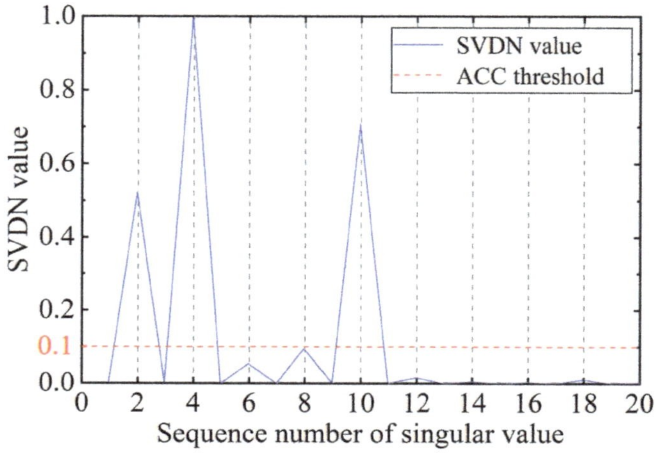

Figure 2. Change in SVDN values of Hankel matrix constructed by $X_1(t)$.

The flowchart of this new algorithm is shown in Figure 3, and the process is described as follows:

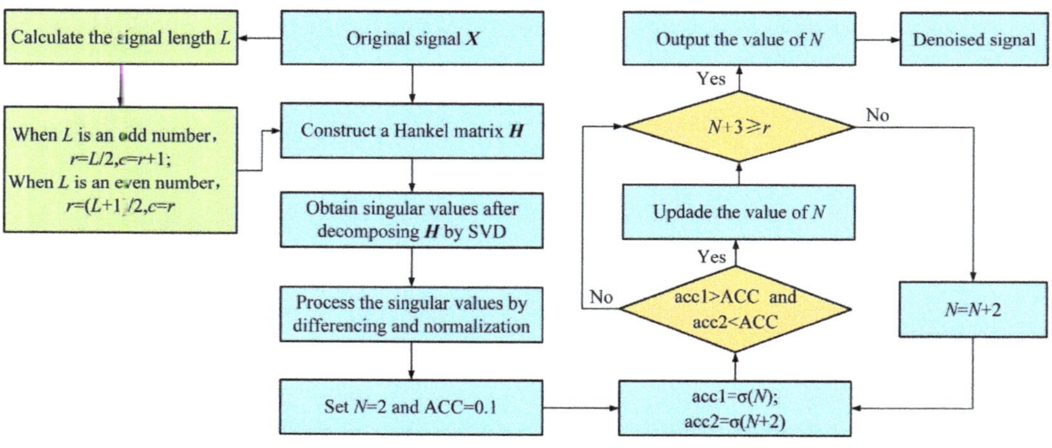

Where σ(*i*) is the *i* th singular value of matrix ***H***.

Figure 3. Flowchart of the ASVD algorithm.

Step 1. Compute the signal length L of the one-dimensional signal $\boldsymbol{X}$, then construct a Hankel matrix $\boldsymbol{H}$ according to Equations (1) and (2);

Step 2. Decompose Hankel matrix $\boldsymbol{H}$ via SVD and record singular values from large to small as $\sigma(1)$,$\sigma(2)$, $\cdots$,$\sigma(r)$.

Step 3. Process the singular values by differencing and normalization.

Step 4. Set the initial value of the iteration variable $N = 2$ and the hyperparameter ACC = 0.1.

Step 5. Assign singular values to acc1 and acc2, that is, acc1 = $\sigma(N)$ and acc2 = $\sigma(N+2)$.

Step 6. Determine whether the condition "acc1 > ACC and acc2 < ACC" is met. When the condition is met, record and update the variable N; otherwise, proceed to the next step.

Step 7. Decide whether the condition "$N + 3 \geq r$" is met. When the condition is met, perform the next step; otherwise, set $N = N + 2$ and jump to step 5.

Step 8. Set the singular values whose sequence numbers are greater than N to 0, reconstruct the Hankel matrix, and use the diagonal averaging method to obtain the denoised signal.

2.2. ICEEMDAN Method

Aiming at the problem of weak residual noise and false modes in CEEMDAN, ICEEMDAN improves CEEMDAN by changing the way white noise is added and introducing the concept of local means. Let $W(i)$ ($i = 1, 2, \cdots, n$) represent white noise with zero mean unit variance; let $< \cdot >$ represent the operator that averages the different signals; and let $E_k(\cdot)$ represent the kth IMF component after performing EMD on the bracketed signal. $\beta_0, \cdots, \beta_{k-1}$ are the parameters controlling the magnitude of the added noise. Let $M(\cdot)$ represent the local means for the bracketed signals, that is, $M(S) = S - E_1(S)$. The flowchart of the ICEEMDAN algorithm is shown in Figure 4, and the decomposition steps are presented as follows:

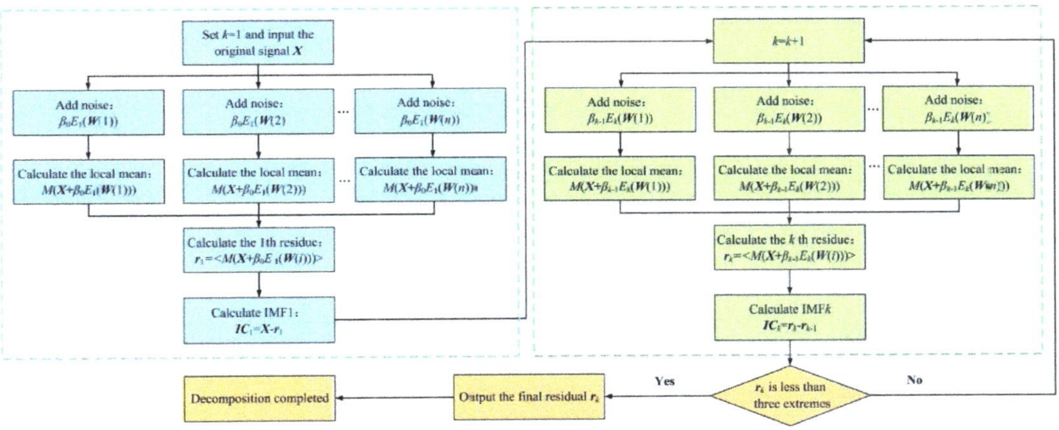

Figure 4. Flowchart of ICEEMDAN algorithm.

Step 1. Add n groups of noise $\beta_0 E_1(W(i))$ to the original signal X to obtain the noise-added signal $x(i) = x + \beta_0 E_1(W(i))$, and perform EMD on the noise-added signal to obtain n local means $M(x(i))$. Then, average the n local means to obtain the first-stage residual value r_1:

$$r_1 = < M(x(i)) > \quad (6)$$

Step 2. Compute the IMF value IC_1 for the first stage:

$$IC_1 = x - r_1 \quad (7)$$

Step 3. Calculate the local means of $r_1 + \beta_1 E_2(w(i))$ and average them to get the second stage residual value:

$$r_2 = < M(r_1 + \beta_1 E_2(W(i))) > \quad (8)$$

Step 4. Calculate the IMF value IC_2 for stage 2:

$$IC_2 = r_1 - r_2 \quad (9)$$

Step 5. For $k = 3, \cdots, K$, compute the residual for the kth stage:

$$r_k = < M(r_{k-1} + \beta_{k-1}E_k(W(i))) > \tag{10}$$

Step 6. Calculate the kth IMF value:

$$IC_k = r_{k-1} - r_k \tag{11}$$

Step 7. Repeat Step 5 to 6 until the signal is less than three extremes, then the last signal that cannot be decomposed is the RES, and ICEEMDAN is complete.

2.3. Combined ASVD–ICEEMDAN Noise Reduction

In the measured dynamic floodgate vibration signal, the floodgates' characteristic information is usually overwhelmed by the background white noise as well as the low-frequency water flow noise. Aiming at these two noise characteristics, the ASVD–ICEEMDAN method is proposed for combined noise reduction. The flow chart is shown in Figure 5, and the noise reduction steps are as follows:

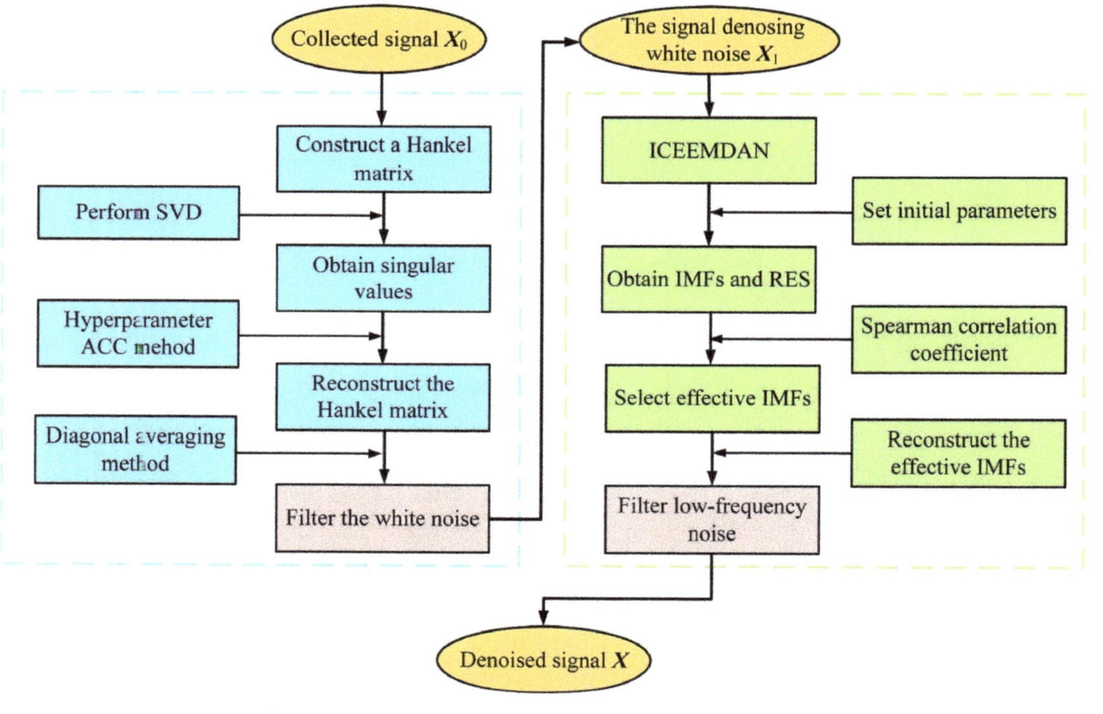

Figure 5. Flowchart of the ASVD–ICEEMDAN noise reduction method.

Step 1. The original signal X_0 is constructed as a Hankel matrix and then decomposed via SVD to obtain the singular values.

Step 2. The improved hyperparameter ACC method is used to automatically search ESVs. Then, the new signal X_1 can be reconstructed via the ASVD algorithm.

Step 3. Decompose X_1 via ICEEMDAN to yield several IMF components whose center frequencies are arranged from high to low.

Step 4. Select the effective IMF components based on the Spearman correlation coefficient [34–36] as in Equation (12)

$$r_s = 1 - \frac{6\sum_{i=1}^{n} d_i^2}{L(L^2-1)} \quad (12)$$

where d_i is the level difference between elements $X(i)$ and $Y(i)$, L is the length of the signal. The value of r_s is between -1 and 1. When the absolute value of r_s approaches 1, it indicates a strong correlation between the two signals; when the absolute value of r_s approaches 0, it indicates a weak correlation between signals. In addition, if the absolute values of r_s of the signal before decomposition and the IMF component are greater than 0.4 [36], the IMF component can be regarded as an effective IMF component.

Step 5. Reconstruct the effective IMF components to obtain the denoised signal X.

3. Numerical Simulation

To clarify the denoising performance of the ASVD–ICEEMDAN method, a pure simulation signal $S_2(t)$ was constructed in MATLAB R2019a. Furthermore, it is assumed that the collected signal $X_2(t)$ is mixed with low-frequency noise (LN) and white noise (WN), and the function expression is shown in Equation (13):

$$\begin{cases} S_2(t) = 6e^{-\frac{t}{4}}\sin(3\pi t) + 8e^{-\frac{t}{3}}\sin(8\pi t) + 9e^{-\frac{t}{3}}\sin(20\pi t) \\ Noise_2(t) = 4e^{-\frac{t}{4}}\sin(\pi t) + P \times std(S_2) \times randn[size(S_2)] \\ X_2(t) = S_2(t) + Noise_2(t) \end{cases} \quad (13)$$

where P is the noise level and $std(S_2)$ is the standard deviation of S_2; $randn[size(S_2)]$ represents a white noise with a zero mean, a standard deviation of 1 and a normal distribution. The sampling frequency is 50 Hz, the total sampling time is 10 s, and the noise level P is 50%. The time waveform and power spectrum curves of pure signals and noisy signals are shown in Figure 6.

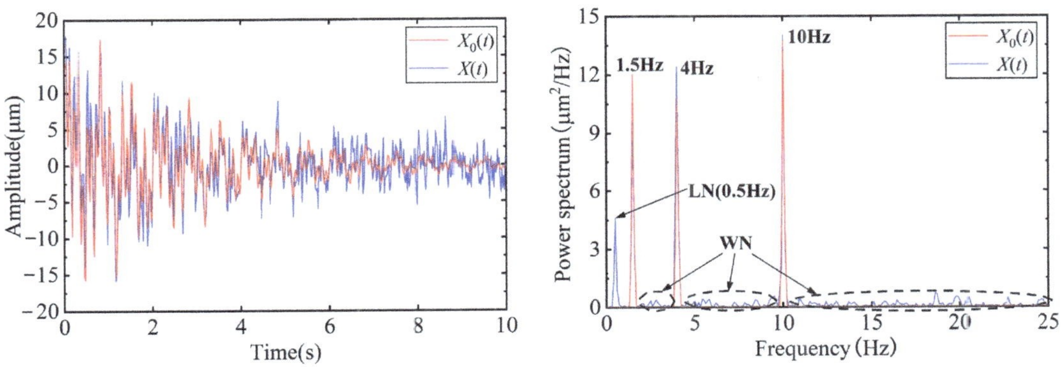

Figure 6. Time waveform and power spectrum curves of pure signals and noisy signals.

After constructing a Hankel matrix for the simulation signal, the SVD operation of the Hankel matrix is performed to calculate its SVDN values. The top 20 values are shown in Table 1, and the line graph is plotted in Figure 7.

Table 1. Top 20 singular values and SVDN values.

Sequence Number	1	2	3	4	5	6	7	8	9	10
Singular value	332.520	327.151	306.594	299.108	260.133	245.444	185.043	156.517	58.924	58.764
SVDN value	0.055	0.211	0.077	0.399	0.151	0.619	0.292	1.000	0.002	0.029
Sequence number	11	12	13	14	15	16	17	18	19	20
Singular value	55.940	55.672	55.038	54.870	52.886	52.711	51.616	51.524	51.410	51.344
SVDN value	0.003	0.006	0.002	0.020	0.002	0.011	0.001	0.001	0.001	0.008

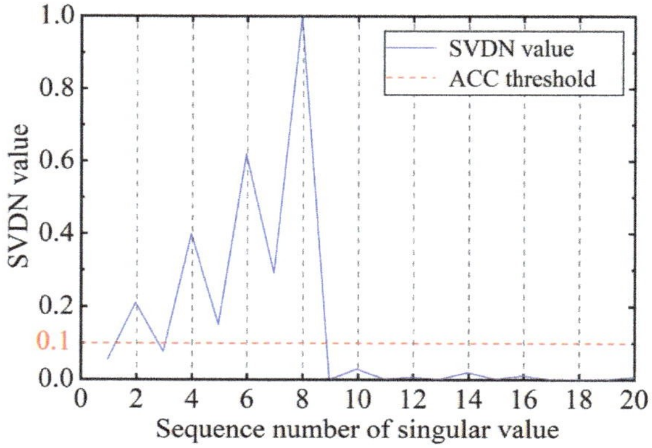

Figure 7. Change in SVDN values of Hankel matrix constructed by $X_2(t)$.

According to Figure 7, the number of ESVs is computed as 8. Next, the signal is reconstructed, and the noise reduction result is shown in Figure 8.

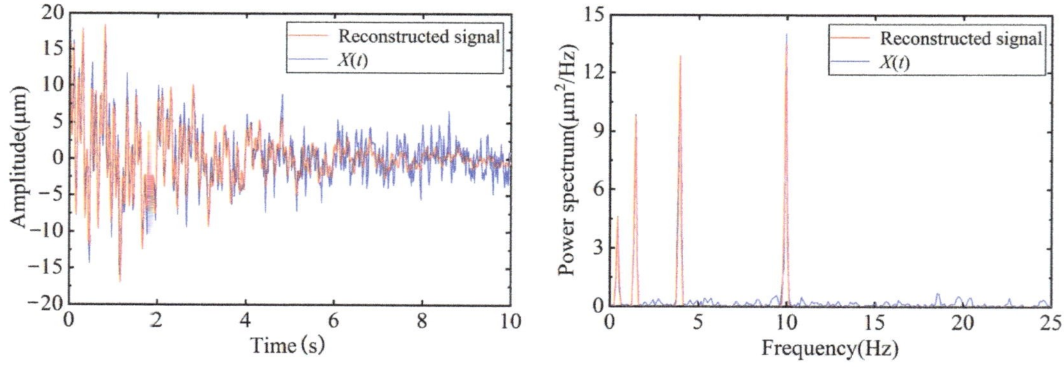

Figure 8. Time waveform and power spectrum curves of signal denoised via the ASVD method.

From Figure 8, it can be observed that the signal denoised by ASVD is smoother, and most of the background white noise in the signal has been filtered out after ASVD filtering. However, low-frequency strong noise at 0.5 Hz still exists in the filtered signal. Next, the CEEMDAN and ICEEMDAN methods are used to decompose the pre-filtered signal, respectively. To ensure that the decomposition results are comparable, the same initial parameters were used in these two methods: the white noise amplitude is set to 0.2, the average number of signals is 100, and the maximum iterations number is 1000. The decomposition results are shown in Figure 9.

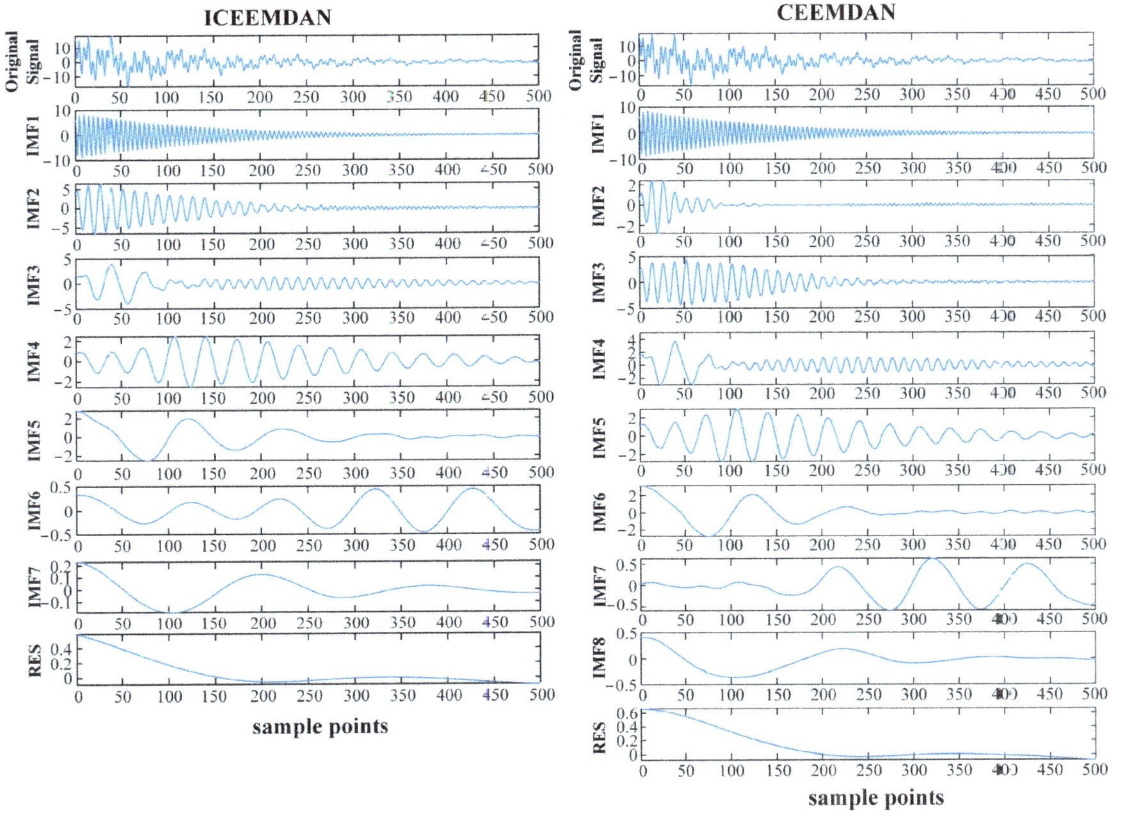

Figure 9. IMFs and RES decomposed via ICEEMDAN and CEEMDAN.

In Figure 9, the signal is decomposed into eight IMFs and a RES by the CEEMDAN method, while only seven IMFs and a RES are generated via ICEEMDAN, which proves that ICEEMDAN can reduce the number of invalid components and inhibit the generation of mode mixing phenomena to a certain extent compared to CEEMDAN.

In addition, to calculate the residual noise of the two decomposition methods, all IMFs and the RESs are summed and reconstructed, and the signal deviation, *RRSE*, between the original signal and the reconstructed signal is calculated via Equation (14):

$$RRSE = \frac{\|S_1 - S_2\|}{\|S_2\|} \tag{14}$$

where S_1 is a vector that sums all IMFs and RESs decomposed by ICEEMDAN or CEEMDAN and S_2 is the original signal vector. The *RRSE* value of the CEEMDAN method is 1.90×10^{-16}, while that of the ICEEMDAN method is only 1.39×10^{-16}, indicating that the signal decomposed by the ICEEMDAN method has lower residual noise.

To select the effective IMFs to reduce noise, the absolute values of r_s of IMF1-IMF7 and RES were calculated and are shown in Table 2. The absolute values of r_s of IMF1-IMF4 are all greater than 0.4, while the remaining components are less than 0.4. Therefore, IMF1-IMF4 were selected for reconstruction, and the signal denoised by ASVD–ICEEMDAN is shown in Figure 10.

Table 2. The absolute values of r_s of IMF1-IMF7 and RES.

IMF Component	Absolute Values of r_s	IMF Component	Absolute Values of r_s
IMF1	0.536	IMF5	0.397
IMF2	0.578	IMF6	0.303
IMF3	0.498	IMF7	0.094
IMF4	0.458	RES	0.076

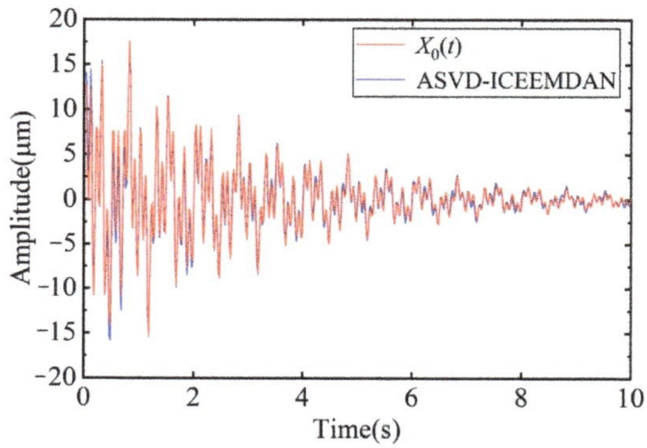

Figure 10. Noise reduction result.

The noise is almost completely filtered out, and the waveforms of the denoised signal and the pure signal are rather similar, as shown in Figure 10. To verify the denoising performance of the combined ASVD–ICEEMDAN method compared with other methods, the signal-to-noise ratio (*SNR*) and the root mean square error (*RMSE*) were calculated via Equations (15) and (16):

$$SNR = 10\log\left[\frac{\sum_{t=1}^{T} X_0^2(t)}{\sum_{t=1}^{T} (X(t) - X_0(t))^2}\right] \tag{15}$$

$$RMSE = \sqrt{\frac{1}{T}\sum_{t=1}^{T}[X_0(t) - X(t)]^2} \tag{16}$$

where T is the sampling number. The calculation results are shown in Table 3. The signal filtered by the ASVD–ICEEMDAN noise reduction method presents the highest SNR and the lowest RMSE, proving the superiority of the combined noise reduction method. The simulation data and MATLAB R2019a script are available at: https://github.com/RabbitInTheNorth/Noise-reduction-simulation.git (accessed on 9 December 2023).

Table 3. Comparison of evaluation indexes of noise reduction performance of different methods.

Noise Reduction Method	Index	
	SNR/dB	*RMSE*
Original signal	4.417	2.286
Moving average	5.414	2.038
IIR digital filters	5.544	2.008
ASVD	9.387	1.290
ICEEMDAN	5.711	1.969
ASVD-CEEMDAN	16.104	0.595
ASVD–ICEEMDAN	16.237	0.586

4. Engineering Example

The studied floodgate is located in the upper reaches of the Hanjiang River, and is a part of a large-scale hydropower station. The hub project is divided into two parts by a longitudinal pier in the middle of the river. The left side is the powerhouse dam section, and the right side is the flood discharge dam section, where there are six holes. From left to right, the six holes are floodgates 1#~4#, a vertical ship lift (also used as a floodgate), and a surface hole in the right auxiliary dam. The actual project is shown in Figure 11a.

Figure 11. Pictures of the project and sensor arrangement: (**a**) top view of the project, (**b**) measuring point layout of hole 2#, (**c**) data acquisition instrument, (**d**) dynamic displacement sensor.

During flood discharge operations in the past, the staff of the project found that there was an obvious vibration phenomenon at the top of the floodgate pier, especially in hole 2#. To study the influence of flood discharge on the pier structure, a vibration test was carried out under various working conditions. The sampling frequency of this test was 50 Hz and the sampling time was 81.92 s. The dynamic displacement sensor layout diagram is shown in Figure 11, and the parameters of the sensors are shown in Table 4.

Table 4. Performance table of DP seismic low frequency vibration sensors.

Sensor Number	Model	Range of Frequency Response	Sensitivity
1~10	DPS-0.5-15-H	0.5~200 Hz	5 mv/μm

The collected vibration signal was processed by the ASVD–ICEEMDAN method introduced in Section 3. Due to length limitations, only the process of measuring point 10 under working condition 1 is described. The time waveform and power spectrum curves of the vibration signal are shown in Figure 12. In the power spectrum figure, there are three obvious frequency peaks (plotted in orange and red wireframes) and many "small burrs" (plotted in green wireframes). These "small burrs" represent white noise components, which would be filtered by the ASVD method. In addition, the main energy of peak 1 is within 1 Hz [37,38], which represents low-frequency noise and would be filtered by the ICEEMDAN method.

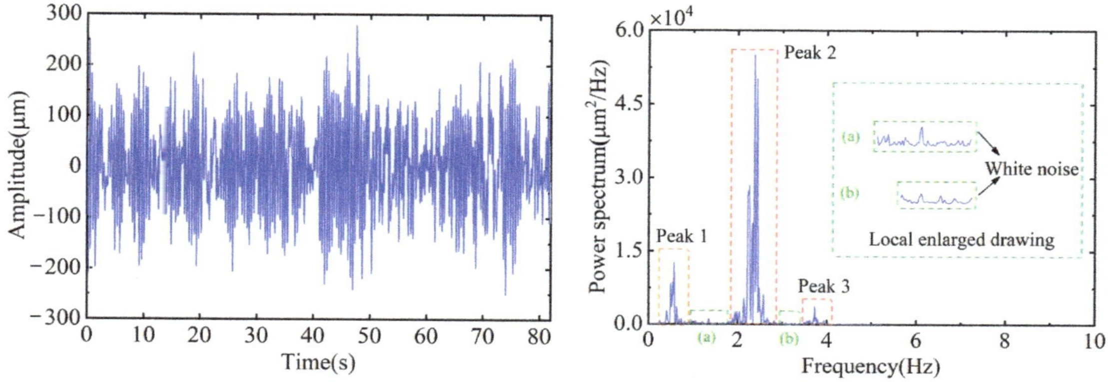

Figure 12. Vibration signal of measuring point 10 under condition 1. (Note: Frame (a) and (b) represent the local enlarged drawing windows.)

After constructing the Hankel matrix of the signal, the ASVD algorithm was performed, determining the number of ESVs, which is 68, as shown in Figure 13. Then, the white-noise-filtered signal was obtained after SVD reconstruction. In order to filter the low-frequency water flow noise, the signal processed by ASVD was decomposed via ICEEMDAN, and the decomposition result is shown in Figure 14. After removing the components whose main frequencies are less than 1 Hz and IMFs whose absolute values of r_s are less than 0.4, the signal was reconstructed. The denoised signals at measuring point 10 under conditions 1 and 2 are shown in Figures 15 and 16, respectively.

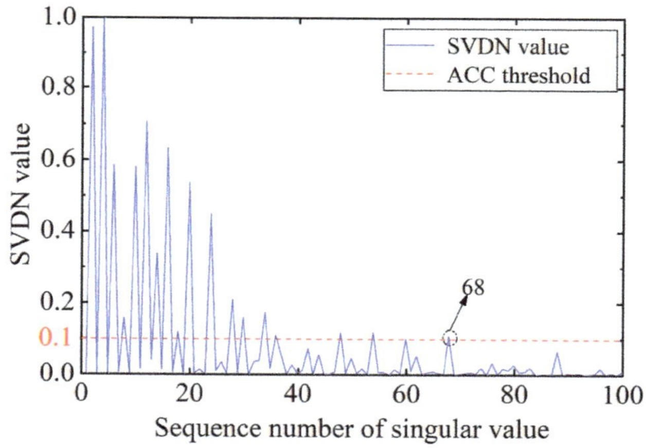

Figure 13. Change in SVDN values of Hankel matrix constructed by vibration signal of measuring point 10 under condition 1.

In Figures 15 and 16, the white noise and the low-frequency water flow noise are considerably filtered out. From the power spectrum curves after windowing, it can clearly be seen that the two-order natural frequencies of the structure under condition 1 are 2.35 Hz and 3.76 Hz, while they are 2.47 Hz and 3.82 Hz under condition 2. The noise reduction results were compared with the results of the eigensystem realization algorithm (ERA) modal identification method [37] in Table 5. The maximum error of the two frequency identification results is only 1.67%, indicating that the method can filter out the noise of the floodgate and restore the structural feature information successfully.

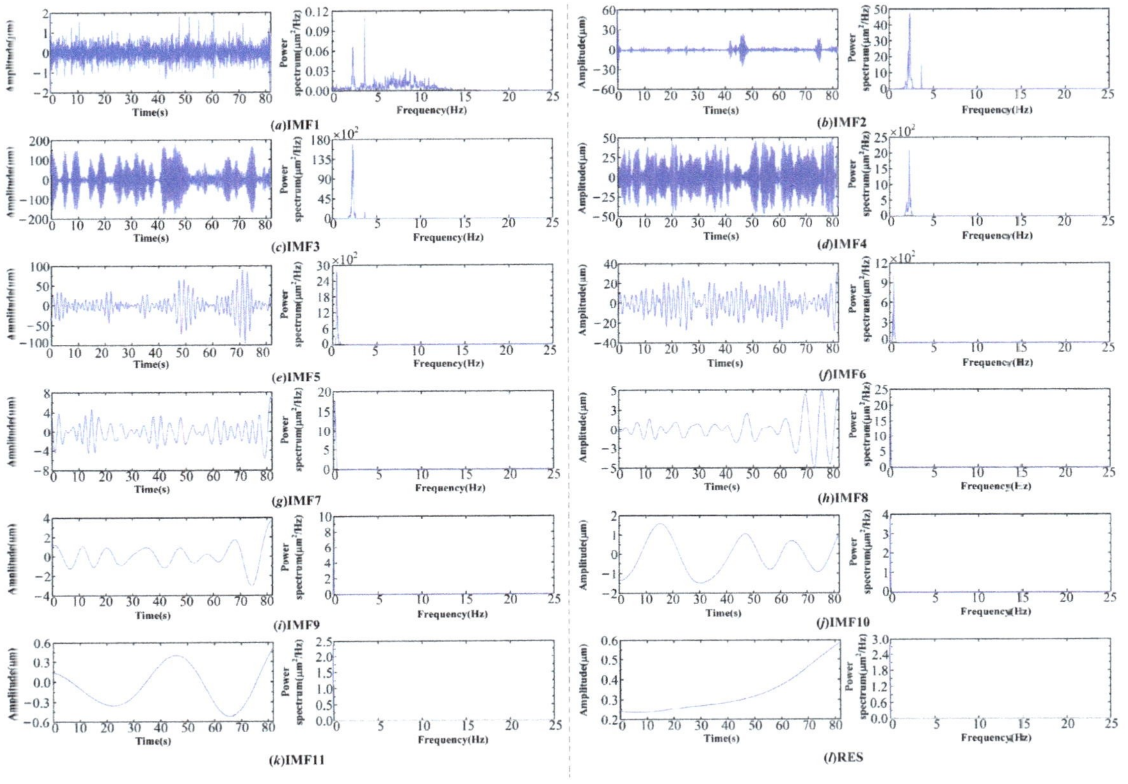

Figure 14. IMFs and RES decomposed via ICEEMDAN in measuring point 10 under condition 1.

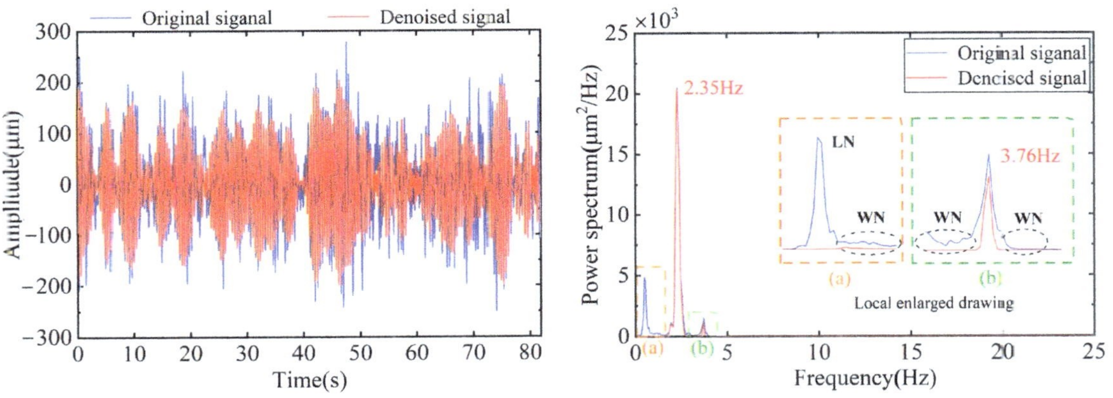

Figure 15. Noise reduction result of measuring point 10 under condition 1. (Note: Frame (a) and (b) represent the local enlarged drawing windows.)

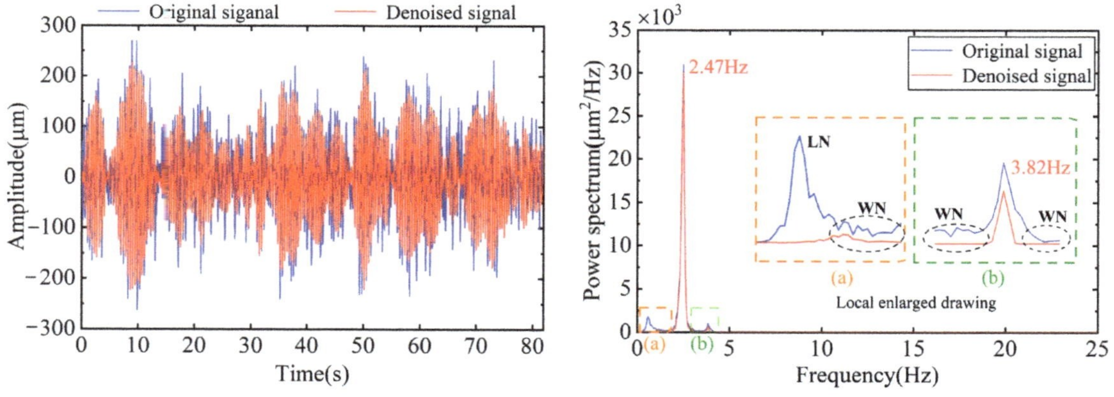

Figure 16. Noise reduction result of measuring point 10 under condition 2. (Note: Frame (a) and (b) represent the local enlarged drawing windows.)

Table 5. Results comparison of ASVD–ICEEMDAN and ERA.

Working Condition	Order	Suggested Method/Hz	ERA/Hz	Relative Error */%
Condition 1	1	2.35	2.39	1.67
	2	3.76	3.77	0.27
Condition 2	1	2.47	2.43	1.65
	2	3.82	3.82	0

Note: * Relative error $= \frac{|\text{Suggested method value} - \text{ERA method value}|}{\text{ERA method value}} \times 100\%$.

5. Conclusions

The ASVD–ICEEMDAN combined noise reduction method is proposed to filter background white noise and low-frequency water flow noise in the vibration signals of a floodgate structure. The following main conclusions are drawn through simulation experiments and an actual engineering case.

(1) An ASVD method is proposed to select effective singular values automatically based on the relationship between singular values and signal components. It avoids the uncertainty of manual selection and improves the efficiency of noise reduction.

(2) The ICEEMDAN algorithm can automatically decompose the signal into several IMF components whose center frequencies are arranged from high to low. Compared to the CEEMDAN algorithm, ICEEMDAN algorithm performs better in terms of suppressing mode mixing and reducing residual noise.

(3) The ASVD–ICEEMDAN method successfully filters out white noise and low-frequency noise in simulated signals, which increases the SNR of the signal (50% noise level) from 4.417 to 16.237 and reduces the RMSE from 2.286 to 0.586. In the engineering case study, the ASVD–ICEEMDAN method effectively filters out the noise and accurately extracts the structural characteristic vibration information, proving it can provide support in operational modal analyses and damage identification in actual structures.

(4) Vibration signals of discharge structures are mainly affected by white noise and low-frequency noise; thus, this method has the potential to be extended to other hydraulic structures under discharge excitation, such as arch dams, gravity dams, and guide walls.

Author Contributions: W.W.: Methodology, Writing—original draft. H.Z.: Validation, Investigation, Data curation, Visualization, Writing—review & editing. Y.C.: Conceptualization, Methodology, Funding acquisition. Y.T.: Resources, Validation, Writing—review & editing. B.L.: Resources, Investigation. H.L.: Funding acquisition, Supervision. F.Y.: Funding acquisition. W.Z.: Conceptualization.

W.H.: Project administration. F.Z.: Investigation. All authors have read and agreed to the published version of the manuscript.

Funding: This work was partially supported by the National Natural Science Foundation of China (grant number 52079061), Jiangxi Province Double Thousand Plan High-end Talent Project of Science and Technology Innovation (grant number jxsq2023201017), Jiangxi Provincial Natural Science Foundation Key Projects (grant number 20232ACB204027), Water Science and Technology Projects of Water Resources Department of Jiangxi Province (grant number 202325ZDKT05).

Data Availability Statement: Due to the safety of the project, the relevant data is confidential and cannot be published in this article.

Conflicts of Interest: The authors declare that they have no known competing financial interests or personal relationships that could have appeared to influence the work reported in this paper.

References

1. Li, H.K.; Zhong, Y.W.; Wei, B.W.; Du, L.; Wang, G. Beat vibration mechanism of a sluice pier under high-speed flood discharge excitation. *J. Low Freq. Noise. Vib. Act. Control* **2020**, *39*, 28–43. [CrossRef]
2. Li, H.K.; Liu, B.; Huang, W.; Liu, H.Y.; Wang, G. Vibration load identification in the time-domain of high arch dam under discharge excitation based on hybrid LSQR algorithm. *Mech. Syst. Signal Process.* **2022**, *177*, 109193. [CrossRef]
3. Liu, B.; Li, H.K.; Wang, G.; Wu, P.Z.; Li, Y.K. Dynamic material parameter inversion of high arch dam under discharge excitation based on the modal parameters and Bayesian optimised deep learning. *Adv. Eng. Inform.* **2023**, *56*, 102016. [CrossRef]
4. Li, H.K.; Wang, G.; Wei, B.W.; Liu, H.Y.; Huang, W. Improved variational mode decomposition method for vibration signal processing of flood discharge structure. *J. Vib. Control* **2022**, *28*, 2556–2569. [CrossRef]
5. Zhang, Y.; Lian, J.J.; Liu, F. An improved filtering method based on EEMD and wavelet-threshold for modal parameter identification of hydraulic structure. *Mech. Syst. Signal Process.* **2016**, *68*, 316–329. [CrossRef]
6. Zhang, J.W.; Jiang, Q.; Zhao, Y.; Zhu, L.H.; Guo, J. De-noising method for vibration signal of flood discharge structure. *J. Vib. Shock* **2015**, *34*, 179–184.
7. Yang, X.; Wang, W. Chaotic signal denoising based on energy selection TQWT and adaptive SVD. *Sci. Rep.* **2023**, *13*, 18873. [CrossRef]
8. Bi, L.; Yu, X.; Jiang, D.R.; Peng, H.L.; Hu, F.; Shi, K.K.; Liang, K.; Ye, X. Denoising Method of Hydro-turbine Vibration Signal Based on Joint WOA-VMD and Improved Wavelet Threshold. In *Journal of Physics: Conference Series*; IOP Publishing Ltd.: Bristol, UK, 2023; Volume 2607, p. 012005.
9. Fattorini, M.; Brandini, C. Observation strategies based on singular value decomposition for ocean analysis and forecast. *Water* **2020**, *12*, 3445. [CrossRef]
10. Zhao, S.X.; Ma, L.S.; Xu, L.Y.; Liu, M.N.; Chen, X.L. A Study of Fault Signal Noise Reduction Based on Improved CEEMDAN-SVD. *Appl. Sci.* **2023**, *13*, 10713. [CrossRef]
11. Zhao, X.Z.; Nie, Z.G.; Ye, B.Y.; Chen, T.J. Number law of effective singular values of signal and its application to feature extraction. *J. Vib. Eng.* **2016**, *29*, 532–541.
12. Koppolu, P.K.; Chemmangat, K. Automatic Selection of IMFs to Denoise the sEMG Signals using EMD. *J. Electromyogr. Kinesiol.* **2023**, *73*, 102834. [CrossRef]
13. Wang, C.G.; Cui, Y.C.; Liu, Y.; Li, K.; Shen, C. High-G MEMS Accelerometer Calibration Denoising Method Based on EMD and Time-Frequency Peak Filtering. *Micromachines* **2023**, *14*, 970. [CrossRef] [PubMed]
14. Chu, T.Y.; Huang, W.C. Application of empirical mode decomposition method to synthesize flow data: A case study of Hushan Reservoir in Taiwan. *Water* **2020**, *12*, 927. [CrossRef]
15. Singh, J.; Darpe, A.K.; Singh, S.P. Bearing damage assessment using Jensen-Rényi Divergence based on EEMD. *Mech. Syst. Signal Process.* **2017**, *87*, 307–339. [CrossRef]
16. Ameri, R.; Hsu, C.C.; Band, S.S.; Zamani, M.; Shu, C.M.; Khorsandroo, S. Forecasting PM 2.5 concentration based on integrating of CEEMDAN decomposition method with SVM and LSTM. *Ecotoxicol. Environ. Saf.* **2023**, *266*, 115572. [CrossRef]
17. Huang, N.E.; Shen, Z.; Long, S.R.; Wu, M.C.; Shih, H.H.; Zheng, Q.; Yen, N.C.; Tung, C.C.; Liu, H.H. The empirical mode decomposition and the Hilbert spectrum for nonlinear and non-stationary time series analysis. *Proc. R. Soc. Lond. Ser. A Math. Phys. Eng. Sci.* **1998**, *454*, 903–995. [CrossRef]
18. Sun, M.; Zhang, X.M.; Huang, Q.; Wu, L. Study on suppression of EMD endpoint effect of blasting seismic wave signal. *Eng. Blasting* **2023**, *29*, 138–143.
19. Zhang, S.H.; Gao, W.X.; Liu, J.C.; Li, X.S.; Hu, Y.; Zhang, X.J. Extraction and analysis of principal components of blasting vibration signal based on EMD decomposition. *Eng. Blasting* **2022**, *28*, 8–14.
20. Zheng, J.D.; Cheng, J.S.; Yang, Y. Partly ensemble empirical mode decomposition: An improved noise-assisted method for eliminating mode mixing. *Signal Process.* **2014**, *96*, 362–374. [CrossRef]
21. Wu, Z.; Huang, N.E. Ensemble empirical mode decomposition: A noise-assisted data analysis method. *Adv. Adapt. Data Anal.* **2009**, *1*, 1–41. [CrossRef]

22. Li, X.J.; Wang, K.; Wang, G.B.; Li, Q. Denoising of the Base Vibration Signal based on EEMD. *Appl. Mech. Mater.* **2012**, *143*, 689–693. [CrossRef]
23. Torres, M.E.; Colominas, M.A.; Schlotthauer, G.; Flandrin, P. A complete ensemble empirical mode decomposition with adaptive noise. In Proceedings of the 2011 IEEE International Conference on Acoustics, Speech and Signal Processing (ICASSP), Prague, Czech Republic, 22–27 May 2011; pp. 4144–4147.
24. Peng, Y.X.; Liu, Y.S.; Zhang, C.; Wu, L. A novel denoising model of underwater drilling and blasting vibration signal based on CEEMDAN. *Arab. J. Sci. Eng.* **2021**, *46*, 4857–4865. [CrossRef]
25. Colominas, M.A.; Schlotthauer, G.; Torres, M.E. Improved complete ensemble EMD: A suitable tool for biomedical signal processing. *Biomed. Signal Process. Control* **2014**, *14*, 19–29. [CrossRef]
26. Yuan, C.Q.; Hu, Z.W.; Liu, Y.; He, S.Y.; Du, J.S. Application of ICEEMDAN to noise reduction of near-seafloor geomagnetic field survey data. *J. Appl. Geophys.* **2023**, *209*, 104933. [CrossRef]
27. Zhang, J.W.; Hou, G.; Bao, Z.L.; Zhang, Y.N.; Ma, Y. A signal de-noising method for vibration signals from flood discharge structures based on CEEMDAN and SVD. *J. Vib. Shock* **2017**, *36*, 138–143.
28. Zhang, J.W.; Jiang, Q.; Ma, B.; Zhao, Y.; Zhu, L.H. Signal de-noising method for vibration signal of flood discharge structure based on combined wavelet and EMD. *J. Vib. Control* **2017**, *23*, 2401–2417. [CrossRef]
29. Li, H.K.; Wu, P.Z.; Huang, W.; Liu, H.Y. Joint noise reduction method of hydraulic structure vibration signals based on CEEMDAN and improved wavelet threshold. *J. Nanchang Univ. (Eng. Technol.)* **2022**, *44*, 1–8.
30. Lv, Y.L.; Lang, R.L.; Liang, J.C. Decision of threshold for singular value decomposition filter based on SNR's empirical value. *Appl. Res. Comput.* **2009**, *26*, 3253–3255.
31. Zhang, Y.; Xing, P.F.; Wang, M.J.; Zhou, Y.N.; Zhou, X.D.; Wei, Y.W. Pipeline Signal Complementary Denoising Method Based on GAHE-VMD and SVD-SCEC. *Noise Vib. Control* **2023**, *43*, 122–129.
32. Wang, Q.H.; Wang, L.J.; Yu, H.T.; Wang, D.; Nandi, A.K. Utilizing SVD and VMD for denoising non-stationary signals of roller bearings. *Sensors* **2021**, *22*, 195. [CrossRef]
33. Li, J.; Chen, Z.; Li, S. Selection of the number of effective singular values for noise reduction. *Mech. Syst. Signal Process.* **2023**, *191*, 110175. [CrossRef]
34. Wu, Q.; Wang, K.W.; Zhao, H.; Shi, W.W. Snow depth retrieval using GPS signal-to-noise ratio data based on improved complete ensemble empirical mode decomposition. *GPS Solut.* **2023**, *27*, 201. [CrossRef]
35. Gao, Y.; Hang, Y.; Yang, M. A cooling load prediction method using improved CEEMDAN and Markov Chains correction. *J. Build. Eng.* **2021**, *42*, 103041. [CrossRef]
36. Luo, C.L.; Liu, Q.H.; Li, W.G. EWT-SSA joint denoising and its application in the analysis of hob vibration signal. *China Meas. Test* **2022**, *48*, 109–116.
37. Lian, J.J.; Li, H.K.; Zhang, J.W. ERA modal identification method for hydraulic structures based on order determination and noise reduction of singular entropy. *Sci. China Ser. E Technol. Sci.* **2009**, *52*, 400–412. [CrossRef]
38. Guo, Y.J.; You, Z.Z.; Wei, B.W. Working Mode Identification Method for High Arch Dam Discharge Structure Based on Improved Wavelet Threshold–EMD and RDT Algorithm. *Water* **2022**, *14*, 3735. [CrossRef]

water MDPI

Article

Study on the Evolution of Risk Contagion in Urban River Ecological Management Projects Based on SEIRS

Junke Xu [1,2,3,*], Jiwei Zhu [1,2,3] and Jiancang Xie [1,2]

1 State Key Laboratory of Eco-Hydraulics in Northwest Arid Region, Xi'an University of Technology, Xi'an 710048, China
2 Research Center of Eco-Hydraulics and Sustainable Development, The New Style Think Tank of Shaanxi Universities, Xi'an 710048, China
3 Department of Engineering Management, Xi'an University of Technology, Xi'an 710048, China
* Correspondence: 1180711014@stu.xaut.edu.cn

Abstract: The risk transmission mechanisms of urban river ecological management engineering projects are examined in this study. Using the Susceptible Exposed Infectious Recovered Susceptible (SEIRS) model for risk transmission, a model of risk propagation delay for urban river ecological management engineering projects on scale-free networks is developed, which takes into account the effects of risk propagation and delay. We conducted a steady-state analysis of the model and obtained the basic reproduction number R. When R > 1, the equilibrium point of risk outbreak is stable, and when R < 1, the equilibrium point of risk disappearance is stable. Numerical simulations of the model were conducted using the MATLAB2022b to reveal the dynamic propagation patterns of risk in urban river ecological management engineering projects. The research results show that the steady-state density of the infected nodes in the network increases with the increase in the effective propagation rate and the propagation delay time; the propagation delay reduces the risk propagation threshold in the network and accelerates the occurrence of the equilibrium state of risk outbreak. There is a correlation between the transmission rate of latent nodes and the transmission rate of infected nodes, and the effective transmission rate of latent nodes has a greater influence on risk propagation. The spread of risk in the network can be effectively controlled and mitigated with targeted immunity for susceptible nodes. This article, based on the theory of complex networks and the mean-field theory, takes into account the propagation delay and spreading of latent nodes. Building a D-SEIRS model for risk propagation broadens the research perspective on urban river ecological management risk propagation.

Keywords: complex networks; ecological management engineering; risk; SRIES model; urban rivers

check for updates

Citation: Xu, J.; Zhu, J.; Xie, J. Study on the Evolution of Risk Contagion in Urban River Ecological Management Projects Based on SEIRS. *Water* **2023**, *15*, 2622. https://doi.org/10.3390/w15142622

Academic Editor: Achim A. Beylich

Received: 26 June 2023
Revised: 16 July 2023
Accepted: 17 July 2023
Published: 19 July 2023

1. Introduction

The urban river ecological governance project is a relatively complex project. When construction projects are developed, designed, constructed, and accepted, risks are present at every stage. The urban river ecological management project's risk network is a complex system with complex interactions between the risk factors, forming a complex network structure [1]. The interaction path between the risk factors provides a path for the spread of risk in the network. When a risk factor presents a risk state, it may spread the risk to its associated risk factors through interaction [2]. This will lead to the spread of risks across the entire network and may have a chain reaction and amplification effect, leading to the outbreak of risks in projects. It is therefore necessary to study the mechanism of risk propagation in urban river ecological management projects so that project risks can be managed and high-quality urban river ecological management projects can be developed and constructed by the departments involved in the project in a quality manner.

Transmission dynamics is a theoretical approach to the quantitative study of infectious diseases. It is possible to study the transmission dynamics of complex networks using mathematical models of epidemic transmission. At present, research on risk transmission is attracting extensive attention from scholars in different fields, including information transmission, internet rumor transmission, financial market risk transmission, corporate risk transmission, rumor transmission risk, and public opinion transmission risk, in addition to infectious diseases. Wang et al. [3] proposed the SIRaRu model and demonstrated that there is a diffusion threshold for all of them in the network and that the network topology has a significant effect on rumor propagation. Tian et al. [4] designed the super SIC model and its evolutionary law based on an opinion super network containing social, environmental, psychological, and opinion sub-networks, drawing on the modeling ideas of the SIR model and introducing rumor clarifiers. Jeon et al. [5] found that multinational banks can transfer financial risk from their parent companies to their foreign subsidiaries through their internal funding markets. Mingyuan et al. [6] used the SIR model as the basic prototype to construct a transmission model SIR-C applicable to unsafe behaviors of workers in construction and explored the transmission characteristics and intervention effects of unsafe behaviors. A model of time-encroaching behavior has been proposed by He et al. [7], which discusses the propagation of time-encroaching behavior in BA scale-free networks, ER random networks, NW small-world networks, and WS small-world networks. This scholars' research has also linked the spread of rumors to the topological nature of social networks. An analysis of rumor transmission on small-world networks was conducted by Zanette [8]. Rumors are spread through pairwise interactions between the purveyor and others through the crowd. When interacting with another spreader or choker, the ignorant person may become infected and spread the rumor. SIR reflects the interactions between rumors and has made advances and applications in the field of information dissemination [9,10]. Since the SIR model was proposed by Kermack and McKendrick in 1927 in their study of the Black Death epidemic in London and its optimization in 1932 with the SIS model [11], researchers have proposed models such as SIRS [12] and SEIR [13]. Lu Miao et al. [14] used a clustering algorithm to cluster the data in the key nodes, based on which an SIS model was constructed and through which the simulation of the evolution of public opinion in group social networks was completed. The results of the research show that the method accurately obtains the number of opinion propagations, search indexes, and high accuracy evolutionary simulations. By constructing a SEIS model with a defined latent period, Li et al. [15] demonstrated that the latent endemic equilibrium point is local and progressive. Yu [16] and Khalkho [17] established the SEIRS model, which is infectious and rehabilitative during both the incubation and infection periods. An infectious disease model was developed by Deng and colleagues [18] to study the transmission mechanism of online mass events.

Domestic and foreign scholars have made a large number of contributions to the study of risk transfer, as well as infectious diseases. Literature analysis shows that only a few scholars have applied the infectious disease model to the risk transfer of construction projects, especially for urban river ecological management. The mechanism of the transmission of infectious diseases is that the source of the disease enters the susceptible population through a certain pathway, which in turn spreads among the population, and the infected person shows different physiological responses depending on his or her immunity [19]. Risk transmission in urban river ecological management projects is similar to infectious disease transmission. During the project construction process, the project risks are easily transmitted to the downstream participants, there is a certain latent period before the project risks occur, and they have been in a hidden state before the risks are revealed. A complex set of internal and external influencing factors, coupled with a dynamic transmission process, constantly changes the node state, making the project risk transmission appear complex. The risk contagion SEIRS model constructed by Xiao Qin et al.'s [20] study on the risk propagation mechanism of amphibious seaplane take-off and landing safety has implications for the research conducted in this paper. Some of the existing studies on infectious disease models have been carried out from the perspectives of both the infectiousness of latency and the delayed nature of risk transmission, respectively. In urban river ecological management projects, the impact of infectious latency and delayed risk transmission on the systemic risk transmission dynamics has not yet been examined.

This article begins by studying the risk evolution in urban river ecological management projects. Based on the existing research, it considers the internal mechanism and influencing factors of risk transmission, as well as the dynamic evolution process of risk. Using principles from infectious disease dynamics, a project risk transmission model based on SEIR is established. By solving the transmission threshold, the analysis and simulation examine the impact of the risk delay time and risk infection rate on the project risk transmission process. This model provides a basis for the effective supervision and control of project participants and theoretical support for indirect supervision and mid-term and post-event supervision. Figure 1 depicts the study's general structure.

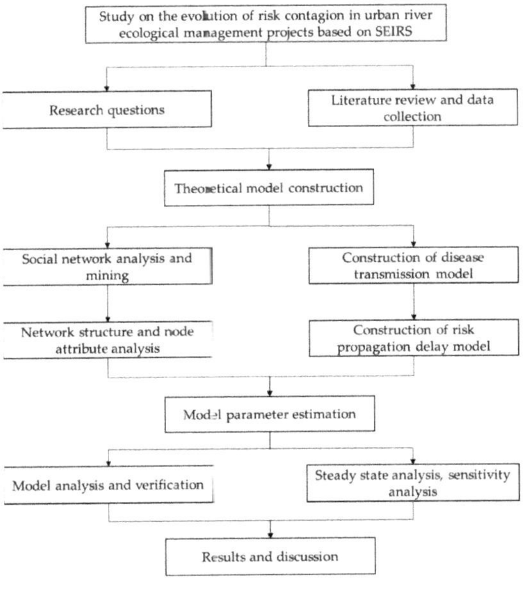

Figure 1. The overall layout of this study.

The rest of the paper if organized as follows. The proposed complex network model is constructed in Section 2 and its characteristics are analyzed. In Section 3, we construct an SEIE-based risk contagion model for urban river ecological management projects. In Section 4, the numerical simulations are presented. In Section 5, we discuss the topic. Section 6 concludes our study.

2. Complex Network of Risks

2.1. Risk Identification

Projects to manage urban rivers ecologically are complex and require high levels of technology; therefore, the risk points are not only diverse, but the factors are often coupled. As a result, project implementation risks are also usually uncontrollable and can lead to large economic losses. The data selected for this paper are mainly obtained from the following sources.

(1) The CNKI and VIP databases are searched for relevant academic papers published between 2011–2022 [21–23], from which all the possible risks in urban river ecological management projects are summarized [24,25].

(2) The list of common problems in the inspection of water conservancy project construction management (2020, Ministry of Water Resources).

(3) Accident investigation report of the safety production management platform of construction units.

Based on the literature and social research, this paper establishes a risk factor system for river ecological management projects from four aspects: project concept [26,27], project decision [28,29], project preparation [30,31], and project implementation [31,32]. Table 1 displays the findings, which reveal that there are 53 risk factors overall and 4 risk stages in the index system.

Table 1. River ecological management project risk factors.

Stage	Risk 1 Level	Risk 2 Level
1 Project concept stage	1 Political Risk	A1 Policy risk A2 Legal and regulatory risks
	2 Economic Risks	A3 Inflation risks A4 Risk of interest rate changes A5 Financing risk
	3 Natural environmental risks	A6 Hydrological and geological risks A7 Risk of meteorological conditions A8 Ecological environment risk
	4 Social Risks	A9 Sociocultural risk A10 Resident negotiated land acquisition risk A11 Social security situation A12 Public opinion
2 Project decision stage	5 Project decision risk	A13 Project approval risk A14 Basic acceptance risk before implementation A15 Risk of decision-making error A16 Risk of land change A17 Risk of incomplete collection of basic data
3 Project preparation phase	6 Bidding risks	A18 Risk of document loss A19 Risk of improper competition A20 Information leakage risk A21 Bid evaluation risk A22 Normative risk of bidding process
	7 Plan and design risks	A23 Risk of qualification of design unit A24 design schedule lag A25 There are defects, errors, omissions, and frequent changes in the design plan A26 Survey accuracy risk
	8 Prepare for risks before construction	A27 Construction site layout and technical preparation risk A28 Project contract risks A29 Risk of insufficient supply of substances (materials) and materials A30 Risk of illegal start
4 Project implementation phase	9 Construction personnel risk	A31 Technical water risk A32 Weak security awareness A33 Employee qualification risk A34 Risk of construction personnel slowing down
	10 Construction technical risks	A35 (construction) drawings improper design risk A36 Engineering and technical risks A37 Construction machinery and equipment condition risk A38 Cross operation condition risk A39 Risk of construction accidents

Table 1. *Cont.*

Stage	Risk 1 Level	Risk 2 Level
4 Project implementation phase	11 Construction management risks	A40 Safety management risks A41 Coordination risks of participating parties (including technical disclosure) A42 Rationality of construction organization design A43 Plan Adjustment and engineering change risk A44 Contract management and enforcement risks A45 Risk of organizational structure setup confusion A46 Manage permission risk
	12 Construction duration factor risk	A47 Certification period A48 Construction period A49 Risk of construction delay
	13 Completion acceptance risk	A50 Risk of file transfer not in place A51 Quality assessment risk A52 Audit risk A53 Risk of cost overruns

2.2. *Complex Network Construction and Characterization*

2.2.1. Construction of Risk Networks

There are nodes and edges in complex networks that represent the influences and their interrelationships. In this way, a complex network can be constructed objectively. Empirical network construction, time series network construction, and correlation coefficient network construction are common methods of network construction [33]. A questionnaire and other forms of research and judgment are used in the empirical network-building method to rate the influencing factors in Table 1. Experts make judgments based on their experience of the influencing factors. If the experts consider them to be relevant, then they are linked; if not, then they are not. In this study, empirical network-building is used.

The columns of the risk factor relationship data matrix are the emitters (causes) and the rows of the matrix are the affected parties (effectors); Relationships are indicated by "1" when they exist, and by "0" when they do not exist. Let there be n risk nodes in risk element set A: $A_h = (R_1, R_2, \ldots, R_h)$ is a set of risk factors for the row; $A_m = (R_1, R_2, \ldots, R_m)$ is for the set of risk elements, and b_{ij} is the binary relational data. The number of rows in the matrix is i and the number of columns is $j, i = 1, 2, 3, \ldots, n, j = 1, 2, 3, \ldots, n$.

$b_{ij} = 1$. That is, the risk element in row i has an effect on the risk element in column j.

$b_{ij} = 1$ means that the risk element in row i does not affect the risk element in column j.

The expert scoring method was used to determine the risk adjacencies for the urban river ecological management project (Table 2).

Table 2. Risk factor adjacency matrix.

	A1	A2	A3	A4	A5	A6	A7	...	A47	A48	A49	A50	A51	A52	A53
A1	0	1	0	1	0	0	0	...	0	0	0	0	0	0	0
A2	0	0	0	0	0	0	0	...	0	0	0	0	0	0	0
A3	1	0	0	1	0	0	0	...	0	0	0	0	0	0	1
A4	0	0	0	0	1	0	0	...	0	0	0	0	0	0	1
A5	0	0	0	0	0	0	0	...	0	0	0	0	0	0	0
A6	0	0	0	0	0	0	0	...	0	0	0	0	0	0	0
A7	0	0	0	0	0	0	0	...	0	0	0	0	0	0	0
⋮	⋮	⋮	⋮	⋮	⋮	⋮	⋮	0	⋮	⋮	⋮	⋮	⋮	⋮	⋮
A47	0	0	0	0	0	0	0	...	0	0	0	0	0	0	0
A48	0	0	0	0	0	0	0	...	1	0	1	0	0	0	0
A49	0	0	0	0	0	0	0	...	1	0	0	0	0	0	1
A50	0	0	0	0	0	0	0	...	0	0	0	0	0	0	0

Table 2. *Cont.*

	A1	A2	A3	A4	A5	A6	A7	...	A47	A48	A49	A50	A51	A52	A53
A51	0	0	0	0	0	0	0	...	0	0	0	0	0	0	0
A52	0	0	0	0	0	0	0	...	0	0	0	0	0	0	0
A53	0	0	1	0	0	0	0	...	0	0	0	0	0	0	0

2.2.2. Network Characteristics and Network Visualization

The MATLAB2022b is used to analyze a large amount of network parameter data for projects, which identifies the key risk factors, as well as the overall characteristics of the network. The results are presented in Table 3.

Table 3. Overall network parameter characteristics.

Parameter Name	Overall Network	Parameter Names	Overall Network
Number of nodes	53	Network diameter	7
Number of network edges	255	Network average aggregation coefficient	0.2977
Network density	0.0925	Intermediation centrality	0.0331
Network average path	2.5287	Approach centrality	0.3015
Network average	9.6226	Global network efficiency	0.5281

Based on the adjacency matrix, MATLAB2022b is used to generate a risk network topology diagram for the urban river ecological management project, as shown in Figure 2.

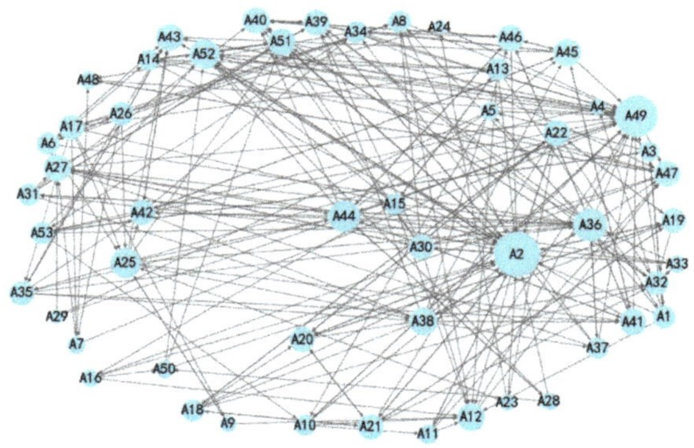

Figure 2. Risk network relationship diagram.

Degree describes the centrality of nodes in a network and is a simple, but important, concept. The degree indicates how many other nodes a node is connected to, and the degree indicates how influential that node is. Based on the directed network definition, degrees can be categorized into the following three types: in-degree, out-degree, and degree. The higher the incidence value of a node, the more vulnerable it is to external influences. The out-degree value can be represented by the number of neighboring edges that the node connects outwards; the larger the out-degree value of the node, the more likely it is to affect other nodes. The sum of the in-degree and out-degree is the degree of the node. The degree value distribution of each risk node in the urban river ecological management project was calculated and obtained.

According to the degree ranking chart in Figure 3, the nodes with larger degrees are listed in order as: A2—laws and regulations risk; A49—schedule delay risk; A36—engineering technology risk; A44—contract management and implementation risk. The risk of each of these on the other risk factors have a greater impact.

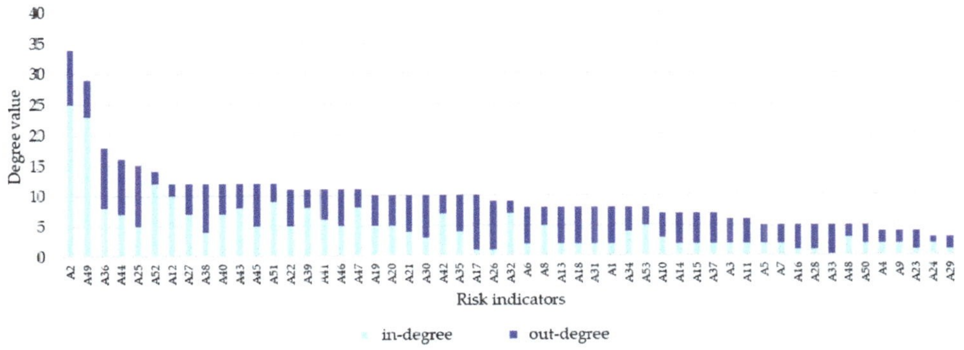

Figure 3. Ranking graph of node degree values in risk evolution network.

2.3. Analysis of Risk Propagation and Delay Effects

2.3.1. Propagation Effects

The risks associated with urban river ecological management projects can propagate from one node to downstream nodes, leading to cascading effects and amplification, ultimately impacting the overall construction project quality. Therefore, effective project risk management should take into account the risk transfer effects. An analysis of the topological relationships of urban river ecological management project risks has been conducted previously. Risk evolution in urban river ecological management projects exhibits characteristics of biological contagion, and the transmission behavior between risk factors is similar to virus diffusion. Firstly, the transmission environment is similar. Viruses are transmitted in social networks, with "people" as nodes, and the transmission channel is the contact between people. Moreover, in the process of interaction, the risk factors may make the risk nodes—which were in a stable state—become potential risk outbreaks or risk outbreak nodes. Secondly, the propagation process is similar. The propagation of a virus is carried out through the virus body to its neighbors, who then continue to propagate to their neighboring nodes, and do not spread across nodes. Risk propagation in urban river ecological management projects is also the propagation of initial risks to its neighboring nodes' risk factors, which eventually leads to outbreaks. Thirdly, the results of propagation are similar. There are similarities between the propagation characteristics and evolutionary laws of urban river ecological management projects and the spread of viruses on social networks, which makes it possible to apply the propagation model of complex networks to the evolution of risk in urban river ecological management projects.

The SEIR model is frequently used to explain how infectious diseases spread. In construction projects, this model can be used to describe the process of risk propagation. The SEIR model consists of four stages: susceptible, exposed, infectious, and recovered. In construction projects, these states can be interpreted as follows: susceptible—that is, not affected by the risk but potentially susceptible; exposed—exposed to the risk but negative effects have not yet occurred; infectious—negative effects of the risk have already appeared, such as personal injury or property damage; recovered—having recovered from the effects of the risk and no longer affected by it.

Based on the SEIR model, risk propagation in construction projects can be described as follows: Initial risk—there may be potential risks in construction projects, such as engineering quality and safety management. These risks can be considered as "infectious sources" and become the infectious state. Risk propagation—once the initial risk has a negative impact, such as an engineering accident or quality issue, these negative impacts may spread to other personnel or stakeholders, such as project participants, government regulatory departments, media, etc. These personnel or stakeholders can be considered as "susceptible" and potentially exposed to the risk of infection. Risk exposure—once other personnel or stakeholders come into contact with the negative impact, such as learning about the occurrence of engineering accidents or quality issues, they are in an exposed state. At this stage, they have not yet suffered direct negative impacts but may face future risks. Risk impact—once other personnel or stakeholders are negatively impacted, such as suffering from the effects of engineering accidents or quality issues, they are in the infectious state. At this stage, they may face negative impacts such as personal injury and property damage. Risk management—once the infectious parties are affected by the risks and recover, they enter the recovered state. At this stage, risk management measures can help prevent future risks.

2.3.2. Delay Effects

The risks associated with urban river ecological management projects can occur at any node in the complex network. Risk generation in urban river ecological management projects is possible, but the interference of other factors can delay it. The delayed effect of risk in urban river ecological management projects in this paper refers to when the risk occurs at one or some nodes in the network and is not manifested because the risk outbreak threshold of other nodes has not been reached or the relevant units have taken temporary control measures on these nodes so that they are concealed until the risk outbreak. Risks at certain nodes do not directly characterize themselves or lead to the outbreak of risks at other nodes, but rather accumulate and eventually lead to the creation of risks. This shows that risk has a delayed effect.

3. SEIRS-Based Risk Contagion Model for Urban River Ecological Management

3.1. Model Assumptions

Based on the transmission principle of infectious diseases and the risk scale-free network topology characteristics of urban river ecological governance projects, the following assumptions are made for this study.

Hypothesis 1. *The nodes in the scale-free network of urban river ecological management projects are divided into four categories: susceptible class S, latent class E (already infected with the risk but not manifested, but with the ability to transmit the risk), infected class I, and immune class R (the risk is eliminated and has some ability to resist the risk, but cannot always resist and may still become susceptible); denoting* $S_k(t)$, $E_k(t)$, $I_k(t)$, $R_k(t)$ *as the densities of the four classes of individuals in nodes of degree k at time t, and meets* $S_k(t) + E_k(t) + I_k(t) + R_k(t) = 1$, $0 \leq S_k(t), E_k(t), I_k(t), R_k(t) \leq 1$.

Hypothesis 2. β, ρ, ε, γ, μ, v, *respectively, denote the probability of conversion of the susceptible state to the latent state, the probability of conversion of the latent state to the infected state, the probability of conversion of the infected state to the immune state, the probability of change from the immune state to the susceptible state, the probability of self-healing of the latent state to the susceptible state, and the probability of self-healing of the infected state to the susceptible state; all of the above parameters being constants between 0 and 1.*

Hypothesis 3. h_1 *denotes the infection rate of latent class nodes,* h_2 *denotes the infection rate of infected class nodes,* $\Theta_1(t)$ *and* $\Theta_2(t)$ *denote the probability of association of susceptible nodes with latent class nodes and the probability of association of susceptible nodes with infected class nodes at moment t, respectively; the above parameters all take values between 0 and 1.*

Hypothesis 4. *T denotes the delay time of risk contagion in the risk-scale-free networks of the urban river ecological management project, and it is assumed that the contagion delay time of latent class nodes and infected class nodes in the network is the same.*

3.2. Construction Based on the SEIRS Model

Based on the above assumptions, the risk propagation SEIRS process for urban river ecological management projects is shown in Figure 4. A set of differential equations for the risk contagion delay model for urban river ecological management projects in a scale-free network model is constructed based on the mean-field theory, as follows.

$$\begin{cases} \frac{dS_k(t)}{dt} = -\beta k[h_1\Theta_1(t) + h_2\Theta_2(t)]S_k(t) + \mu E_{k,T}(t) + \nu I_{k,T}(t) + \gamma R_k(t), \\ \frac{dE_{k,0}(t)}{dt} = \beta k[h_1\Theta_1(t) + h_2\Theta_2(t)]S_k(t) - (\mu+\rho)E_{k,0}, \\ \frac{dE_{k,1}(t)}{dt} = -(\mu+\rho)E_{k,1}(t) + (\mu+\rho)E_{k,0}, \\ \cdots \\ \frac{dE_{k,T}(t)}{dt} = -(\mu+\rho)E_{k,T}(t) + (\mu+\rho)E_{k,T-1}(t), \\ \frac{dI_{k,0}(t)}{dt} = -(\nu+\varepsilon)I_{k,0}(t) + \rho E_{k,T}(t), \\ \frac{dI_{k,1}(t)}{dt} = -(\nu+\varepsilon)I_{k,1}(t) + (\nu+\varepsilon)I_{k,0}(t), \\ \cdots \\ \frac{dI_{k,T}(t)}{dt} = -(\nu+\varepsilon)I_{k,T}(t) + (\nu+\varepsilon)I_{k,T-1}(t), \\ \frac{dR_k(t)}{dt} = \varepsilon I_{k,T}(t) - \gamma R_k(t), \end{cases} \tag{1}$$

where $E_{k,\tau}(t), I_{k,\tau}(t)$ denote the latent and infected nodes of degree k, respectively, at $t-\tau$ moments and satisfies $E_k(t) = \sum_{\tau=0}^{T} E_{k,\tau}$, $I_k(t) = \sum_{\tau=0}^{T} I_{k,\tau}$. $\Theta_1(t) = \sum_{j=1}^{n} \frac{jP(j)E_j(t)}{\langle k\rangle}$, $\Theta_2(t) = \sum_{j=1}^{n} \frac{jP(j)I_j(t)}{\langle k\rangle}$ Letting the right-hand side of Equation (1) equal zero shows that $E_{k,0} = E_{k,1} = \cdots = E_{k,T}$, $I_{k,0} = I_{k,1} = \cdots = I_{k,T}$, so that $h = h_1\Theta_1(t) + h_2\Theta_2(t)$, simplifying Equation (1) to:

$$\begin{cases} \dot{S}_k = -\beta hkS_k + \frac{\mu}{T+1}E_k + \frac{\nu}{T+1}I_k + \gamma R_k, \\ \dot{E}_k = \beta hkS_k - \frac{\mu+\rho}{T+1}E_k, \\ \dot{I}_k = \frac{\rho}{T+1}E_k - \frac{\nu+\varepsilon}{T+1}I_k, \\ \dot{R}_k = \frac{\varepsilon}{T+1}I_k - \gamma R_k. \end{cases} \tag{2}$$

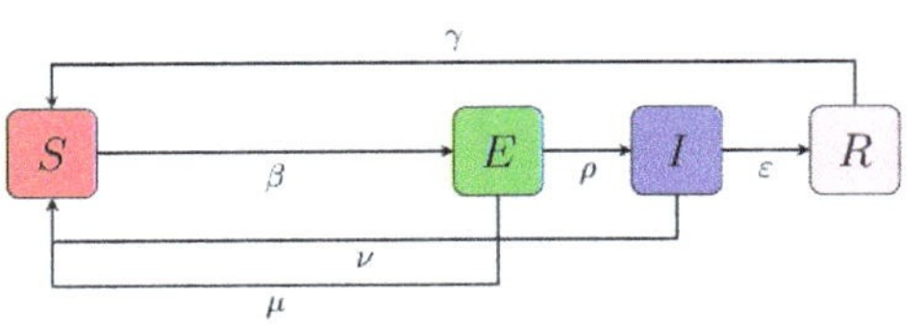

Figure 4. SEIRS model.

3.3. Immunization Strategy Construction

Assuming that the immunization ratio of susceptible nodes is a, the immunization strategy is modeled as in Figure 5, $a \in [0,1]$; with the other parameters defined as before, the differential equation for the post-immunization system can be expressed as:

$$\begin{cases} \dot{S}_k = -\beta hkS_k - aS_k + \frac{\mu}{T+1}E_k + \frac{\nu}{T+1}I_k + \gamma R_k, \\ \dot{E}_k = \beta hkS_k - \frac{\mu+\rho}{T+1}E_k, \\ \dot{I}_k = \frac{\rho}{T+1}E_k - \frac{\nu+\varepsilon}{T+1}I_k, \\ \dot{R}_k = aS_k + \frac{\varepsilon}{T+1}I_k - \gamma R_k, \end{cases} \tag{3}$$

where $h = h_1\Theta_1 + h_2\Theta_2$. According to the analysis process of the unimmunized contagion model, the effective contagion rate and steady-state density of the immunized network can be found in the same way, without going over the solution process here; the contagion threshold h_c and the steady-state density of infected nodes I_1 of the immunized network are as follows:

$$h_c = \frac{(a+\gamma)\langle k \rangle}{\gamma(T+1)\langle k^2 \rangle}, \tag{4}$$

$$I_1 = \frac{2\rho\left\{m\gamma\beta(T+1)[h_1(\varepsilon+\nu)+h_2\rho]\left(e^{\frac{(\nu+\varepsilon)(\mu+\rho)(a+\gamma)}{m\beta\gamma(T+1)[h_1(\varepsilon+\nu)+h_2\rho]}}-1\right)-(\nu+\varepsilon)(\mu+\rho)(a+\gamma)\right\}}{m\beta[h_1(\varepsilon+\nu)+h_2\rho][(\rho+\varepsilon+\nu)(T+1)\gamma+\varepsilon\rho]\left(e^{\frac{(\nu+\varepsilon)(\mu+\rho)(a+\gamma)}{m\beta\gamma(T+1)[h_1(\varepsilon+\nu)+h_2\rho]}}-1\right)^2}. \tag{5}$$

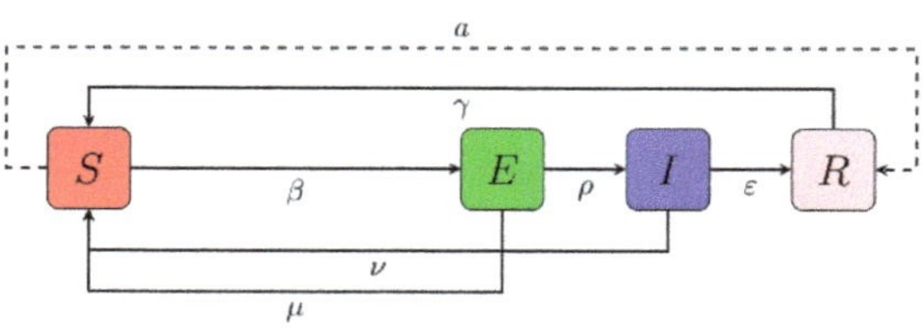

Figure 5. SEIRS model target immunity model.

4. Numerical Simulation

4.1. Initial Model Parameter Setting

Based on the previous assumptions, there is β probability of the conversion of the susceptible state to the latent state, ρ probability of the conversion of the latent state to the infected state, ε probability of the conversion of the infected state to the immune state, γ probability of the conversion of the immune state to the susceptible state, μ probability of the conversion of the latent state to the susceptible state, ν probability of the conversion of the susceptible state to the infected state. Combining the actual situation of the urban river ecological management project with expert opinions and research [34], the above parameters are assigned the values: 0.4, 0.6, 0.2, 0.1, 0.1, 0.2.

4.2. Model Dynamics Simulation

Considering the characteristics of urban river ecological management projects, project risks will persist. In this regard, we conducted dynamic analysis of the risk network for $N = 1500$, $N = 50$, and $N = 10$.

The parameters are as follows:

$$\beta = 0.4, \rho = 0.6, \varepsilon = 0.2, \gamma = 0.1, \mu = 0.1, \nu = 0.2, T = 5, h_1 = 0.1, h_2 = 0.3, m = 3$$

From Figure 6a–c, it can be observed that in cases where the network size is 1500, 50, and 10, respectively, the instantaneous density of different-scale network models eventually reaches a steady-state density (i.e., parallel to the time axis). Moreover, based on the simulations conducted above, it can be concluded that as the evolution time of the risk network increases, the nodes tend to balance within their respective groups. Under these parameter settings, the equilibrium achieved is a balance of risk outbreak rather than the disappearance of network risks. This validates the correctness of the stability analysis of the equilibrium points in Appendix A.1.1; namely, if network risks are not effectively controlled, risks will persist. The theoretical analyses further support the need for timely risk control and management by project stakeholders involved in urban river ecological management projects. Only by doing so can the continuous presence of risks on the network be prevented, thereby avoiding potential project losses.

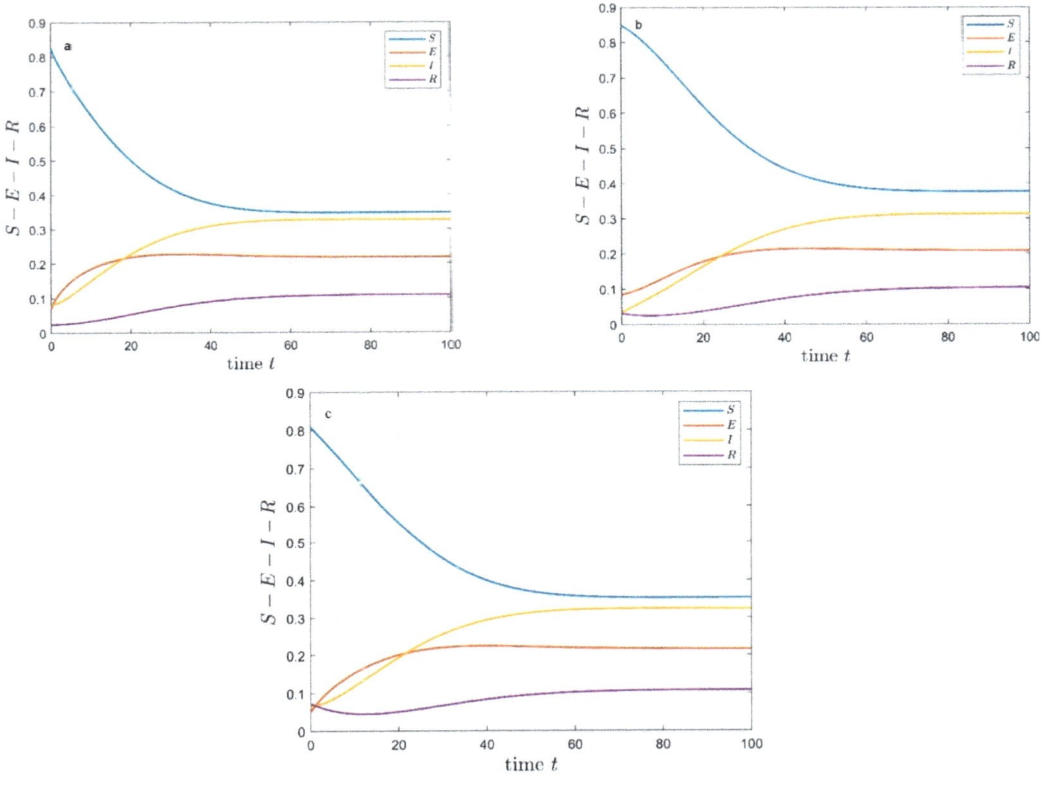

Figure 6. Dynamic evolution of the network model over time: (**a**) Dynamic evolution of the risk network for N = 1500. (**b**) Dynamic evolution of the risk network for N = 50. (**c**) Dynamic evolution of the risk network for N = 10.

4.3. Dynamic Simulation Analysis of the SEIRS Model

4.3.1. Effect of Delay Time and Network Size on Propagation Thresholds

The above analysis shows that $\lambda = \left(\frac{h_1(\nu+\varepsilon)}{\rho} + h_2\right)\frac{\beta\rho}{(\nu+\epsilon)(\mu+\rho)}$, i.e., the effective contagion rate of risk is influenced by the probability of $\beta, \rho, \varepsilon, \gamma, \mu, \nu$. The infection rates are h_1 and h_2 for latent and infected nodes.

In this study, the parameters are set as constants to analyze the pattern of the effective propagation rate with the network size. From $\lambda_c = \frac{\langle k \rangle}{(T+1)\langle k^2 \rangle}$, the propagation threshold of the security risk evolution network is related to the average degree $\langle k \rangle$, $\langle k^2 \rangle$ and delay time T of the network. When the size of the network is large enough, the average degree of the network $\langle k \rangle \approx 2m$, $\langle k^2 \rangle \approx 2m^2 \ln K_c/m$, $K_c \approx mN^{1/2}$, where K_c is the network, the maximum value of neutrality, m, is the minimum number of connected edges in the network, and N is the total number of nodes in the network. The transformation $\lambda_c = \frac{2}{m(T+1)\ln N}$.

It is clear from the previous analysis, and $\langle k \rangle \approx 9$, that the minimum connected edge $m = 3$ in this paper. The relationships between the propagation threshold λ_c, T, and N are shown in Figure 7.

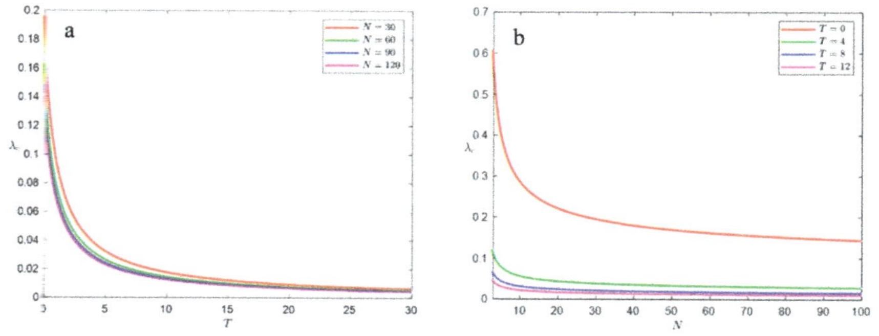

Figure 7. Propagation threshold λ_c Relationship between, T, and N. (**a**) A function of delay time for the contagion threshold; (**b**) Contagion threshold as a function of network size.

Figure 7a analyzes the variation pattern of the risk propagation thresholds in the network with the risk contagion delay time for the sizes of the different networks. In networks of a certain size, the contagion threshold decreases with the increasing delay time, and when the delay time is sufficiently long, the propagation threshold approaches zero, indicating that risk contagion delays in the network eventually lead to increased risk contagion. When the delay time is certain, the propagation threshold varies according to the size of the network. The larger the network size, the lower the propagation threshold, which is in line with the propagation characteristics of scale-free networks, and it can also be obtained that the propagation threshold is almost the same for different sizes of networks when the delay time is long enough. Urban river ecological management projects involve a long project cycle and many stakeholders; therefore, in the case of a certain scale of project risks, the project risks can be detected and dealt with on time according to the law of the change of the contagion threshold with the delay time in order to avoid the project risks from not being dealt with on time, resulting in a lower risk contagion threshold and causing uncontrollable project risk contagion.

According to Figure 7b, the variation patterns of the contagion threshold are shown for different delay times and network sizes in the urban river ecological management project risk network. The risk network's contagion threshold tends to decrease as its network size increases, and the threshold with a time delay is lower than the threshold without a delay. Based on the trends, the contagion threshold of the network is almost zero when there is a time delay and the network is large enough; however, when there is no time delay and the network is large enough, the contagion threshold drops. However, the structural characteristics of the network prevent it from dropping to zero. When the network is large enough, the contagion thresholds for different delay times converge to the same value.

Figure 7a,b show that the marginal change in the project risk contagion threshold with the delay time is greater than the marginal change with the network size, indicating that the risk contagion threshold is more sensitive to the risk delay time, i.e., risk delay affects the contagion threshold more than the network size. Therefore, in the construction process of urban river ecological management projects, risks are found to be dealt with promptly to avoid the contagion of the project risks caused by the failure to deal with them promptly.

4.3.2. Effect of Delay Time on Steady-State Density

When the network size $N = 1500$, the maximum value of the contagion threshold of the network without delay $\lambda_c(\max) = 0.0912$, when the effective contagion rate of risk $\lambda = \frac{h_1}{2} + \frac{3h_2}{13}$. The relationship between λ and h_1 and h_2 is shown in Figure 8, where the intersecting line segments are $\frac{h_1}{2} + \frac{3h_2}{13} = 0.0912$.

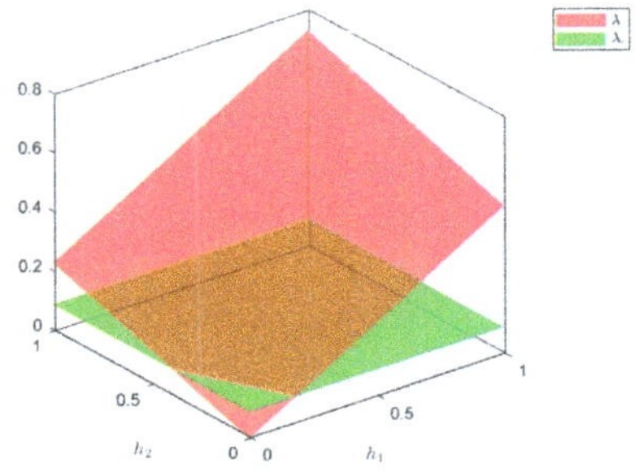

Figure 8. The variation pattern of effective transmission rate λ with h_1 and h_2.

In two contexts, with $h_1 = 0.01, h_2 = 0.03$, and $h_1 = 0.1, h_2 = 0.3$, i.e., $\lambda < \lambda_c(\max)$ and $\lambda > \lambda_c(\max)$.

A MATLAB2022b simulation was used to analyze the variation law of the steady-state density of various nodes in the risk network of engineering projects with a delay time. Figure 8 shows the specific results.

Figure 9a analyzes the network's steady-state density with a delay time when $h_1 = 0.01$, $h_2 = 0.03$, i.e., when the initial effective infection rate is less than the infection threshold. The density of the susceptible nodes in the network at the initial moment (zero contagion delay time) is 1, and the steady-state density of the latent, infectious, and immune nodes is 0. As the delay time of transmission increases, the transmission threshold of the risk gradually decreases, and it can be seen from Figure 9a that when the delay time is less than 5, the steady-state density of various nodes does not change, indicating that the effective transmission rate of the risk is always less than the transmission threshold. When the delay time is longer than 5, the steady-state density of the susceptible nodes decreases rapidly, and when the delay time is long enough, the density drops to 0. The steady-state densities of the latent and infected nodes show an increasing trend and eventually equalize, and the steady-state density of the latent nodes is close to 0.6, the steady-state density of the infected nodes is close to 0.3, and the steady-state density of the immune nodes changes less but shows an overall trend of first increasing and then decreasing to 0. The steady-state density of the immune nodes is less variable but shows a trend of increasing and then decreasing until it reaches zero. This indicates that when the delay time is greater than 5, the risk starts to explode in the network as the effective contagion rate of the risk is greater than the contagion threshold.

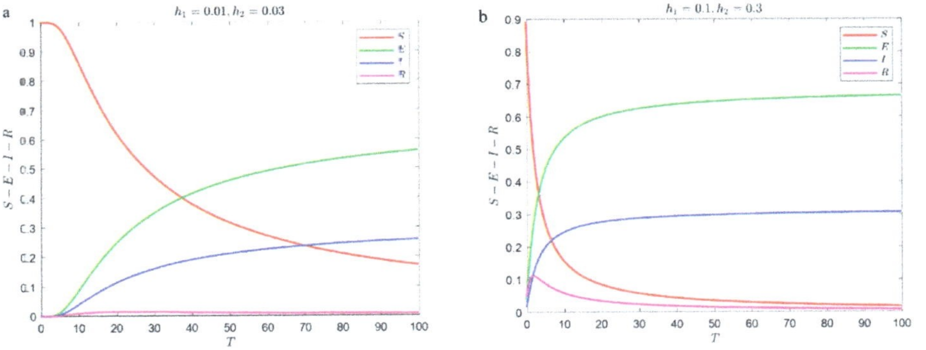

Figure 9. Variation of steady-state density with delay time T. (**a**) $h_1 = 0.01, h_2 = 0.03$ Variation of steady-state density with delay time T; (**b**) $h_1 = 0.1, h_2 = 0.3$ Variation of steady-state density with delay time T.

Figure 9b analyzes how the steady-state density of the network varies with the contagion delay time when $h_1 = 0.1$ and $h_2 = 0.3$, i.e., when the initial effective contagion rate of risk in the network is greater than the contagion threshold. The steady-state density of the four types of nodes in the network is approximated to 0.9, 0.03, 0.02, and 0.05 at the initial moment; when the risk delay effect appears, the risk will be transmitted in the network regardless of the delay time, resulting in a rapid decrease in the steady-state density of the susceptible nodes until it reaches 0. The steady-state densities of the latent and infected nodes increased to 0.68 and 0.31, respectively, and the immune nodes increased and then decreased rapidly until they approached 0.

In cases where the effective rate of risk transmission is less than the transmission threshold, a longer transmission delay is required to bring the steady-state density of each node into equilibrium, and when the effective rate of risk transmission is greater than the transmission threshold, a shorter transmission delay is required. In cases where the effective transmission rate is less than the transmission threshold, risk transmission delays play a greater role than in cases where the effective transmission rate is greater. Different values of the risk transmission rate for latent and infected nodes affect the steady-state density at the initial moment, but not at equilibrium. The results provide some theoretical guidance for the development of risk contagion control strategies for urban river ecological management projects.

4.3.3. Infection Rate Effects on Steady-State Density

The relationship between the infected nodes, latent nodes, and h_1, h_2, as well as the results of the analysis of the above steady-state density over time, show that when the density of infection is relatively small, the delay time exceeds 5 before infection occurs. Therefore, in this paper, assuming the delay time $T = 5$ for risk contagion, a MATLAB simulation was used to analyze the trend of infected nodes and latent nodes with h_1, h_2. Figure 10 illustrates the specific results.

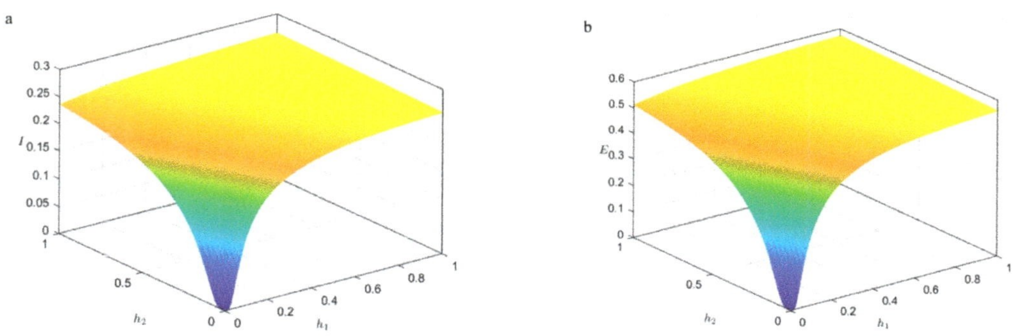

Figure 10. The relationship between infection node, latent node and h_1, h_2. (**a**) Steady-state density of infected nodes as a function of h_1 and h_2; (**b**) Steady-state density of latent nodes as a function of h_1 and h_2.

From Figure 10a, it can be seen that when the values of h_1 and h_2 are small, there is no significant change in the steady-state density of the infected nodes and because the delay time is fixed at this time, which indicates that the risk transmission threshold is certain. As a result, the value of the effective infection rate is small and lower than the propagation threshold, and the risk cannot spread in the network, so the steady-state density of the infected nodes will remain at zero. As h_1 and h_2 increase, the effective infection rate gradually increases until it exceeds the propagation threshold, and the steady-state density shows a rapidly increasing trend. The marginal change of the steady-state density of the infected nodes concerning h_1 is greater than that concerning h_2; that is, the steady-state density of the infected nodes is more sensitive to h_1. This result is because h_1 has a greater impact on the infection rate in the expression of the effective infection rate.

Based on Figure 10b, it can be observed that when h_1 and h_2 have relatively small values, the steady-state density of the latent nodes does not show significant changes. As h_1 and h_2 gradually increase, the pattern of change in the steady-state density of the latent nodes follows the same trend as the variation of the infected nodes in Figure 10a. Therefore, it can be concluded that the steady-state density of the latent nodes is more sensitive to h_1. The reason for this result is that h_1 has a greater impact on the transmission rate in the expression of the effective transmission rate.

To further explore the relationship between the steady-state density of the infected and latent nodes with h_1, h_2, and T. Taking the delay time $T = 0, 10, 20$, the variation law of the steady-state density of the infected and latent nodes with h_2 for fixed $h_1 = 0.01$, and the variation law of the steady-state density of the infected and latent nodes with h_1 for fixed $h_2 = 0.03$, respectively, were analyzed. In Figures 11 and 12, the specific results are shown.

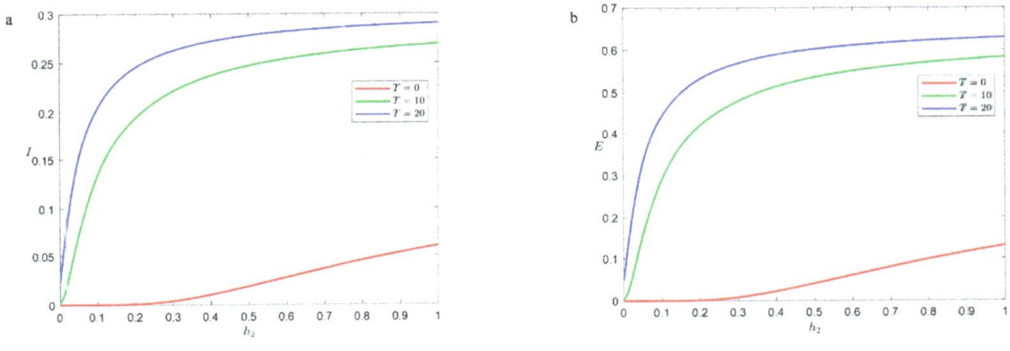

Figure 11. Steady-state density of infected and latent nodes with h_2 for $h_1 = 0.01$. (**a**) The law of steady-state density of infected nodes changing with h_2;(**b**) The law of steady-state density of Latent Nodes changing with h_2.

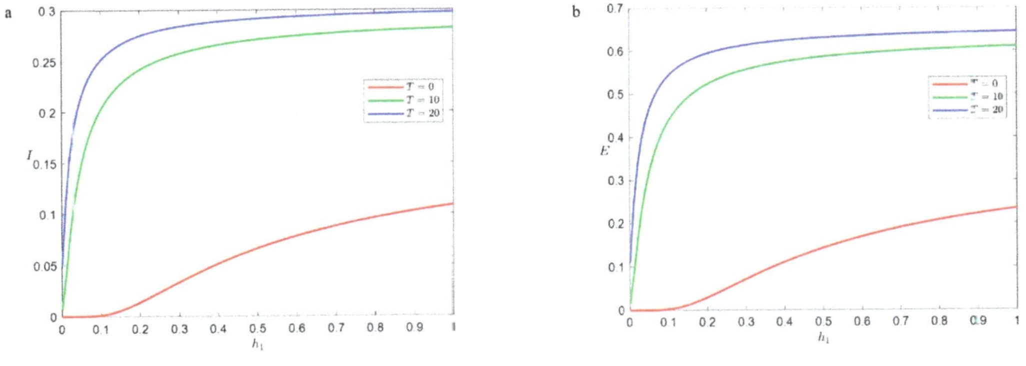

Figure 12. Steady-state density of infected and latent nodes with h_1 for $h_2 = 0.03$. (**a**) The law of steady-state density of infected nodes changing with h_1;(**b**) The law of steady-state density of Latent Nodes changing with h_1.

Figure 11a,b show that when $h_1 = 0.01$, the steady-state densities of the infected node I and the latent node E increase with the increase in h_2. By comparing the cases of the risk delay being 0, 10, and 20, it is found that the project risk delay leads to an increase in the steady-state densities of node I and node E. This, in turn, accelerates the spread of the project risk within the network. In addition, as the time units of the delay increase, the growth rate of their steady-state densities also increases.

Figure 12a,b show that when $h_2 = 0.03$, the steady-state densities of the infected node I and the latent node E increase with the increase in h_1. By comparing the cases of the risk delay being 0, 10, and 20, it is found that the changing pattern of the steady-state densities of the infected node I and the latent node E concerning h_1 is consistent with the pattern shown in Figure 10. This implies that increasing the value of h_1 in the model leads to an increase in the steady-state densities of the infected node and latent node, indicating that more individuals are infected or in a latent state in the system.

By comparing the results of Figures 11 and 12, it is observed that when parameter h_1 is fixed, the equilibrium steady-state density of the infected node I and the latent node E is smaller than the result with a fixed parameter h_2. This indicates that increasing parameter h_1 is more likely to lead to a risk outbreak in urban river ecological management projects. The equilibrium steady-state density of both the infected node I and the latent node E increases as parameters h_1 and h_2 increase. When one parameter is fixed and the other parameter is increased, it also leads to an increase in the steady-state density, and the combined effect of increasing both parameters is more pronounced. Furthermore, the latent node E exhibits a growth rate faster than the infected node I, and the maximum steady-state density of the latent node is greater than that of the infected node. This suggests that in the risk network of urban river ecological management projects, the contagiousness of the risk factors in the latent period plays a significant role in the entire risk-spreading process, and the transmission delay accelerates the propagation of risks within the network. Therefore, when devising risk control strategies for urban river ecological management projects, stakeholders should pay attention to the contagiousness of the latent nodes and the delay in risk propagation.

4.3.4. Sensitivity Analysis

Sensitivity analysis is a method used to assess the sensitivity of the model outputs to changes in the input parameters. By systematically varying the key parameters within the model and observing the resulting changes in the output, we can gain insights into the model's sensitivity to different parameters, thus enhancing our understanding of the model's behavior and predictive capabilities. In sensitivity analysis, the effects of the model outputs can be observed by changing the values of parameters such as $m, T, \beta, \rho, \varepsilon, \gamma, \nu, \mu, h_1, h_2$.

Non-normalized sensitivity formula:

$$D_x^T = \frac{\partial T}{\partial x} \tag{6}$$

Normalized sensitivity formula:

$$F_x^T = \frac{\partial T}{\partial x} \cdot \frac{x}{T} \tag{7}$$

where T is a variable, and x is a variable. In this article, T takes the values of S, E, I, R, and x takes the values of $m, T, \beta, \rho, \varepsilon, \gamma, \nu, \mu, h_1$, and h_2. Based on the two formulas above and taking parameters 3, 5, 0.4, 0.6, 0.2, 0.1, 0.2, 0.1, 0.1, and 0.3, we can obtain the results shown in the graph.

According to the non-normalized analysis in Figure 13a, the parameters $m, T, \beta, \rho, \varepsilon, \gamma, \nu, \mu, h_1, h_2$ have relatively small numerical values overall, but still have some impact on the nodes S, E, I, R in the model. According to the analysis results in Figure 13b, for node S, the parameters $m, T, \beta, \rho, \varepsilon, \gamma, \nu, \mu, h_1, h_2$ have little impact on it. For the latent node E, the parameters $m, T, \beta, \rho, \varepsilon, \nu, \mu$ have a negative influence on E, while h_1 and h_2 have a positive impact on E. Additionally, as h_1 and h_2 increase, the risk of the latent node E also continues to increase. For the infected node I, the parameters $m, T, \beta, \varepsilon, \nu, \mu$ have a negative impact on I, while h_1 and h_2 have a positive impact on I. Similarly, as h_1 and h_2 increase, the risk of the infected node I also continues to increase. Therefore, in the risk evolution study of urban river ecosystem management projects, it is crucial to strengthen the control of h_1 and h_2 as they play a significant role in controlling the spread of risks.

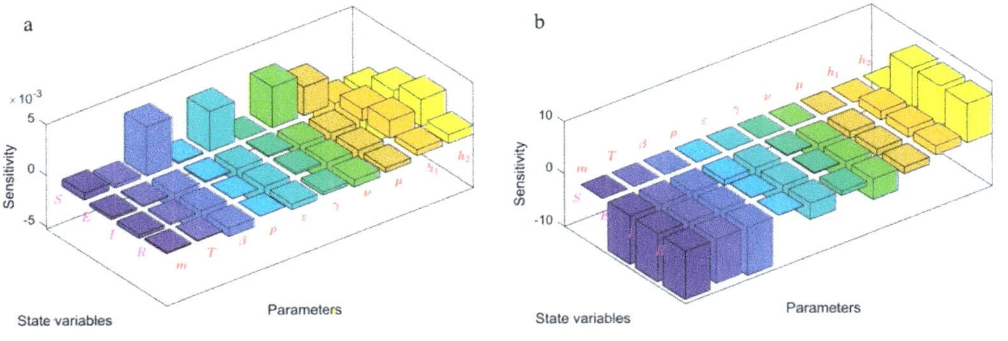

Figure 13. The sensitivity of model parameters and model states. (**a**) Non-normalized sensitivity; (**b**) Normalized sensitivity.

4.4. Immunization Strategy

Comparing the effective transmission rate after immunization with the effective transmission rate and transmission threshold before immunization, we can see that $h_c > \lambda_c$, indicating that the transmission threshold of the risk transmission model after immunization is greater than that before immunization, and the spread of risk in the network can be controlled to a certain extent, effectively avoiding the risk of transmission.

The steady-state density versus T and a was obtained by substituting $\beta = 0.4, \rho = 0.6, \varepsilon = 0.2, \gamma = 0.1, \mu = 0.1, \nu = 0.2, m = 3, h_1 = 0.01,$ and $h_2 = 0.03$ into the expression for the steady-state density of the infected node after immunization. Analyze the relationship between I and a when $T = 0, 10, 20, 30, 40, 50$; I_1 and T when $a = 0, 0.2, 0.4, 0.6, 0.8, 1$. This study normalized the immune density of the infected nodes after immunization to ensure a convincing analysis. I in the figure indicates the steady-state density of the infected nodes before immunization, i.e., the maximum value of the steady-state density of the infected nodes, and the results of the analysis are shown in Figure 14.

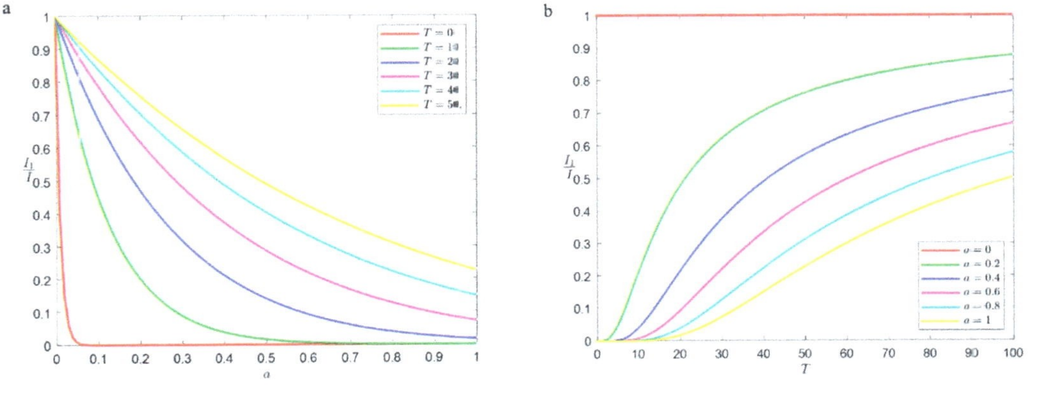

Figure 14. Relationship between relative steady-state density of risk infection nodes and immune probability and delay time. (**a**) Relative value of steady-state density of risk-infected nodes versus probability of immunity; (**b**) Relative value of steady-state density of risk-infected nodes versus delay time.

As can be seen from Figure 14a, the steady-state density decreases as the probability of the immunization of the susceptible nodes increases, suggesting that enhanced immunization is effective in controlling contagion in the risk network of urban river ecological management projects. It was observed that the relative value of the steady-state density of the network is 1 when the immunity probability of the susceptible nodes is zero, at which point the steady-state density is at its maximum. The relative value of the steady-state density is equal to the marginal change in the immunity probability of the susceptible node, indicating that the immunity probability of the susceptible node plays an important role in the relative value of the steady-state density of the infected node.

As can be seen from Figure 14b, when the immunization probability of the susceptible nodes is 0, the relative steady-state density of the infected nodes remains at 1, which is the same as before immunization. When immunizing susceptible nodes with a certain probability, the higher the immunization probability, the smaller the relative value of the steady-state density of the infected nodes, and the better the immunization effect when the delay time is held constant. For a given probability of immunization, the relative value of the steady-state density first remains constant at 0 and then increases with the increasing delay time, i.e., the immunization becomes less effective. The reason for this phenomenon is that after immunizing the nodes in the network, the contagion threshold of the risk increases. When the delay time is short, the contagion threshold of the risk is still greater than the effective contagion rate of the network and the risk cannot spread in the network, so the relative value of the steady-state density remains 0. However, as the delay time continues to increase, the contagion threshold decreases further until it is less than the effective contagion rate, at which point the risk will spread in the network, leading to an increase in the relative value of the steady-state density. The results show that the immunization of the susceptible nodes can effectively inhibit the spread of risk; in addition, the effectiveness of immunization can be enhanced by controlling the delay time of the risk in the network.

The above analysis shows that the immunization of the susceptible nodes can effectively control the propagation of risks in the risk network of urban river ecological management projects, and controlling the delay time of risks in the network can strengthen the immunization effect of the network against risks. Therefore, in the practical work of risk management for urban river ecological management projects, project participants can combine the results of the analysis of the topology of the scale-free network for the evolution of risk in urban river ecological management projects and target immunization at the nodes with the highest degree values.

5. Discussion

The research results show that the model accurately describes the law of risk propagation in urban river ecological management projects and explains the trend in delay propagation. By adjusting the values of various parameters in the model and analyzing the influence of their changes on the proportion of various nodes, the propagation law of risk in urban river ecological management projects can be derived, which provides a theoretical basis for preventing and controlling the propagation of project risks. For the application of the research results in the construction of urban river ecological management engineering projects, for the nodes of the class of easily infected risk factors, the supervision of the project needs to be strengthened so that the factors affecting the risks in urban river ecological management engineering projects are kept within a controllable range, and the probability of the risk factors that may affect urban river ecological management engineering projects being transformed into latent or infected states is minimized as far as possible. For latent category risk factors, through strengthening the risk early warning, the timely understanding of the risk changes when the early warning signals promptly invokes the emergency plan control to reduce the risk value of the risk factor to a manageable range or directly eliminate. The most direct management tools are used to transform the risk factor into an immune node for the infection category while avoiding the probability of exposure to this risk factor with the susceptible and latent risk factors and reducing the number of risk factors transformed into the infection status. The contagiousness of the risk also needs to be considered along with the latency of the risk, and when adopting an immunization strategy in a network, the most effective risk control strategy needs to be developed in conjunction with the latency of the risk.

Applying the SEIR model to urban river revitalization projects enables comprehensive and objective risk assessment, revealing the mechanisms and patterns of risk transmission. Establishing a project risk transmission model allows for in-depth research on the interactions and influences among various stakeholders, providing theoretical support for risk management and control. In practical applications, the introduction of the SEIR model can assist managers in effectively addressing the issues of risk transmission, thereby improving the smooth progress and quality outcomes of the project. Specifically, it provides the following areas of assistance: (1) Enhanced project risk management: The application of the SEIR model helps to elevate the level of project risk management. It allows for a comprehensive and objective assessment of project risks, enabling the timely identification of and response to potential risks. (2) Improved project schedule and risk control: By analyzing the risk transmission pathways, speeds, and their impact on the project schedule and costs, the SEIR model enhances the project's ability to manage progress and mitigate risks. It aids in developing effective risk management plans and facilitates timely adjustments in the schedule and resource allocation. (3) Strengthened coordination and collaboration among project stakeholders: The introduction of the SEIR model fosters better coordination and collaboration among project participants. It helps in predicting risk trends, analyzing the interactions and influences among stakeholders, optimizing resource allocation, and enhancing the overall coordination and cooperation. This ultimately improves the overall project outcomes and quality levels.

Although this study focuses on risk propagation in urban river ecological management projects, the analytical models and methods used can be applied to risk propagation research in other domains. Similar SEIRS models and network analysis methods can be employed to analyze and understand various risk propagation phenomena, such as the spread of infectious diseases, information dissemination, and influence propagation in social networks. By adjusting the model parameters and network structures, this approach can be extended to other domains that exhibit characteristics of risk propagation, including financial risk propagation, supply chain risk propagation, and aviation safety risk propagation. The flexibility of these models and methods allows for their application in diverse contexts and fields, thereby offering the potential for a comprehensive understanding and management of risk propagation processes.

6. Conclusions

Based on the complex network theory and mean-field theory, this paper develops a D-SEIRS model to predict risk propagation in urban river ecological management projects. This model takes into account both the propagation latency and latent node propagation, thus providing a new perspective on risk propagation. Using a complex network of risks for urban river ecological management, the network's overall characteristics and key risk factors in project risk transfer can be determined. By analyzing the network topologically, we can identify the risk nodes with high degree values. These risks, including A2, A49, and A36, significantly affect the other risk variables.

The results of the network stability analysis revealed that when the basic reproduction number is less than 1, there exists a risk avoidance equilibrium point for urban river ecological management project risks. There is a global asymptotically stable equilibrium point at this point. The model exhibits a risk outbreak equilibrium point, which is locally asymptotically stable, when the basic reproduction number is greater than 1. The derivation of the steady-state density demonstrates that, under specific parameter settings, the steady-state densities of various nodes in the risk network of urban river ecological management projects are influenced by the contagion rates (h_1 and h_2) and the delay times.

The simulation results indicate that in a scale-free network of project risks, the presence of risk is persistent, and the delay in risk propagation leads to a lower risk propagation threshold within the network, thereby accelerating the spread of the risk. Additionally, the decrease in the risk transmission threshold within the network, caused by the delay in risk propagation, facilitates the diffusion of network risks and the emergence of a balanced state of risk outbreak within the network. Furthermore, the steady-state densities of both the infected nodes (*I*) and latent nodes € in the risk network increase with higher effective transmission rates and longer propagation delay times. Moreover, the transmission rate of the latent nodes has a greater impact on the steady-state density of the risk nodes. According to the simulations involving the immunization of the susceptible risk nodes in the network, strengthening the immunity of the susceptible nodes can effectively control risks in the urban river ecological management network.

The study acknowledges the specific characteristics and limitations influenced by external conditions in the evolution of risks within urban river ecological management projects. In risk propagation theory, it is typically assumed that all risk nodes have the same attributes, meaning they are subject to the same probability of change due to the risk factors. However, in practical situations, the conditions influencing the risk factors tend to be complex and diverse. The next research focus is on integrating multidisciplinary knowledge and leveraging big data analysis techniques to construct more reasonable analytical models. These models will consider the heterogeneity of the nodes and the external conditions, ultimately enhancing the effectiveness and accuracy of project risk management.

Author Contributions: Conceptualization, J.Z.; Methodology, J.X. (Junke Xu); Software, J.X. (Junke Xu); Formal analysis, J.X. (Junke Xu); Investigation, J.X. (Junke Xu); Writing—original draft, J.X. (Junke Xu); Visualization, J.X. (Junke Xu); Supervision, J.X. (Jiancang Xie). All authors have read and agreed to the published version of the manuscript.

Funding: This research was supported by the Jinghe New City Jinghe Flood Control and Ecological Management Engineering Consulting Research Project (No. 107-441220098).

Data Availability Statement: The data used to support the findings of this study are available from the corresponding author upon request.

Conflicts of Interest: The authors declare no conflict of interest.

Appendix A

Appendix A.1. Equilibrium Point Stability and Steady-State Density Analysis

Appendix A.1.1. Equilibrium Point and Stability of the Model

Risk Aversion Balance Point and Stability

To obtain the equilibrium point, let the right-hand side of the equation in the Equation (A2) equal 0, and then according to $S_k + E_k + I_k + R_k = 1$. One can obtain the risk-averse equilibrium point $(1, 0, 0, 0)$ and the unique risk-burst equilibrium point $(S_k^*, E_k^*, I_k^*, R_k^*)$, where:

$$S_k^* = \frac{\gamma(\nu+\varepsilon)(\mu+\rho)}{\beta hk[(\rho+\varepsilon+\nu)(T+1)\gamma+\rho\varepsilon]+\gamma(\nu+\varepsilon)(\mu+\rho)},$$

$$E_k^* = \frac{\gamma\beta hk(\nu+\varepsilon)(T+1)}{\beta hk[(\rho+\varepsilon+\nu)(T+1)\gamma+\rho\varepsilon]+\gamma(\nu+\varepsilon)(\mu+\rho)},$$

$$I_k^* = \frac{\gamma\beta hk\rho(T+1)}{\beta hk[(\rho+\varepsilon+\nu)(T+1)\gamma+\rho\varepsilon]+\gamma(\nu+\varepsilon)(\mu+\rho)},$$

$$R_k^* = \frac{hk\rho\beta\varepsilon}{\beta hk[(\rho+\varepsilon+\nu)(T+1)\gamma+\rho\varepsilon]+\gamma(\nu+\varepsilon)(\mu+\rho)}.$$

Substituting E_k^* and I_k^* into h gives:

$$h = \left(\frac{h_1(\nu+\varepsilon)}{\rho} + h_2\right)\sum_{j=1}^{n}\frac{jP(j)}{\langle k\rangle}\frac{\gamma\beta hj\rho(T+1)}{\beta hj[(\rho+\varepsilon+\nu)(T+1)\gamma+\rho\varepsilon]+\gamma(\nu+\varepsilon)(\mu+\rho)} \triangleq f(h). \tag{A1}$$

Let the function:

$$F(h) = f(h) - h. \tag{A2}$$

It follows that the equation $F(h) = 0$ has a banal solution $h = 0$, i.e., the risk-free equilibrium point $S_k = 1, E_k = I_k = R_k = 0$. At this point, the risk-averse equilibrium point is $(E, I, R) = (0, 0, 0)$, whose Jacobi matrix is:

$$J_0 = \begin{bmatrix} -\frac{\mu+\rho}{T+1} & 0 & 0 \\ \frac{\rho}{T+1} & -\frac{\nu+\varepsilon}{T+1} & 0 \\ 0 & \frac{\varepsilon}{T+1} & -\gamma \end{bmatrix}. \tag{A3}$$

It is known that the characteristic roots are all negative, $-\frac{\mu+\rho}{T+1}, -\frac{\nu+\varepsilon}{T+1}, -\gamma$. Thus, according to the theory of differential equations, it is known that there is a risk-averse equilibrium point $(1, 0, 0, 0)$ of the system and that the equilibrium point is locally asymptotically stable.

Risk Outbreak Equilibrium and Stability Analysis

Substituting $h = 1$ into Equation (A2) gives:

$$F(1) < 1.$$

Derivation of Equation (A2) concerning h gives:

$$\begin{aligned} F'(h) &= f'(h) - 1 \\ &= \left(\frac{h_1(\nu+\varepsilon)}{\rho} + h_2\right)\sum_{j=1}^{n}\frac{jP(j)}{\langle k\rangle}\frac{\gamma^2\beta j\rho(T+1)(\nu+\varepsilon)(\mu+\rho)}{\{\beta hj[(\rho+\varepsilon+\nu)(T+1)\gamma+\rho\varepsilon]+\gamma(\nu+\varepsilon)(\mu+\rho)\}^2} - 1. \end{aligned}$$

The second derivative of Equation (A2) for h gives:

$$F''(h) < 0.$$

Thus, to obtain the equation $F(h) = 0$ to have a non-trivial solution on $0 < h < 1$, i.e., the risk burst equilibrium point $(E, I, R) = (E^*, I^*, R^*)$, it must satisfy:

$$F'(0) > 0.$$

Hence, we have:

$$\begin{aligned} \left.\frac{\mathrm{d}F}{\mathrm{d}h}\right|_{h=0} &= \left(\frac{h_1(\nu+\varepsilon)}{\rho} + h_2\right)\sum_{j=1}^{n}\frac{j^2P(j)}{\langle k\rangle}\frac{\beta\rho(T+1)}{(\nu+\varepsilon)(\mu+\rho)} - 1 \\ &= \frac{\langle k^2\rangle}{\langle k\langle}\left(\frac{h_1(\nu+\varepsilon)}{\rho} + h_2\right)\frac{\beta\rho(T+1)}{(\nu+\varepsilon)(\mu+\rho)} - 1 > 0 \end{aligned}.$$

The critical conditions are:

$$\frac{\langle k^2\rangle}{\langle k\rangle}\left(\frac{h_1(\nu+\varepsilon)}{\rho} + h_2\right)\frac{\beta\rho(T+1)}{(\nu+\varepsilon)(\mu+\rho)} = 1. \tag{A4}$$

Let $R_0 = \frac{\langle k^2\rangle}{\langle k\rangle}\left(\frac{h_1(\nu+\varepsilon)}{\rho} + h_2\right)\frac{\beta\rho(T+1)}{(\nu+\varepsilon)(\mu+\rho)}$, then R_0 is called the fundamental regeneration number, and when $R_0 < 1$, the risk vanishes and there exists a risk-averse equilibrium point $(1, 0, 0, 0)$ and that equilibrium point is locally asymptotically stable. When $R_0 > 1$, the risk will be stable after some control and there exists a unique equilibrium point of risk outbreak $(S_k^*, E_k^*, I_k^*, R_k^*)$. Substituting the fundamental regenerative number R_0 into the expression for h^* gives $S^* = \frac{1}{R}I^*$, $E^* = \frac{\nu+\varepsilon}{\rho}I^*$, $R^* = \frac{\varepsilon}{(T+1)\gamma}I^*$, at which point the Jacobi matrix of the equilibrium point of the risk outbreak is:

$$J^* = \begin{bmatrix} -\frac{(\nu+\varepsilon)(\mu+\rho)R_0}{(T+1)}I^* - \frac{\mu+\rho}{T+1} & \frac{(\nu+\varepsilon)(\mu+\rho)R_0}{(T+1)}(S^* - I^*) & -\frac{(\nu+\varepsilon)(\mu+\rho)R_0}{(T+1)}I^* \\ \frac{\rho}{T+1} & -\frac{\nu+\varepsilon}{T+1} & 0 \\ 0 & \frac{\varepsilon}{T+1} & -\gamma \end{bmatrix}.$$

Its characteristic equation can be obtained as:

$$x^3 + ax^2 + bx + c = 0, \tag{A5}$$

of which:

$$\begin{aligned} a = & \frac{(\nu+\varepsilon)(\mu+\rho)R}{(T+1)}I^* + \frac{\mu+\rho}{T+1} + \frac{\nu+\varepsilon}{T+1} + \gamma > 0, \\ b = & \left[\frac{(\nu+\varepsilon)(\mu+\rho)R}{(T+1)}I^* + \frac{\mu+\rho}{T+1}\right]\frac{\nu+\varepsilon}{T+1} + \frac{\nu+\varepsilon}{T+1}\gamma + \left[\frac{(\nu+\varepsilon)(\mu+\rho)R}{(T+1)}I^* + \frac{\mu+\rho}{T+1}\right]\gamma \\ & -\frac{(\nu+\varepsilon)(\mu+\rho)R}{(T+1)}(S^* - I^*)\frac{\rho}{T+1} > 0, \\ c = & -\frac{(\nu+\varepsilon)(\mu+\rho)R}{(T+1)}(S^* - I^*)\frac{\rho}{T+1}\gamma + \left[\frac{(\nu+\varepsilon)(\mu+\rho)R}{(T+1)}I^* + \frac{\mu+\rho}{T+1}\right]\frac{\varepsilon}{T+1}\gamma \\ & + \frac{(\nu+\varepsilon)(\mu+\rho)R}{(T+1)}I^*\frac{\rho}{T+1}\frac{\varepsilon}{T+1} > 0. \end{aligned}$$

Therefore, we have:

$$ab - c > 0.$$

According to the Routh–Hurwitz stability criterion, the real part of all the characteristic roots of the characteristic Equation (A5) are all negative, and the theory of differential equations shows that there is a risk outbreak equilibrium point (S^*, E^*, I^*, R^*) in the system and that the equilibrium point is locally asymptotically stable. Let $\lambda = \frac{\beta[h_1(\nu+\varepsilon)+h_2\rho]}{(\nu+\varepsilon)(\mu+\rho)}$, $\lambda_c = \frac{\langle k \rangle}{(T+1)\langle k^2 \rangle}$, where λ is the effective propagation rate of risk in the network and λ_c is the risk contagion threshold. It follows that when $n \to +\infty$, $\langle k^2 \rangle \to +\infty$, $\lambda_c \to 0$, i.e., a very small contagion rate of risk in a scale-free network can also make risk persistent.

Appendix A.2. Steady-State Density Analysis of the Model

In the scale-free grid, the average degree and degree distribution of the network satisfies:

$$P(k) = 2m^2k^{-3}, \langle k \rangle = \int_m^{+\infty} kP(k) = 2m, \tag{A6}$$

where m is the minimum number of connections in the network, substituting Equation (A6) into $h = h_1\Theta_1(t) + h_2\Theta_2(t)$.

$$\begin{aligned} h &= \frac{\left(\frac{h_1(\nu+\varepsilon)}{\rho} + h_2\right)}{2m} \sum_{k=1}^{n} \frac{2m^2k^{-1}\gamma\beta h\rho(T+1)}{\beta hk[(\rho+\varepsilon+\nu)(T+1)\gamma+\rho\varepsilon]+\gamma(\nu+\varepsilon)(\mu+\rho)} \\ &= \left(\frac{h_1(\nu+\varepsilon)}{\rho} + h_2\right)m\gamma\beta h\rho(T+1) \times \int_m^{+\infty} \frac{1}{\{\beta hk[(\rho+\varepsilon+\nu)(T+1)\gamma+\rho\varepsilon]+\gamma(\nu+\varepsilon)(\mu+\rho)\}k}dk. \end{aligned} \tag{A7}$$

Integrate to obtain:

$$h = \frac{\gamma(\nu+\varepsilon)(\mu+\rho)}{m\beta[(\rho+\varepsilon+\nu)(T+1)\gamma+\rho\varepsilon]\left(e^{\frac{(\nu+\varepsilon)(\mu+\rho)}{[h_1(\nu+\varepsilon)+h_2\rho]m\beta(T+1)}} - 1\right)}. \tag{A8}$$

The proportion of infected nodes in the whole network is $I = \sum P(k) I_k(t)$, where $I_k(t)$ denotes the proportion of nodes of degree k in the steady-state infected state, combined with Equations (A6)–(A8), we can obtain I as in Equation (A9). According to Equation (A9), we can find out the steady-state density of easily infected nodes, latent nodes, and immune nodes, see Equation (A10).

$$
\begin{aligned}
I &= \sum 2m^2k^{-3}\frac{\gamma\beta hk\rho(T+1)}{\beta hk[(\rho+\varepsilon+\nu)(T+1)\gamma+\rho\varepsilon]+\gamma(\nu+\varepsilon)(\mu+\rho)} \\
&= 2m^2\gamma\beta h\rho(T+1)\int_m^{+\infty}\frac{1}{\{\beta hk[(\rho+\varepsilon+\nu)(T+1)\gamma+\rho\varepsilon]+\gamma(\nu+\varepsilon)(\mu+\rho)\}k^2}\,\mathrm{d}k \\
&= 2m^2\beta h\gamma\rho(T+1) \\
&\quad\times\left(-\frac{h[(\rho+\varepsilon+\nu)(T+1)\gamma+\rho\varepsilon]}{m\gamma^2(T+1)(\nu+\varepsilon)(\mu+\rho)[h_1(\nu+\varepsilon)+h_2\rho]}+\int_m^{+\infty}\frac{1}{\gamma(\nu+\varepsilon)(\mu+\rho)k^2}\mathrm{d}k\right) \\
&= 2m^2\beta h\gamma\rho(T+1) \\
&\quad\times\left(-\frac{h[(\rho+\varepsilon+\nu)(T+1)\gamma+\rho\varepsilon]}{m\gamma^2(T+1)(\nu+\varepsilon)(\mu+\rho)[h_1(\nu+\varepsilon)+h_2\rho]}+\frac{1}{\gamma(\nu+\varepsilon)(\mu+\rho)m}\right) \\
&= 2m\beta h\rho\left(-\frac{h[(\rho+\varepsilon+\nu)(T+1)\gamma+\rho\varepsilon]}{\gamma(\nu+\varepsilon)(\mu+\rho)[h_1(\nu+\varepsilon)+h_2\rho]}+\frac{(T+1)}{(\nu+\varepsilon)(\mu+\rho)}\right) \\
&= \frac{2\rho\gamma\left\{m\beta(T+1)[h_1(\nu+\varepsilon)+h_2\rho]\left(e^{\frac{(\nu+\varepsilon)(\mu+\rho)}{[h_1(\nu+\varepsilon)+h_2\rho]m\beta(T+1)}}-1\right)-(\nu+\varepsilon)(\mu+\rho)\right\}}{m\beta[h_1(\nu+\varepsilon)+h_2\rho][(\rho+\varepsilon+\nu)(T+1)\gamma+\rho\varepsilon]\left(e^{\frac{(\nu+\varepsilon)(\mu+\rho)}{[h_1(\nu+\varepsilon)+h_2\rho]m\beta(T+1)}}-1\right)^2}
\end{aligned}
\quad (A9)
$$

Similarly, we can obtain:

$$
\begin{cases}
S = \dfrac{\frac{2(\nu+\varepsilon)(\mu+\rho)}{m\beta(T+1)[h_1(\nu+\varepsilon)+h_2\rho]}+\left(e^{\frac{(\nu+\varepsilon)(\mu+\rho)}{[h_1(\nu+\varepsilon)+h_2\rho]m\beta(T+1)}}-1\right)\left(e^{\frac{(\nu+\varepsilon)(\mu+\rho)}{[h_1(\nu+\varepsilon)+h_2\rho]m\beta(T+1)}}-3\right)}{\left(e^{\frac{(\nu+\varepsilon)(\mu+\rho)}{[h_1(\nu+\varepsilon)+h_2\rho]m\beta(T+1)}}-1\right)^2}, \\
E = \dfrac{2\gamma(\nu+\varepsilon)\left\{m\beta(T+1)[h_1(\nu+\varepsilon)+h_2\rho]\left(e^{\frac{(\nu+\varepsilon)(\mu+\rho)}{[h_1(\nu+\varepsilon)+h_2\rho]m\beta(T+1)}}-1\right)-(\nu+\varepsilon)(\mu+\rho)\right\}}{m\beta[h_1(\nu+\varepsilon)+h_2\rho][(\rho+\varepsilon+\nu)(T+1)\gamma+\rho\varepsilon]\left(e^{\frac{(\nu+\varepsilon)(\mu+\rho)}{[h_1(\nu+\varepsilon)+h_2\rho]m\beta(T+1)}}-1\right)^2}, \\
R = \dfrac{2\rho\varepsilon\left\{m\beta(T+1)[h_1(\nu+\varepsilon)+h_2\rho]\left(e^{\frac{(\nu+\varepsilon)(\mu+\rho)}{[h_1(\nu+\varepsilon)+h_2\rho]m\beta(T+1)}}-1\right)-(\nu+\varepsilon)(\mu+\rho)\right\}}{m\beta(T+1)[h_1(\nu+\varepsilon)+h_2\rho][(\rho+\varepsilon+\nu)(T+1)\gamma+\rho\varepsilon]\left(e^{\frac{(\nu+\varepsilon)(\mu+\rho)}{[h_1(\nu+\varepsilon)+h_2\rho]m\beta(T+1)}}-1\right)^2}.
\end{cases}
\quad (A10)
$$

References

1. Xin, T.X.; Fan, L. Study on human-induced risk infection process in airport Flight area based on SEIRS model. *Ind. Eng.* **2016**, *19*, 56–63.
2. Deng, G.; Lu, F.; Wu, L.; Xu, C. Social network analysis of virtual water trade among major countries in the world. *Sci. Total Environ.* **2021**, *753*, 142043. [CrossRef] [PubMed]
3. Wang, J.; Zhao, L.; Huang, R. SIRaRu rumor spreading model in complex networks. *Phys. A Stat. Mech. Appl.* **2014**, *398*, 43–55. [CrossRef]
4. Tian, R.-Y.; Zhang, X.-F.; Liu, Y.-J. SSIC model: A multi-layer model for intervention of online rumors spreading. *Phys. A Stat. Mech. Its Appl.* **2015**, *427*, 181–191. [CrossRef]
5. Jeon, B.N.; Olivero, M.P.; Wu, J. *Multinational Banking and Financial Contagion: Evidence from Foreign Bank Subsidiaries*; SSRN: Rochester, NY, USA, 2013.
6. Mingyuan, Z.; Qin, X. Propagation model and intervention effect analysis of Unsafe Behavior of construction workers. *China Saf. Sci. J.* **2021**, *31*, 1–9. [CrossRef]
7. Jianjia, H.; Jusheng, L.; Fuyuan, X. Dynamic Modeling and Simulation of time Encroachment Propagation based on SIR Model. *Appl. Res. Comput.* **2018**, *35*, 1360–1364.
8. Zanette, D.H. Dynamics of rumor propagation on small-world networks. *Phys. Rev. E* **2002**, *65*, 041908. [CrossRef]

9. Ikeda, K.; Okada, Y.; Toriumi, F.; Sakaki, T.; Kazama, K.; Noda, I.; Shinoda, K.; Suwa, H.; Kurihara, S. Multi-agent information diffusion model for twitter. In Proceedings of the 2014 IEEE/WIC/ACM International Joint Conferences on Web Intelligence (WI) and Intelligent Agent Technologies (IAT), Warsaw, Poland, 11–14 August 2014; pp. 21–26.
10. Zhu, K.; Ying, L. Information Source Detection in the SIR Model: A Sample Path Based Approach. *IEEE/ACM Trans. Netw.* **2016**, *24*, 408–421. [CrossRef]
11. Kermack, W.O. McKendrick, A.G. Contributions to the mathematical theory of epidemics-II. The problem of endemicity. *Bull. Math. Biol.* **1991**, *53*, 57–87.
12. Xue, M.; Xu, F. Yang, Y.; Tao, Z.; Chen, Y.; Wang, S.; Yin, J.; Min, M.; Shi, D.; Yao, C. Diagnosis of sepsis with inflammatory biomarkers, cytokines, endothelial functional markers from SIRS patients. *Medicine* **2022**, *101*, e28681. [CrossRef]
13. Chernov, A.A.; Shemendyuk, A.A.; Kelbert, M.Y. Fair insurance premium rate in connected SEIR model under epidemic outbreak. *Math. Model. Nat. Phenom.* **2021**, *16*, 34. [CrossRef]
14. Lu, M.; Men, K.; Ma, Y.H. Evolution simulation of public opinion in group Social networks based on SIS model. *J. Jilin Univ. (Inf. Sci. Ed.)* **2023**, *41*, 106–111. [CrossRef]
15. Jianquan, L.; Zhien, M. Analysis of two types of SEIS infectious disease models with definite incubation period. *Syst. Sci. Math.* **2006**, *26*, 9.
16. Yu, Z.; Arif, R.; Fahmy, M.A.; Sohail, A. Self organizing maps for the parametric analysis of COVID-19 SEIRS delayed model. *Chaos Solitons Fractals* **2021**, *150*, 111202. [CrossRef]
17. Khalkho, B.R.; Kurrey, R.; Deb, M.K.; Karbhal, I.; Sahu, B.; Sinha, S.; Sahu, Y.K.; Jain, V.K. A simple and convenient dry-state SEIRS method for glutathione detection based on citrate functionalized silver nanoparticles in human biological fluids. *New J. Chem.* **2021**, *45*, 1339–1354. [CrossRef]
18. Chunlin, D.; Zhen, H.; Liu, Y. Research on the transmission and prevention of mass incidents based on SIS model. *Chin. J. Inf.* **2016**, *35*, 79–84+90.
19. Hallam, T.G.; Zhien, M. Persistence in population models with demographic fluctuations. *J. Math. Biol.* **1986**, *24*, 327–339. [CrossRef]
20. Qin, X.; Fan, L. Safety Risk communication mechanism of amphibious seaplanes. *Traffic Inf. Saf.* **2022**, *40*, 1–9.
21. Zhang, H.Y.; Liu, Y.H. A Study on Risk System of Water Conservancy and Hydropower Engineering. *Adv. Mater. Res.* **2012**, *459*, 450–453. [CrossRef]
22. He, Z.; Huang, D.; Zhang, C.; Fang, J. Toward a stakeholder perspective on social stability risk of large hydraulic engineering projects in China: A social network analysis. *Sustainability* **2018**, *10*, 1223. [CrossRef]
23. Yang, Z.; Feng, L.J. Application of Improved LEC Method in the Construction of Water Conservancy Project Hazard Evaluation. *Adv. Mater. Res.* **2014**, *1028*, 297–299. [CrossRef]
24. Meng, J.; Zhang, J. Study on risk assessment of EPC water conservancy project based on entropy weight set pair analysis. *IOP Conf. Ser. Earth Environ. Sci.* **2021**, *804*, 022045. [CrossRef]
25. Xin, W.; Wang, Y. Research on influencing factors of reservoir construction risk based on interpretative structural modeling. *World J. Eng. Technol* **2021**, *9*, 727–736. [CrossRef]
26. Sun, X. Research on risk management of engineering project. In Proceedings of the 2016 International Conference on Logistics, Informatics and Service Sciences (LISS), Sydney, Australia, 24–27 July 2016; pp. 1–5.
27. Chen, J. Risk management of large water conservancy projects. *J. Yangtze River Sci. Res. Inst.* **2012**, *29*, 15.
28. Liu, X.; Xu, C.; Zhang, Z.; Zhang, Y.; Wu, S. Study on the role of eco-environmental impact assessment in performance management assessment of large water conservancy projects. *Ekoloji* **2019**, *28*, 1165–1172.
29. Zeng, J.; An, M.; Smith, N.J. Application of a fuzzy based decision making methodology to construction project risk assessment. *Int. J. Proj. Manag.* **2007**, *25*, 589–600. [CrossRef]
30. Arabi, S.; Eshtehardian, E.; Shafiei, I. Using Bayesian networks for selecting risk-response strategies in construction projects. *J. Constr. Eng. Manag.* **2022**, *148*, 04022067. [CrossRef]
31. El-Sayegh, S.M.; Manjikian, S.; Ibrahim, A.; Abouelyousr, A.; Jabbour, R. Risk identification and assessment in sustainable construction projects in the UAE. *Int. J. Constr. Manag.* **2021**, *21*, 327–336. [CrossRef]
32. Perry, J.G.; Hayes, R. Risk and its management in construction projects. *Proc. Inst. Civ. Eng.* **1985**, *78*, 499–521. [CrossRef]
33. Heng, Y.; Wu, M.; Wen, X.; Bedon, C. Identifying Key Risk Factors in Air Traffic Controller Workload by SEIR Model. *Math. Probl. Eng.* **2022**, *2022*, 7600754. [CrossRef]
34. Yuting, W.; Chun, F.; Ke, Z.; Ming, L.; Song, X. Research on SEIRS Based Quality Risk Transmission Model and Simulation of Construction Projects. *Oper. Res. Manag.* **2020**, *29*, 214–221.

water

MDPI

Article

Deep-Learning-Enhanced CT Image Analysis for Predicting Hydraulic Conductivity of Coarse-Grained Soils

Jiayi Peng [1,2], Zhenzhong Shen [1,2,*], Wenbing Zhang [3] and Wen Song [4]

1 College of Water Conservancy and Hydropower Engineering, Hohai University, Nanjing 210098, China; pengjiayiphd@hhu.edu.cn
2 State Key Laboratory of Hydrology-Water Resources and Hydraulic Engineering, Hohai University, Nanjing 210098, China
3 College of Ocean Science and Engineering, Shanghai Maritime University, Shanghai 201306, China; zhangwb@shmtu.edu.cn
4 Hunan Institude of Water Resources and Hydropower Research, Changsha 410000, China; songwen0879@163.com
* Correspondence: zhzhshen@hhu.edu.cn

Abstract: Permeability characteristics in coarse-grained soil is pivotal for enhancing the understanding of its seepage behavior and effectively managing it, directly impacting the design, construction, and operational safety of embankment dams. Furthermore, these insights bridge diverse disciplines, including hydrogeology, civil engineering, and environmental science, broadening their application and relevance. In this novel research, we leverage a Convolutional Neural Network (CNN) model to achieve the accurate segmentation of coarse-grained soil CT images, surpassing traditional methods in precision and opening new avenues in soil granulometric analysis. The three-dimensional (3D) models reconstructed from the segmented images attest to the effectiveness of our CNN model, highlighting its potential for automation and precision in soil-particle analysis. Our study uncovers and validates new empirical formulae for the ideal particle size and the discount factor in coarse-grained soils. The robust linear correlation underlying these formulae deepens our understanding of soil granulometric characteristics and predicts their hydraulic behavior under varying gradients. This advancement holds immense value for soil-related engineering and hydraulic applications. Furthermore, our findings underscore the significant influence of granular composition, particularly the concentration of fine particles, on the tortuosity of water-flow paths and the discount factor. The practical implications extend to multiple fields, including water conservancy and geotechnical engineering. Altogether, our research represents a significant step in soil hydrodynamics research, where the CNN model's application unveils key insights into soil granulometry and hydraulic conductivity, laying a strong foundation for future research and applications.

Keywords: coarse-grained soil; hydraulic conductivity; computed tomography image segmentation; convolutional neural network; deep learning

Citation: Peng, J.; Shen, Z.; Zhang, W.; Song, W. Deep-Learning-Enhanced CT Image Analysis for Predicting Hydraulic Conductivity of Coarse-Grained Soils. *Water* **2023**, *15*, 2623. https://doi.org/10.3390/w15142623

Academic Editors: Jie Yang, Lin Cheng and Chunhui Ma

Received: 2 July 2023
Revised: 17 July 2023
Accepted: 18 July 2023
Published: 19 July 2023

1. Introduction

Coarse-grained soil, defined by particle sizes from 0.075 to 60 mm, is a prevalent component in natural environments and engineering sites. Its utility as a fundamental constituent in embankment dams, forming key parts like rockfill zones, inverted filters, and cushion layers, is particularly noteworthy. The complex granular composition and variation in particle sizes make coarse-grained soil an interesting yet challenging subject for understanding permeability characteristics. Furthermore, the hydraulic behavior of coarse-grained soil greatly affects the stability and safety of structures built upon it, especially in water conservancy projects. The soil's permeability substantially dictates its stability, load-bearing capability, and seepage deformation—vital engineering attributes. Consequently, it is of paramount importance in the design, construction, and safety during the operation of embankment dams. Deepening our understanding of the permeability attributes of coarse-grained soil and accurately forecasting its hydraulic conductivity, a metric for seepage capacity, are essential for comprehending and manipulating its seepage behavior. The implications of these findings extend into various disciplines, including hydrogeology, civil engineering, and environmental science.

Hydraulic conductivity is traditionally determined via experimental testing or numerical simulations. Empirical tests, conducted both in laboratories and in field conditions, offer valuable insights, for instance, the mixed-method [1] investigation of Nam et al. into the impact of test conditions on the hydraulic conductivity of natural coarse-grained soils. However, the accuracy of such empirical tests is susceptible to factors like specimen preparation and boundary conditions [2–8], which demand rigorous operational protocols and equipment. In contrast, numerical simulations leverage finite or discrete elements to model real-world coarse-grained soils [9–12]. However, these models often rely on parameters that are experimentally elusive, undermining their applicability and limiting experimental validation.

Computed tomography (CT) image analysis offers a promising hybrid approach, combining empirical data with simulation [13–17]. This technique's efficacy hinges on accurate CT image segmentation, a process where conventional methods such as the adaptive threshold [18,19], hysteresis threshold [20], and watershed segmentation [21,22] methods often fall short when applied to the bulk structures of coarse-grained soils. Coarse-grained soils necessitate CT image analysis that segments each soil particle in adjacent stacks, crucial for extracting parameters like particle size and shape factor, which subsequently inform the study of their impact on hydraulic conductivity. Consequently, a need exists for more refined methods that can accurately and effectively perform this complex analysis.

The advent of deep learning models, particularly convolutional neural networks (CNNs) [23,24], offers a potential solution. By categorizing or labeling each pixel in an image, these models facilitate segmentation into discrete regions or objects. With their inherent ability to learn intricate features directly from data, CNNs can efficiently manage high-resolution, multi-scale, and multi-class images. While these models have gained traction in studying the engineering properties of hydraulic and geotechnical materials, their use has primarily been on solid materials, such as concrete [25–32] and rock cores [33–37]. Comparatively, coarse-grained soils present more complex imaging tasks, necessitating extensive research into the application of deep learning models to soil CT images, accounting for variations in particle size distributions, shapes, and materials. This is paramount in tailoring the model to the unique requirements of coarse-grained soil CT image segmentation.

This paper addresses these research gaps through comprehensive CT scanning and infiltration tests on coarse-grained soils of varying grain size distributions and porosity. We study the hydraulic conductivity of blasted coarse-grained soils from the Yalong River Lianghekou Hydropower Station construction site in China. An initial coarse-grained soil hydraulic conductivity calculation model is introduced, derived from a simplified cohesionless soil model and Poiseuille's law. We propose a CNN-based method for CT image segmentation, focusing on the unique characteristics of coarse-grained soils. The model employs a U-net architecture and is guided by a novel loss function incorporating both grain size and shape perception terms. We verify the accuracy of our segmentation method against a gold standard image and the known grain size distribution of the coarse-grained-soil CT image. The results are utilized to devise a prediction formula for hydraulic conductivity based on CT image analysis. The presented approach offers an efficient, precise, and intelligent method for studying the permeability characteristics of coarse-grained soils, promising to significantly impact the optimization of water conservancy project design and construction.

2. Materials and Methods

2.1. Hydraulic Conductivity Calculation Model for Coarse-Grained Soil Based on Equivalent Simplified Model and Poiseuille's Law

A simplified equivalent model of coarse-grained soil is depicted in Figure 1, which idealizes the soil as a uniform sphere and abstracts seepage channels as parallel capillaries. Assuming the total volume and total surface area of the pore channels of the simplified model are equal to those of coarse-grained soil, the diameter of the pore channels (d_0) is given by [38]:

$$d_0 = \frac{1}{\beta}\frac{2}{3}\frac{n}{1-n}d_e \quad (1)$$

where d_e represents the diameter of ideal soil particles; n is the porosity; and β is the particle shape correction factor of coarse-grained soil.

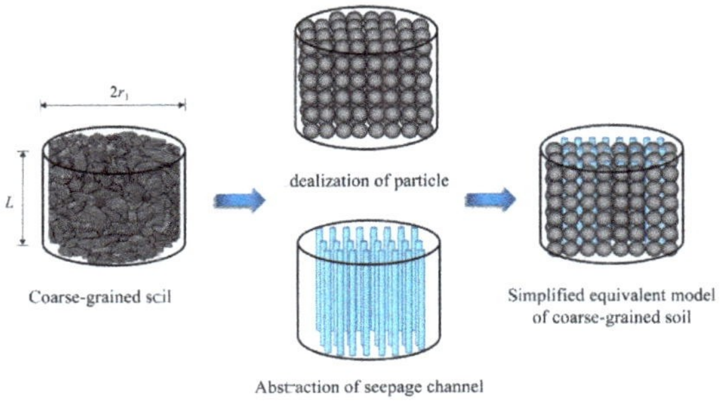

Figure 1. Simplified equivalent model of coarse-grained soil.

As depicted in Figure 2, water flow in the single pore pipe of the simplified model adheres to Poiseuille's law [39], stating that flow resistance in the pore pipe is inversely proportional to the fourth power of the pipe radius and proportional to the length of the pipe and the liquid viscosity coefficient. The average flow velocity of the single pore pipe (V) is then:

$$V = \frac{Q}{A_0} = \frac{gJ}{8\mu}r_0^2 \quad (2)$$

where Q is the pore pipe's flow rate, A_0 is the pore pipe's cross-section, g is the acceleration of gravity, J is the hydraulic gradient, μ is the coefficient of water movement viscosity, and r_0 is the radius of the pore pipe.

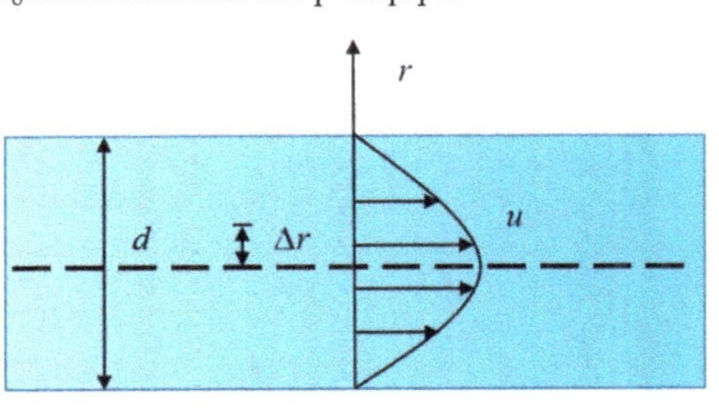

Figure 2. Schematic diagram of the flow velocity distribution in the pore tube of the simplified equivalent model of coarse-grained soil.

For a coarse-grained soil overflow section A with N_0 pore ducts, the actual overflow area is N_0A_0, which, when expressed in terms of porosity, can be equated as

$$nA = NA_0 \quad (3)$$

$$n\pi r_1^2 = N_0\pi r_0^2 \tag{4}$$

where r_1 is the radius of the coarse-grained soil sample (Figure 1 in Section 2.1).

For a coarse-grained soil simplified equivalent model comprising an N_e ideal soil particles of diameter d_e, both the mass of the coarse-grained soil specimen (M_1; Equation (5)) and the total mass of the ideal soil of the simplified equivalent model (M_e; Equation (6)) are equal.

$$M_1 = \rho_d \pi r_1^2 L \tag{5}$$

$$M_e = N_e\rho_e \cdot \frac{4}{3}\pi\left(\frac{d_e}{2}\right)^3 = \frac{\pi}{6}d_e^3\rho_e N_e \tag{6}$$

where ρ_d and ρ_e are the dry density of coarse-grained soil and the density of the ideal soil of the simplified equivalent model, respectively, and L is the stacking height of coarse-grained soil.

The discount factor α represents the degree of loss in the actual seepage channel of coarse-grained soil caused by pore connectivity and tortuosity, which increases with the complexity of the pore structure of coarse-grained soil and decreases with an increase in pore connectivity. The total surface area of the pore channel of the simplified equivalent model of coarse-grained soil is defined as:

$$S_e = 2\pi r_0^2 L N_0 = \alpha \cdot 4\pi\left(\frac{d_e}{2}\right)^2 N_1 \tag{7}$$

Combining Equations (2)–(7), we obtain the discount factor as

$$\alpha = \frac{nd_e}{3(n-1)}\sqrt{\frac{ngJ}{8\mu V}} \tag{8}$$

Inserting Equation (8) into Darcy's law [40] $V = KJ$ and simplifying it, we find the equation for the hydraulic conductivity (K) of coarse-grained soil:

$$K = \frac{g}{72\mu\alpha^2}\frac{n^3}{(1-n)^2}d_e^2 \tag{9}$$

The difficulty in using Equation (9) to calculate the hydraulic conductivity of coarse-grained soils is establishing the equations for d_e and α, which are related to equivalent particle size and particle shape and pore structure, respectively. The determination of d_e and α through tests requires a combined CT scan and constant-head permeability test for coarse-grained soils. By analyzing CT images, we can obtain the particle size, volume, and equivalent volume sphere diameter of each particle in the specimen. The hydraulic conductivity of the specimen can be obtained by substituting the hydraulic conductivity of the specimen and d_e into Equation (9). For some coarse-grained soil specimens with a different particle size distribution and porosity, a series of discount factors can be obtained by the above method. Further details about the combined CT scan and permeability test for coarse-grained soils and the method of CT image analysis for coarse-grained soils are outlined in the following section.

2.2. Materials

The coarse-grained soil used for the experiment was sourced from blasted tuff at the construction site of the Lianghekou Rockfill Dam, with a grain size ranging between 1 and 20 mm (Figure 3). Located in Sichuan Province, Southwest China, the 295 m high Lianghekou Rockfill Dam serves as the largest hydropower project in the Tibetan region.

(a) (b)

Figure 3. Materials. (**a**) Materials used for the tests. (**b**) The construction site of the Lianghekou core-wall rockfill dam.

2.3. Laboratory Test

A combination of CT scanning and constant-head permeability tests were carried out on the coarse-grained soil samples. Initially, the specimens were prepared inside a resin-made permeameter, conforming to the test numbers and porosities outlined in Table 1, and following the particle-size distribution curves detailed in Figure 4. The particle-size distribution curves of each specimen (Figure 4) aligned with the original grading of the material from the Lianghekou core-wall rockfill dam quarry, with specimen S1 and S7 representing the upper and lower envelopes, respectively. Table 1 lists the minimum porosity of each specimen under the specified particle size distribution. The specimens, each cylindrical with a diameter of 10 cm and a filling height of 8.5 cm, totaled 12 in number.

Table 1. Coarse-grained soil sample IDs and porosities.

Sample ID	**S1**	**S2**	**S3**	**S4**	**S5**	**S6**
Porosity	40%	30%	40%	40%	38%	35%
Sample ID	**S7**	**S8**	**S9**	**S10**	**S11**	**S12**
Porosity	38%	35%	35%	32%	30%	35%

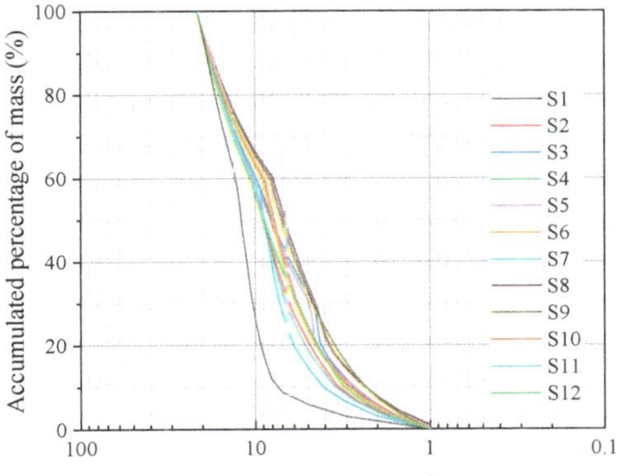

Figure 4. Particle size distribution of coarse-grained soil specimens.

The non-uniformity coefficient (C_u) and the curvature coefficient (C_c) of the particle size distribution curve serve as key parameters in evaluating soil grading. A soil grade is considered good when $C_u \geq 5$ and C_c lies between 1 and 3. Conversely, if the soil grading is poor, it can cause pipe surge phenomena due to the lack of intermediate-sized soil particles.

The formulae to calculate C_u and C_c are given by Equations (10) and (11), respectively, with the computation results for each specimen listed in Table 2.

$$C_u = \frac{d_{60}}{d_{10}} \tag{10}$$

$$C_c = \frac{d_{30}^2}{d_{10}d_{60}} \tag{11}$$

where, d_{10}, d_{30} and d_{60} represent the particle sizes corresponding to the 10%, 30%, and 60% mass accumulation percentages in the particle size distribution curve of the coarse-grained soil.

Table 2. Non-uniformity coefficient and curvature coefficient of particle size distribution of coarse-grained soil specimens.

Sample ID	S1	S2	S3	S4	S5	S6
C_u	1.65	3.05	3.58	3.66	3.68	3.05
C_c	1.07	1.20	0.80	1.06	0.89	1.29
Sample ID	**S7**	**S8**	**S9**	**S10**	**S11**	**S12**
C_u	2.38	3.36	3.52	3.11	2.93	3.45
C_c	1.34	1.11	1.14	1.31	1.27	0.98

All specimens, as per Table 2, displayed C_u values less than 5 and C_c values between 0.80 and 1.34, indicating poor grading across all specimens.

Specimens were prepared in the permeameter and scanned using a diondo d2 industrial high-resolution nano-focus CT machine from Germany. The machine has a spatial resolution of 96 μm and scans in a bottom-up direction, creating 1333 scanned slices for each coarse-grained soil specimen. Following this, a constant-head permeability test was conducted on the specimens, following these steps:

1. A settlement measurement device was installed at the top of the coarse-grained soil specimens to prevent seepage deformation during the test.
2. Aerated water was used to negate the impact of air bubbles on the percolation volume.
3. Before the test, the specimen was saturated with bottom-up exhaust under a lower head and soaked for over 8 h to eliminate the influence of non-saturation on the permeability test results.
4. The test head from the starting hydraulic slope dropped from 0.05 to 0.30, loaded step by step, with each head level loaded for 20 min before measuring the overflow in the permeameter and recording it.
5. The next level of head was loaded only when the overflow in the unit time remained unchanged. This process continued until the test concluded.

2.4. Coarse-Grained Soil CT Image Segmentation Method Based on Convolutional Neural Network

In this research, a convolutional neural network-based CT image segmentation method for coarse-grained soil is proposed, implemented using a Python program that we wrote ourselves. The details are as follows:

2.4.1. U-Net Structured Convolutional Neural Network

Ronneberger et al. [41] proposed a novel Convolutional Neural Network (CNN) named U-Net, a significant advancement in robust and efficient image segmentation. The unique U-shaped structure of U-Net, reminiscent of an autoencoder but exhibiting unique characteristics, divides the network into two separate segments. The schematic diagram of U-Net with an input image resolution of 800 × 800 is shown in Figure 5.

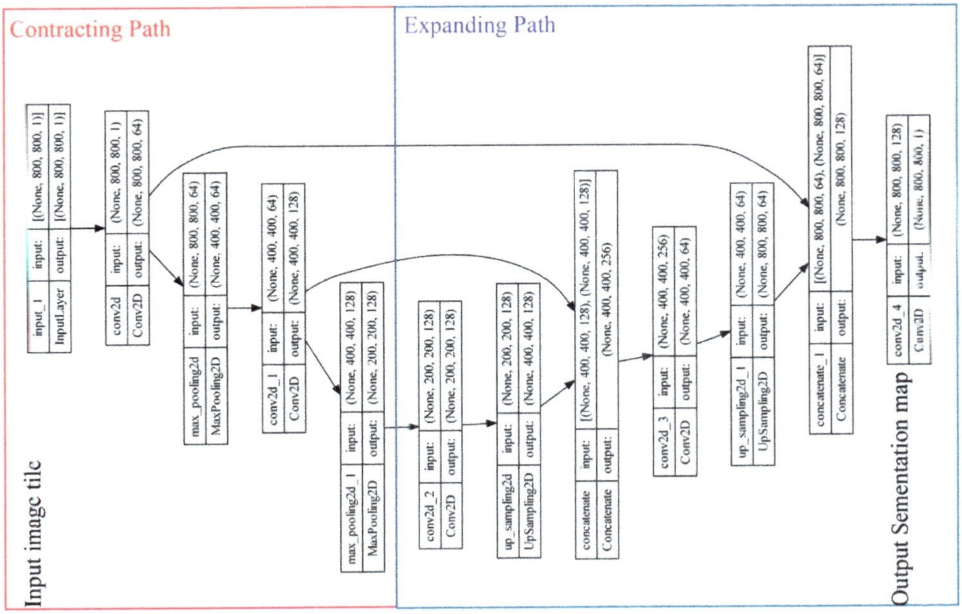

Figure 5. Convolutional neural network with U-net structure (input image resolution of 800 × 800).

The first segment, known as the Contracting or Downsampling path (forming the left side of the 'U'), comprises dual 3 × 3 convolutions, each succeeded by a rectified linear unit (ReLU), and a subsequent 2 × 2 max pooling operation with a stride of 2 for downsampling Notably, each step in this downsampling operation doubles the number of feature channels Conversely, the Expansive or Upsampling path (making up the right side of the 'U') begins with an upsampling of the feature map, followed by a 2 × 2 convolution, known as an "up-convolution". This up-convolution is concatenated with the corresponding feature map from the downsampling path and is succeeded by dual 3 × 3 convolutions, each followed by a ReLU. This innovative design of U-Net provides a streamlined and efficient process for image segmentation. 'None' in Figure 5 indicates that the image at that layer can have any batch size.

2.4.2. Loss Function

The loss function plays a vital role in CNN-based CT image segmentation. This utility function quantifies the discrepancy between predicted segmentation results and the actual or "ground truth" segmentations. Crucially, it allows for model performance evaluation and guides the optimization process, aiming to minimize this loss function.

The choice of an appropriate loss function is essential. In the context of coarse-grained soil CT image segmentation, attention is given to the shape and size of particles. Therefore, the loss function in this code includes shape perception and particle size perception terms.

The shape perception term uses the image similarity index *IoU* as a judgment metric to compute the shape similarity between the predicted segmentation and the actual segmentation:

$$ShapeLoss(A, B) = -IoU \tag{12}$$

where A and B represent the masks of the binary segmentation results of the actual image and the predicted image, respectively, and IoU is the metric for evaluating the similarity between the actual image and the predicted image. The calculation formula for IoU is:

$$IoU = \frac{|Predicted \cap Ground\ Truth|}{|Predicted \cup Ground\ Truth|} \tag{13}$$

The particle size perception term's loss function employs the radius of the equivalent circle of the target particles in the binary segmentation of the true and predicted images as a judgment metric:

$$DiameterLoss(A, B) = \frac{|EqDiameter(A) - EqDiameter(A)|}{EqDiameter(A)} \tag{14}$$

where $EqDiameter$ represents the equivalent circle radius:

$$EqDiameter = \sqrt{\frac{4|S|}{\pi}} \tag{15}$$

where S is the area of the target particle.

The loss function combines the standard cross-entropy loss with the shape-aware and diameter-aware terms, enabling the loss-functional supervised model to focus on capturing the shape and size of particles by introducing a weighting factor:

$$L(y, \hat{y}, A, B) = CE(y, \hat{y}) + \alpha \cdot Shapeloss(A, B) + \beta \cdot DiameterLoss(A, B) \tag{16}$$

where α and β are weight coefficients that ensure a balance between standard cross-entropy loss and additional geometric constraints, which are adaptively adjusted by model learning. CE is the cross-entropy loss function [42]:

$$CE(y, \hat{y}) = -sum(y \cdot \log(\hat{y})) \tag{17}$$

where y is ground truth image, and $\hat{y}$ is predicted image.

2.4.3. Workflow of Convolutional Neural Network Segmentation Model

The U-Net CNN workflow designed for the segmentation of coarse-soil CT images is illustrated in Figure 6. This workflow includes six primary steps: preprocessing, generating training and validation sets, training the U-Net model, validation, post-processing, and evaluation.

Stage 1: Preprocessing—initially, anisotropic diffusion filtering is employed to denoise all CT images of the coarse-grained soil samples. After this, images are resampled to achieve a pixel resolution of 800 × 800 per CT image, which aligns with computational memory capabilities. Lastly, all CT images are normalized.

Stage 2: Generating training and validation sets—the VGG Image Annotator (VIA) is used to create a ground truth image set via annotations, aiding the CNN in distinguishing between soil particles and pores. The CT images are then divided into a training set (70% of the images), a validation set (15% of the images), and a testing set (remaining 15%), to facilitate model training and performance evaluation.

Stage 3: U-net model training—the training set is used to train the U-net model, aiming to minimize the loss function Equation (16), thereby enabling it to accurately predict the segmentation of coarse soil CT images.

Stage 4: Validation—the model's performance is validated during or after the training process using the validation set to avoid the overfitting of the training data.

Stage 5: Post-processing—post-processing involves assigning a '0' value to pixels corresponding to particle parts and a '1' value to pore parts in the CT image. This process results in a binary image output for the segmentation results.

Stage 6: Evaluation—the model's performance is evaluated using metrics such as Intersection over Union (IoU), Precision, Recall, Accuracy, and Specificity, as detailed in Table 3. If all these parameters exceed 0.95, the training is concluded, and the model, along with its weight coefficients, is saved. If not, the model returns to Stage 3 for retraining.

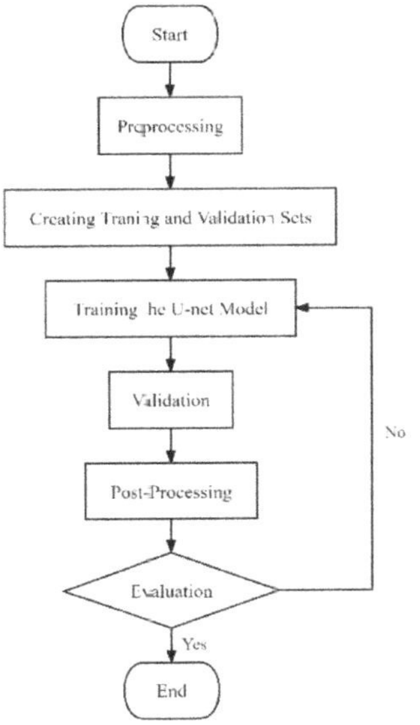

Figure 6. Workflow of coarse-soil CT segmentation using U-net CNN.

Table 3. Metric to evaluate the performance of model.

Metric	Expression [43]	Range
IoU	$IoU = \frac{\lvert Predicted \cap Ground\ Truth \rvert}{\lvert Predicted \cup Ground\ Truth \rvert}$	Metric is between 0 and 1, and the closer it is to 1, the better the model performs.
Precision	$Precision = \frac{True\ Positives}{True\ Positives + False\ Positives}$	
Recall	$Recall = \frac{True\ Positives}{True\ Positives + False\ Negativies}$	
Accuracy	$Accuracy = \frac{True\ Positives + True\ Negativies}{Total\ Predictions}$	
Specificity	$Specificity = \frac{True\ Negativies}{True\ Negatives + False\ Positives}$	

To evaluate our model's effectiveness, we used a set of metrics as detailed in Table 2. Intersection over Union (IoU) is a key parameter, quantifying the overlap between predicted and ground truth areas, thus providing insights into its localization accuracy. Precision measures the model's ability to avoid false positives, highlighting the reliability of the model's positive predictions. Conversely, Recall evaluates the model's ability to identify all positive instances, offering insights into its skill in avoiding false negatives. Accuracy encapsulates overall model performance, defined as the proportion of correct predictions over the entire dataset, although interpretability may be impaired in situations with imbalanced datasets. Lastly, we use Specificity to indicate the model's skill in correctly identifying negative instances, contributing to minimizing false alarms. Taken together, these metrics offer a comprehensive evaluation of the model's performance, illuminating various aspects of its classification competency.

2.5. Geometric Characterization of Coarse Soil Particles via CT Image Analysis

This section delineates the procedure to build three-dimensional (3D) models of soil particles following the segmentation of coarse-grained soil CT images using a U-net structured CNN. Furthermore, it describes the methodology employed to calculate the major axis, minor axis, volume, and isovolumetric sphere diameter of each particle within the 3D model. The computational framework comprises two principal components: 3D modeling and morphological calculations.

Three-dimensional modelling: After segmentation, each binary image slice, where '1' represents soil particles and '0' represents pores, is sequentially stacked to create a three-dimensional voxel model. A voxel, or volume pixel, represents a point within the 3D grid of the object, encompassing spatial relationships and physical quantities. Within this constructed model, each soil particle manifests as a 3D object, discernible by neighboring voxels designated a value of '1'.

Morphological calculations: After the 3D model construction, morphological parameters—major axis, minor axis, volume, and isovolumetric sphere diameter—are calculated for each distinct soil particle. The major and minor axes are determined by evaluating the eigenvalues of the covariance matrix corresponding to the voxel coordinates of the particle—the maximum eigenvalue aligns with the longest axis (major axis), while the minimum aligns with the shortest (minor axis). The volume of the particle is ascertained by counting the number of constituent voxels, and given the isotropic nature of the voxels, this count is multiplied by the voxel volume. The isovolumetric sphere diameter—corresponding to the diameter of a sphere with volume equivalent to the particle—is calculated using Equation (18). These computations enable the extraction of essential morphological parameters from the 3D model, providing critical insights into the physical characteristics of soil particles. The computed parameters for each particle can be exported and stored for subsequent analysis.

$$d_{eq} = \sqrt[3]{\frac{6V_s}{\pi}} \tag{18}$$

where vs. represents the volume of the particle; and d_{eq} is the diameter of a sphere with a volume equivalent to the particle.

In summary, this proposed methodology offers a comprehensive approach, facilitating the construction of a 3D model of soil particles post-image-segmentation, while also enabling the calculation of vital particle characteristics. This paves the way for a more nuanced analysis and interpretation of soil properties.

2.6. Data Analysis

The least squares method is utilized for the linear regression analysis of variables d_e, α and k. Additionally, the strength and direction of the linear correlation are assessed employing the Pearson correlation coefficient.

The least squares method, which is commonly applied for the approximation of solutions in overdetermined systems—where there are more equations than unknowns—operates by minimizing the sum of the squares of the residuals. These residuals represent discrepancies between observed and estimated values, and the method facilitates the optimal linear fit for a given dataset. The method is implemented through the following formula [44]:

$$\beta_1 = \frac{\sum(x_i - \overline{x})(y_i - \overline{y})}{\sum(x_i - \overline{x})^2} \tag{19}$$

$$\beta_0 = \overline{y} - \beta_1\overline{x} \tag{20}$$

where n is the number of observations and i is between 1 and n; x_i is the independent variable; y_i is the dependent variable; $\overline{x}$ is the mean of x; $\overline{y}$ is the mean of y; and β_0 and β_1 are defined as the intercept and slope of the coefficients for the least squares line, respectively. Once β_0 and β_1 are calculated, the linear regression model (Equation (21)) is utilized to predict the value of y based on a given x value.

$$y = \beta_0 + \beta_1 x \tag{21}$$

In parallel, the Pearson correlation coefficient (R) is employed to quantify the degree of the linear relationship between pairs of variables. For any pair of random variables, denoted as (X,Y), with standard deviations σ_X and σ_Y, and expectations $E[X]$ and $E[Y]$, the Pearson correlation coefficient is computed by the following formula [45]:

$$R = \frac{E[(X - E[X])(y - E[Y])}{\sigma_X \sigma_Y} \tag{22}$$

The value of the Pearson correlation coefficient ranges from -1 to 1. A value of 1 implies a strong positive correlation, -1 signifies a strong negative correlation, and 0 indicates no correlation. With this statistical tool, the strength and direction of the linear relationship between d_e, α, and k are assessed. The integration of the least squares method for linear regression and the Pearson correlation coefficient facilitates a robust evaluation and the characterization of d_e, α, and k.

3. Results

3.1. Accuracy Verification of CT Image-Segmentation Program Based on Convolutional Neural Network

3.1.1. Verification of CT Image-Segmentation Accuracy Based on Convolutional Neural Networks

We present a CNN-based approach designed specifically for CT image segmentation of coarse-grained soil. The efficacy of this method is evaluated via five crucial metrics: Intersection over Union (IoU), Recall, Accuracy, Precision, and Specificity. As these metrics approach a value of 1, it indicates optimal prediction and segmentation accuracy. Higher values correspond to more precise outcomes of image segmentation, as detailed in Table 3.

Figure 7 visualizes the segmentation results of our validation set alongside a five-metric evaluation of the ground-truth image. Remarkably, all metrics exceed a value of 0.95 within 300 iterations of the CNN model, demonstrating the robustness of our methodology It shows its capability to accurately segment CT images of coarse-grained soil, thereby meeting the objectives of this study.

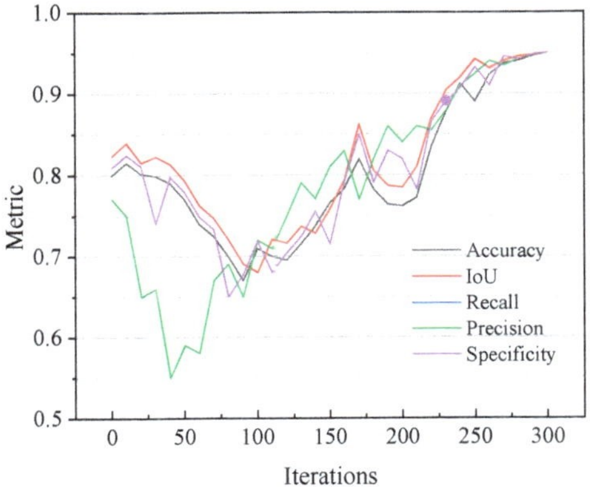

Figure 7. Metrics of test-set CT image segmentation results predicted by the CNN model.

When examining the clarity of segmented images, a pattern emerges: there seems to be a linear correlation between clarity and the number of CNN model iterations, resulting in a large oscillation amplitude. The maximum oscillation amplitude can be attributed to the balance between precision and recall that the CNN model strives to optimize during segmentation.

As the CNN model iteratively learns and adapts, it improves its ability to distinguish between true positive and false positive regions in the image, thereby enhancing image clarity. The increase in clarity over the iterations suggests that the CNN model is not only identifying regions of interest but is also refining the precision of their boundaries. The observed oscillation may serve as an insightful indicator of the model's learning progress, signifying an improvement in both accuracy and refinement in identifying and defining segmented region boundaries.

Our results underscore the effectiveness and potential applicability of our novel CNN-based method. Given its proficiency in the CT image segmentation of coarse-grained soil, it opens up exciting opportunities for future research in this field.

3.1.2. Comparison of Segmentation Results between CNN Model and Traditional Methods for CT Images of Coarse-Grained Soil

Besides quantitative image similarity indicators, we emphasize visual manual inspection to ascertain the accuracy of CT image segmentation. Manual validation was carried out on a randomly selected subset comprising 30% of the test set predicted images after the CNN model training iterations. This was executed alongside the calculation of metrics for all test-set predicted CT images and their corresponding ground truth images. The results affirmed the CNN model's competence in distinguishing coarse-grained soil particles into separate entities, thereby meeting the analytical criteria for this study.

Due to space limitations, we present two randomly selected CT image segmentation results predicted by the CNN model. Initially displayed as a binary black and white image, the CT segmentation result undergoes color representation to enhance the distinction between adjacent particles. Effective segmentation by the CNN model is, thus, manifested if adjacent particles exhibit distinct colors. Notably, the assignment of particle colors is arbitrary and independent of particle size. We also applied adaptive threshold segmentation [46] and marker-based watershed segmentation [47] methods for comparative analysis on identical CT images to validate the superiority of our CNN model-based CT image segmentation approach. Figure 8 showcases the results of CT image segmentation of coarse-grained soil via the CNN model, adaptive threshold segmentation method, and marker-based watershed segmentation method.

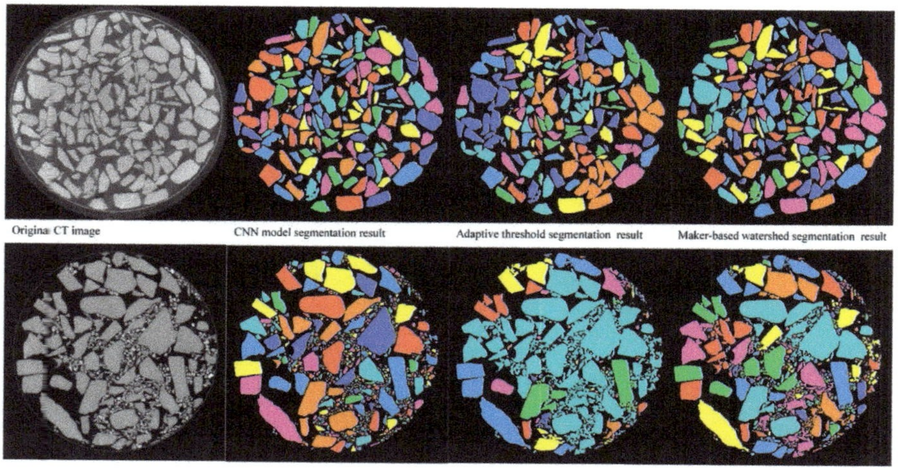

Figure 8. Comparison of results from various CT image segmentation methods.

Figure 8 shows that the CNN model provides the most precise segmentation, with particles distinctly defined. In contrast, the adaptive threshold segmentation method falls short, while the marker-based watershed method yields intermediate results. These differences primarily stem from the varying capabilities of the three methods in handling edge pixels of smaller particles in the images.

The marker-based watershed segmentation algorithm retains a degree of accuracy due to its adherence to the dilation principle, but it becomes less efficient with an increase in fine particles and complex pore structures in the CT image. The simultaneous rise in phase-change areas and the smoothing of pixel intensity distribution leads to the failure of both the adaptive threshold segmentation method and eigenvalue segmentation method, due to the diminished edge contrast of fine particles in the image. Although the marker-based watershed segmentation method outperforms the threshold segmentation method, the accuracy of its segmentation results falls short of the study requirements.

In contrast, our CNN model leverages deep learning, freeing it from reliance on image pixel intensity distribution characteristics for segmentation. Instead, it mimics the process of human truth image segmentation through extensive data training, enabling the accurate prediction of segmentation outcomes for new images. Therefore, the CNN model's segmentation results demonstrate its remarkable efficacy.

3.2. Equivalent Simplified Model of Ideal Particle Diameter in Coarse-Grained Soil

3.2.1. Three-Dimensional Model Reconstruction of Coarse-Grained Soil Based on CT Image Segmentation Results

Upon segmenting the coarse-grained soil CT images from our twelve samples using the CNN model, we reconstructed the segmentation results through voxelization to create corresponding three-dimensional particle models (Figure 9). To confirm the accuracy of these models and further validate the segmentation accuracy of the CNN model, we calculated the non-uniformity and curvature coefficients of the particle size distribution within each model. We then carried out a comparative error analysis against the actual non-uniformity and curvature coefficients of the respective coarse-grained soil samples, as listed in Table 2. An error margin of less than 2% proves that the accuracy of the three-dimensional models meets the research requirements.

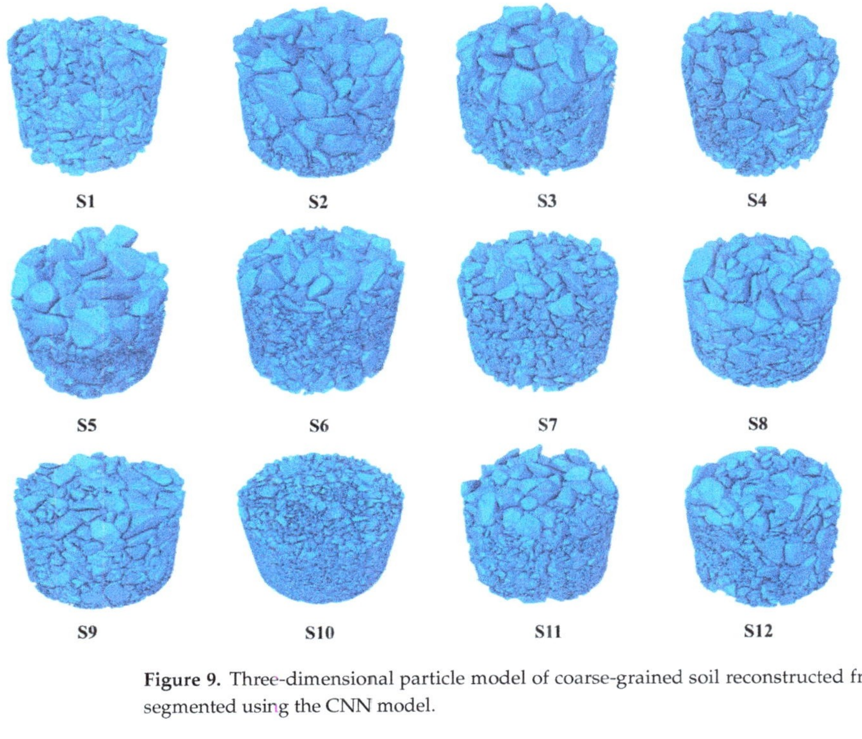

Figure 9. Three-dimensional particle model of coarse-grained soil reconstructed from CT images segmented using the CNN model.

Figure 10 illustrates the relative errors in particle size distribution for all the three-dimensional particle models corresponding to the twelve soil samples. It is apparent that the relative error between the calculated non-uniformity and curvature coefficients and their true values for each sample is under the acceptable 2% threshold. This verifies the high accuracy of the models and not only corroborates the precision of the three-dimensional models but also reinforces the segmentation accuracy of the CNN model. This is a significant outcome because it suggests that the CNN model has successfully learned to recognize and segment different sizes of particles in the soil samples, despite their complex and variable nature. Therefore, these results reinforce the viability and reliability of using deep learning models like CNN for soil-particle segmentation and subsequent 3D modeling in soil studies, thereby broadening the prospects of automating and enhancing precision in such processes.

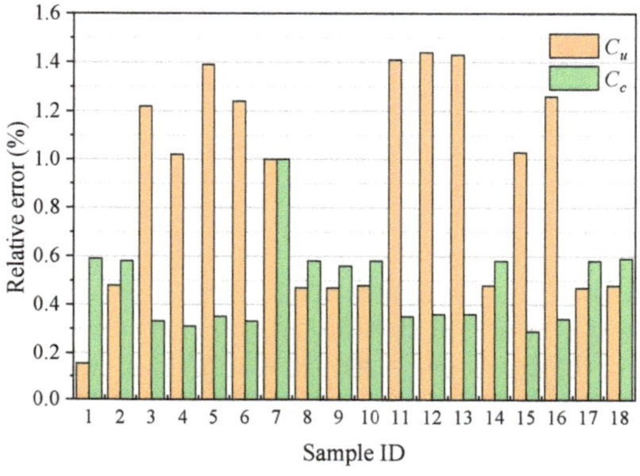

Figure 10. Relative error in the three-dimensional particle models of coarse-grained soil.

3.2.2. Empirical Formula for the Ideal Particle Size of Coarse Soil Particles

An error margin of less than 2% between the 3D models of coarse-grained soil particles, obtained from CT images segmented via the CNN model, and actual specimens, attests to the close alignment of geometric parameters from the 3D models—such as the long axis, short axis, aspect ratio (Φ), and diameter of the equal volume sphere (d_{eq})—with those of real soil particles. This accuracy satisfies the requirements of our study.

Our data processing reveals a compelling linear relationship between the product of d_{eq} and Φ ($d_{eq} \times \Phi$) and the diameter of soil particles. Accordingly, we applied Equations (18)–(22) to regress the particle size of soil particles, as shown in Figure 11. A congruent linear relationship appears across all 12 coarse-grained soil specimens. However, due to space limitations, we only present the linear regression analyses of particle sizes for specimens S5 and S6. Figure 11 presents the Pearson correlation coefficients of $d_{eq} \times \Phi$ and the particle size for all specimens, exceeding 0.95 for all specimens except S1, which displays a smaller particle size range (5–13 mm) compared to the other specimens. This smaller range results in a slightly lower Pearson correlation coefficient of 0.88; yet, this figure still denotes a substantial linear correlation between these variables.

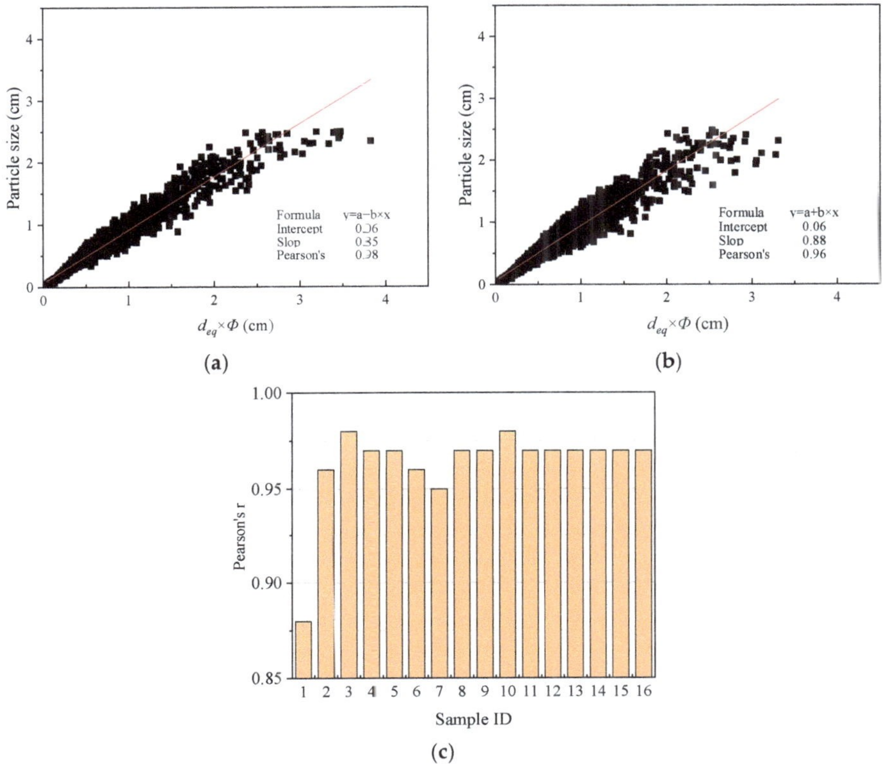

Figure 11. Linear regression results of coarse soil particle size. (**a**) Linear regression results of particle size for S5 sample. (**b**) Linear regression results of particle size for S6 sample. (**c**) Pearson correlation coefficient for linear regression of particle size of all samples.

Kozeny and Terzaghi et al. [48–52], from their research on the permeability tests of cohesionless soils along with theoretical analyses, concluded that the hydraulic conductivity of cohesionless soils should be represented by the particle diameter d_{20}, corresponding to a 20% cumulative mass fraction. This view is further corroborated by several other studies; hence, we incorporated this perspective into our research.

In this regard, we performed a linear regression analysis using d_{20} of the specimen, with the equivalent particle diameter d_{eq20} corresponding to the 20% cumulative mass fraction, and the average aspect ratio of the specimen. These parameters are presented in Table 4. Our results, depicted in Figure 12, establish a strong linear correlation between d_{20} and the product of d_{eq20} and average aspect ratio, $\overline{\Phi}$.

Table 4. Characteristic particle size and shape factor of coarse-grained soil samples.

Sample ID	S1	S2	S3	S4	S5	S6	S7	S8	S9	S10	S11	S12
d_{eq20} (mm)	6.26	2.71	1.72	2.21	2.01	2.46	3.29	1.19	1.24	2.30	2.90	2.30
$\overline{\Phi}$	2.31	1.93	2.03	1.95	2.04	1.86	1.99	1.98	2.03	1.93	1.92	2.00
d_{20} (mm)	10.29	5.20	3.94	4.49	4.16	4.82	6.34	3.66	3.74	4.69	5.48	4.55

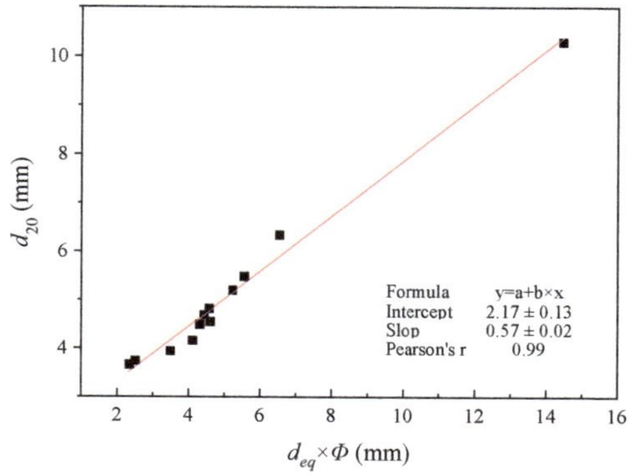

Figure 12. Linear regression results of ideal particle size of coarse-grained soil particles.

The regression line's intercept and slope in Figure 12 suggest that the ideal equation for the particle size of the coarse-grained soil aligns with Equation (23). This finding contributes to a more nuanced understanding of the granulometric characteristics of coarse-grained soils and provides an empirically derived equation for predicting particle size, helping to better inform soil-related engineering and environmental applications.

$$d_e = \frac{d_{20} - 2.17}{0.57\overline{\Phi}} \tag{23}$$

where d_e is the rational particle size of coarse-grained soil, which is equal to the particle size d_{eq20} corresponding to a cumulative mass percentage of 20% in the diameter distribution of equal volume spheres; d_{20} is the particle size corresponding to a cumulative percentage of the coarse soil mass of 20%; and $\overline{\Phi}$ is the average aspect ratio of coarse-grained soil particles.

3.3. *Prediction Formula for Permeability Coefficient of Coarse-Grained Soil Based on CT Image Analysis*

3.3.1. Constant-Head Permeability Test Results

The results from the constant-head permeability tests conducted on coarse-grained soil specimens are displayed in Figure 13. According to Darcy's Law (Equation (24)) [40], a linear relationship exists between the water velocity (V) and the hydraulic gradient (J). This relationship enables us to calculate the hydraulic conductivity of the specimens, which are presented in Table 5.

$$V = KJ \tag{24}$$

where V is the water flow velocity, J is the hydraulic gradient, and K is hydraulic conductivity.

Upon referring to the characteristic particle size of the coarse-grained soil specimens from Table 2 and using our empirically derived Equation (9) for estimating the hydraulic conductivity of coarse-grained soil, we can compute the discount factor (α) for the specimens. These computed values are also enumerated in Table 5. This calculated discount factor and hydraulic conductivity are essential parameters in understanding and predicting the behavior of coarse-grained soil under varying hydraulic gradients.

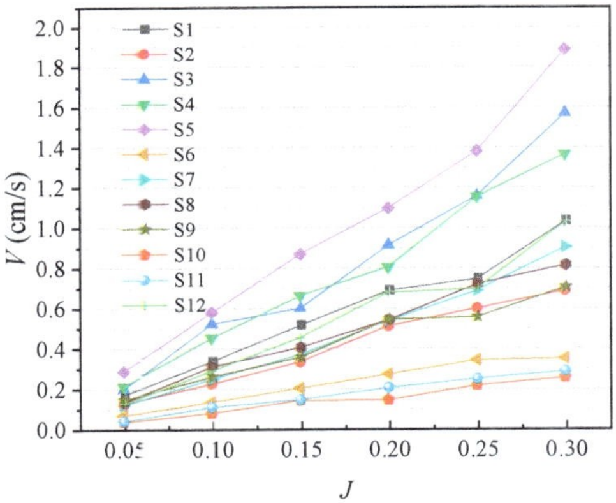

Figure 13. Permeability test results of coarse-grained soil samples (test water temperature at 20 °C).

Table 5. Hydraulic gradient and discount factor of coarse-grained soil samples.

Sample ID	S1	S2	S3	S4	S5	S6	S7	S8	S9	S10	S11	S12
K (cm/s)	3.45	2.30	4.58	4.30	5.81	1.37	2.70	2.72	2.62	0.86	0.99	3.04
α	162.29	48.79	39.42	52.27	36.65	77.88	87.99	26.74	28.39	76.80	79.58	48.88

3.3.2. Empirical Formula for Discount Factor

Tables 4 and 5 indicate a distinct trend: as the ideal particle size of the coarse-grained soil samples increases, there is a concurrent rise in the discount factor. To better define this relationship, we conducted a linear regression analysis using the least squares method. As illustrated in Figure 14, the resulting Pearson correlation coefficient is a robust 0.95, signifying a strong linear correlation between the discount factor and the ideal soil particle size. Therefore, the empirical formula for the discount factor of coarse-grained soil fitted in this article is shown in Equation (25).

$$\alpha = -4.31 + 26.85d_e \tag{25}$$

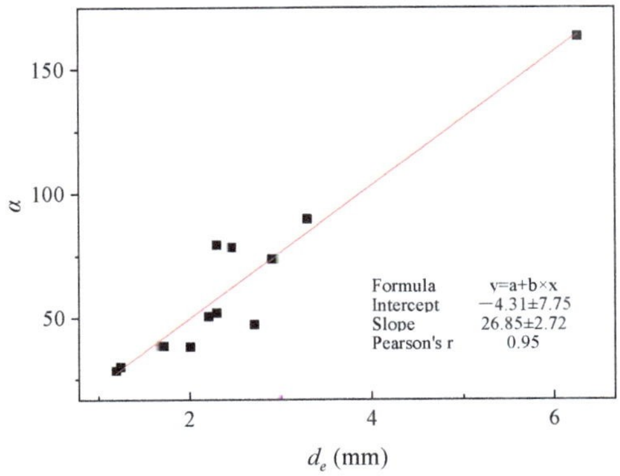

Figure 14. Regression analysis results of reduction coefficient for coarse-grained soil.

This relationship is deeply rooted in the soil's granular composition, with the particle size playing a significant role. A higher concentration of fine particles within a unit volume of coarse-grained soil invariably augments the tortuosity of water-flow paths. This labyrinth-like network intensifies the interactions between the water and particles, causing greater water loss as it traverses the soil, thereby increasing the discount factor—a measure of this water loss. Moreover, particle size and distribution have a substantial impact on the soil's void ratio, which in turn affects hydraulic conductivity. Typically, larger particles lead to a higher void ratio and greater pore spaces. Conversely, the presence of fine particles tends to fill these voids, reducing total pore space and causing an increased tortuosity in water-flow paths. This elevated tortuosity contributes to a higher reduction coefficient, necessitating more energy for water to pass through the soil matrix.

While the primary focus of this study lies in examining the influence of particle size on the discount factor, it is important to acknowledge that the soil's pore structure has a fundamental role in this relationship. Future research will explore in greater depth the specific effects of pore structure parameters on the discount factor, shedding light on the complex interactions between particle size, pore structure, and water movement in coarse-grained soils. Such insights will advance our understanding of soil hydrodynamics and its implications for a variety of fields, including water conservancy and geotechnical engineering.

3.3.3. Prediction Formula and Accuracy Verification of Hydraulic Conductivity of Coarse-Grained Soil

Following the segmentation of coarse-grained soil specimens' CT images via the CNN model, we successfully constructed their respective 3D particle models. An examination of these particle models' parameters allowed us to derive the empirical Equation (23), which represents the ideal particle size of coarse-grained soil. Furthermore, through an analysis of the infiltration tests conducted on coarse-grained soil specimens, we deduced the empirical Equation (25), illustrating the reduction factor. Integrating Equations (23) and (25) into our established model for calculating the hydraulic conductivity of coarse-grained soil (Equation (9)), we derived a predictive formula for coarse-grained soil hydraulic conductivity, based on CT image analysis, as

$$K = \frac{g}{72\mu\alpha^2} \cdot \frac{n^3}{(1-n)^2} \cdot \frac{(d_{20} - 2.17)^2}{0.32\overline{\Phi}^2} \tag{26}$$

among which, the discount factor α is

$$\alpha = -4.31 + \frac{83.91(d_{20} - 2.17)^2}{\overline{\Phi}^2} \tag{27}$$

This section aims to probe the precision and applicability of various models for predicting the hydraulic conductivity of coarse-grained soils. Existing models mainly fall into two categories. The first type comprises empirical formulae based on pore structure and particle characteristics, exemplified by the Kozeny–Carman equation [53]. The second category includes empirical models derived from experimental data and statistical analysis, such as the Terzaghi [52] and Hazen [54] equations. In this study, we compared our derived model for predicting the hydraulic conductivity of coarse-grained soils (based on CT image analysis) against the aforementioned eight formulae. This comparison entailed calculating the hydraulic conductivity for the coarse-grained soil specimens using each formula, and subsequently comparing these values against the actual hydraulic conductivities derived from permeability tests. To assess the predictive performance of these formulae, we utilized the Pearson correlation coefficient as an evaluation index, with possible values ranging from −1 to 1. A coefficient closer to 1 indicates a strong positive correlation, while a value closer to −1 denotes a strong negative correlation. A value near 0 suggests a lack of any significant correlation.

Table 6 lists the Pearson correlation coefficients between the hydraulic conductivity predictions from the eight formulae and the actual hydraulic conductivities of the specimens. As shown in Table 6, our proposed formula for predicting the hydraulic conductivity of coarse-grained soil provides the most accurate predictions. It attains a Pearson correlation coefficient of 0.91 with the actual hydraulic conductivity values.

Table 6. Comparative analysis of the accuracy of various hydraulic conductivity calculation formulae.

Formula	Equation (25)	KC [53]	Terzaghi [52]	Hazen [54]	Safari [55]	Chapuis [56]	Cote [57]
K Pearson's r	0.91	0.70	0.36	-0.01	0.75	0.72	0.28

4. Discussion

Existing methods for determining the hydraulic conductivity of coarse-grained soils fall into two main categories: experimental methods and numerical simulation methods. Experimental methods typically utilize a constant-head test to directly measure hydraulic conductivity [2–8]. However, these methods are often time-consuming and labor-intensive, with the accuracy of the results heavily reliant on sample preparation and boundary conditions. Numerical simulation methods, on the other hand, employ finite elements or discrete elements to mimic actual coarse-grained soils [9–12]. By resolving the fluid flow equation, the hydraulic conductivity can be calculated. Yet, conclusions drawn from these methods often lack experimental validation. A more effective approach to predicting the hydraulic conductivity of coarse-grained soils is to leverage CT image analysis. This method marries experimental results with simulations, thereby offering a more comprehensive and accurate evaluation. The accuracy of such predictions hinges on the precision of the CT image segmentation. However, as coarse-grained soils exhibit bulk structures and their CT image pixel distributions are complex, the accuracy of traditional image-segmentation methods often falls short of the research requirements for coarse-grained soils.

This paper presents a novel method for segmenting CT images of coarse-grained soil using deep learning models. Based on the segmentation results, we created a three-dimensional particle model of coarse-grained soil and extracted the geometric feature parameters of the particles. Subsequently, we fitted a prediction formula for the hydraulic conductivity of coarse-grained soil using CT image analysis. We discovered that the U-Net convolutional neural network, under the supervision of a specially tailored loss function with particle size and shape perception terms, can effectively capture the profiles of coarse soil particles. After adequate training, the network can precisely segment coarse soil. The formula predicting the hydraulic conductivity of coarse-grained soil, based on CT image analysis and fitting, reveals a particular pattern in coarse-grained soil's hydraulic conductivity when the particle size range remains constant. If the particle size distribution is uneven, smaller particles can fill in the pores formed within the large particle skeleton, leading to lower porosity and permeability. For a constant particle size distribution, a larger aspect ratio of coarse-grained soil (indicating more needle-shaped particles) results in a smaller discount factor, representing the loss degree of actual seepage channel due to pore connectivity, and hence lower permeability.

Our results further the research conducted by other scholars through seepage tests [58,59] and discrete element simulations [60–63], microscopically confirming, based on CT images, that the hydraulic conductivity of coarse-grained soil primarily depends on the characteristic particle size d_{20}, porosity, and aspect ratio of the particles. Moreover, our proposed formula predicting the hydraulic conductivity of coarse-grained soil based on CT image analysis is more precise, demonstrating a higher value for engineering applications. To our knowledge, this is the first study predicting the hydraulic conductivity of coarse-grained soil using CT images segmented based on deep learning models. Taking the practical implications of our findings into account, the accurate prediction of hydraulic conductivity of coarse-grained soils using our model presents transformative potential in the field of water conservancy and geotechnical engineering. Firstly, in embankment dam design, our approach offers a more precise and reliable method to analyze the granulometric characteristics of the soil, thereby aiding in choosing the right type of soil and understanding its behavior under various conditions. This ultimately impacts the safety and longevity of the dam structure. Secondly, in the context of geotechnical engineering, the understanding of granular composition and its impact on the tortuosity of water-flow paths can assist in effective ground water management and the design of structures requiring soil as the foundational support. With a clearer understanding of soil's hydraulic behavior, engineers can better anticipate and mitigate potential issues related to soil stability, water seepage and deformation under load [64–66].

However, it is important to note some limitations. Although we have accurately segmented the CT images of coarse-grained soils using U-Net convolutional neural networks, the study of network structure optimization and enhancing learning efficiency remains unexplored. Future research should aim at optimizing the deep learning model for studying coarse-grained soil CT images, with a focus on improving learning efficiency.

5. Conclusions

In this study, we innovatively utilized a Convolutional Neural Network (CNN) model for the accurate segmentation of coarse-grained soil CT images, making significant strides in soil granulometric analysis. Our approach outperforms traditional methods, revealing new empirical formulae to comprehend the granulometric characteristics and hydraulic conductivity of coarse-grained soils, thereby offering fresh insights into soil hydrodynamics. Some valuable conclusion are as follows:

1. The implementation of the CNN model demonstrates unparalleled precision in the segmentation of coarse-grained soil CT images, ascertaining the model's superiority over traditional segmentation methods. The accuracy of the 3D models reconstructed from these segmented images corroborates the effectiveness of this approach and broadens the prospects of automation and precision in soil particle segmentation.
2. We established and validated empirical formulae for the ideal particle size of coarse-grained soil and the discount factor, both predicated on a robust linear correlation found in the study. These novel formulae contribute significantly to understanding the granulometric characteristics of soils and predicting their behavior under various hydraulic gradients, thus providing valuable insights for soil-related engineering and hydraulic applications.
3. Our research underlines the strong influence of the granular composition, especially the concentration of fine particles, on the tortuosity of water flow paths and the discount factor. These findings highlight the potential of the CNN model in soil hydrodynamics research and its implications for a variety of fields, including water conservancy and geotechnical engineering.

Author Contributions: Conceptualization, J.P. and Z.S.; methodology, J.P.; software, J.P.; validation, J.P., W.Z. and Z.S.; formal analysis, J.P.; investigation, J.P. and W.Z.; resources, W.S.; data curation, W.S.; writing—original draft preparation, J.P.; writing—review and editing, J.P.; supervision, Z.S.; project administration, Z.S.; funding acquisition, J.P. and Z.S. All authors have read and agreed to the published version of the manuscript.

Funding: This research was funded by the National Natural Science Foundation of China (grant No. 52179130), the Fundamental Research Funds for the Center Universities (grant No. 2019B70814), and the Postgraduate Research & Practice Innovation Program of Jiangsu Province (grant No. SJKY19_0483).

Data Availability Statement: Not applicable.

Conflicts of Interest: The authors declare no conflict of interest.

References

1. Nam, S.; Gutierrez, M.; Diplas, P.; Petrie, J. Laboratory and in Situ Determination of Hydraulic Conductivity and Their Validity in Transient Seepage Analysis. *Water* **2021**, *13*, 1131. [CrossRef]
2. Shaker, A.A.; Dafalla, M.; Al-Mahbashi, A.M.; Al-Shamrani, M.A. Predicting Hydraulic Conductivity for Flexible Wall Conditions Using Rigid Wall Permeameter. *Water* **2022**, *14*, 286. [CrossRef]

3. Yoon, S.; Lee, S.R.; Kim, Y.T.; Go, G.H. Estimation of Saturated Hydraulic Conductivity of Korean Weathered Granite Soils Using a Regression Analysis. *Geomech. Eng.* **2015**, *9*, 101–113. [CrossRef]
4. Edeh, I.G.; Mašek, O. The Role of Biochar Particle Size and Hydrophobicity in Improving Soil Hydraulic Properties. *Eur. J. Soil Sci.* **2022**, *73*, e13138. [CrossRef]
5. Liu, Z.R.; Cui, Y.J.; Ye, W.M.; Chen, B.; Wang, Q.; Chen, Y.G. Investigation of the Hydro-Mechanical Behaviour of GMZ Bentonite Pellet Mixtures. *Acta Geotech.* **2020**, *15*, 2865–2875. [CrossRef]
6. Mijic, Z.; Dayioglu, A.Y.; Hatipoglu, M.; Aydilek, A.H. Hydraulic and Environmental Impacts of Using Recycled Asphalt Pavement on Highway Shoulders. *Constr. Build. Mater.* **2020**, *234*, 117226. [CrossRef]
7. Hatipoglu, M.; Cetin, B.; Aydilek, A.H. Effects of Fines Content on Hydraulic and Mechanical Performance of Unbound Granular Base Aggregates. *J. Transp. Eng. Part B Pavements* **2020**, *146*, 04019036. [CrossRef]
8. Wang, J.P.; Zhuang, P.Z.; Luan, J.Y.; Liu, T.H.; Tan, Y.R.; Zhang, J. Estimation of Unsaturated Hydraulic Conductivity of Granular Soils from Particle Size Parameters. *Water* **2019**, *11*, 1826. [CrossRef]
9. Wang, T.; Yan, C.; Zheng, Y.; Jiao, Y.-Y.; Zou, J. Numerical Study on the Effect of Meso-Structure on Hydraulic Conductivity of Soil-Rock Mixtures. *Comput. Geotech.* **2022**, *146*, 104726. [CrossRef]
10. Huang, L.; He, R.; Yang, Z.; Tan, P.; Chen, W.; Li, X.; Cao, A. Exploring Hydraulic Fracture Behavior in Glutenite Formation with Strong Heterogeneity and Variable Lithology Based on DEM Simulation. *Eng. Fract. Mech.* **2023**, *278*, 109020. [CrossRef]
11. Tseng, C.-Y.; Ghadiri, M.; Kumar, P.; Meidani, H. Estimation of Hydraulic Conductivity in a Watershed Using Sparse Multi-Source Data via Gaussian Process Regression and Bayesian Experimental Design. *Adv. Water Resour.* **2023**, *178*, 104489. [CrossRef]
12. Luo, M.; Ye, C.; Wang, X.; Huang, E.; Yan, X. Analytical Model of Flow Velocity in Gravel-Bed Streams under the Effect of Gravel Array with Different Densities. *J. Hydrol.* **2022**, *608*, 127581. [CrossRef]
13. An, R.; Wang, Y.; Zhang, X.; Chen, C.; Liu, X.; Cai, S. Quantitative Characterization of Drying-Induced Cracks and Permeability of Granite Residual Soil Using Micron-Sized X-Ray Computed Tomography. *Sci. Total Environ.* **2023**, *876*, 163213. [CrossRef] [PubMed]
14. Lin, S.; Zheng, H.; Zhang, Z.; Li, W. Evaluation of Permeability of Soil & Rock Aggregate Using Meshless Numerical Manifold Method. *Comput. Geotech.* **2022**, *151*, 104953. [CrossRef]
15. Liu, S.; Zhang, L.; Su, X.; Zhao, L.; Wang, Y. Micro-CT Characterization on Pore Structure Evolution of Low-Permeability Sandstone under Acid Treatment. *Appl. Geochem.* **2023**, *152*, 105633. [CrossRef]
16. Ferreira, T.R.; Archilha, N.L.; Cássaro, F.A.M.; Pires, L.F. How Can Pore Characteristics of Soil Aggregates from Contrasting Tillage Systems Affect Their Intrinsic Permeability and Hydraulic Conductivity? *Soil Tillage Res.* **2023**, *230*, 105704. [CrossRef]
17. Zhang, L.; Wang, J. Prediction of the Soil Saturated Hydraulic Conductivity in a Mining Area Based on CT Scanning Technology. *J. Clean. Prod.* **2023**, *383*, 135364. [CrossRef]
18. Liu, Y.; Zhang, C.; Li, C.; Cheng, J.; Zhang, Y.; Xu, H.; Song, T.; Zhao, L.; Chen, X. A Practical PET/CT Data Visualization Method with Dual-Threshold PET Colorization and Image Fusion. *Comput. Biol. Med.* **2020**, *126*, 104050. [CrossRef]
19. Zhang, J.; Yan, C.-H.; Chui, C.-K.; Ong, S.-H. Fast Segmentation of Bone in CT Images Using 3D Adaptive Thresholding. *Comput. Biol. Med.* **2010**, *40*, 231–236. [CrossRef]
20. Mat Jais, I.S.; Liu, X.; An, K.-N.; Tay, S.C. A Method for Carpal Motion Hysteresis Quantification in 4-Dimensional Imaging of the Wrist. *Med. Eng. Phys.* **2014**, *36*, 1699–1703. [CrossRef]
21. Peter, Z.; Bousson, V.; Bergot, C.; Peyrin, F. A Constrained Region Growing Approach Based on Watershed for the Segmentation of Low Contrast Structures in Bone Micro-CT Images. *Pattern Recognit.* **2008**, *41*, 2358–2368. [CrossRef]
22. Cristoforetti, A.; Faes, L.; Ravelli, F.; Centonze, M.; Del Greco, M.; Antolini, R.; Nollo, G. Isolation of the Left Atrial Surface from Cardiac Multi-Detector CT Images Based on Marker Controlled Watershed Segmentation. *Med. Eng. Phys.* **2008**, *30*, 48–58. [CrossRef]
23. Minnema, J.; Wolff, J.; Koivisto, J.; Lucka, F.; Batenburg, K.J.; Forouzanfar, T.; van Eijnatten, M. Comparison of Convolutional Neural Network Training Strategies for Cone-Beam CT Image Segmentation. *Comput. Methods Programs Biomed.* **2021**, *207*, 106192. [CrossRef]
24. Roslin, A.; Marsh, M.; Provencher, B.; Mitchell, T.R.; Onederra, I.A.; Leonardi, C.R. Processing of Micro-CT Images of Granodiorite Rock Samples Using Convolutional Neural Networks (CNN), Part II: Semantic Segmentation Using a 2.5D CNN. *Miner. Eng.* **2023**, *195*, 108027. [CrossRef]
25. Dong, Y.; Su, C.; Qiao, P.; Sun, L. Microstructural Crack Segmentation of Three-Dimensional Concrete Images Based on Deep Convolutional Neural Networks. *Constr. Build. Mater.* **2020**, *253*, 119185. [CrossRef]
26. Nguyen, T.; Truong, T.T.; Nguyen-Thoi, T.; Van Hong Bui, L.; Nguyen, T.-H. Evaluation of Residual Flexural Strength of Corroded Reinforced Concrete Beams Using Convolutional Long Short-Term Memory Neural Networks. *Structures* **2022**, *46*, 899–912. [CrossRef]
27. Mirbod, M.; Shoar, M. Intelligent Concrete Surface Cracks Detection Using Computer Vision, Pattern Recognition, and Artificial Neural Networks. *Procedia Comput. Sci.* **2023**, *217*, 52–61. [CrossRef]
28. Liu, Y.; Zhang, J.; Zhao, T.; Wang, Z.; Wang, Z. Reconstruction of the Meso-Scale Concrete Model Using a Deep Convolutional Generative Adversarial Network (DCGAN). *Constr. Build. Mater.* **2023**, *370*, 130704. [CrossRef]
29. Tian, W.; Cheng, X.; Liu, Q.; Yu, C.; Gao, F.; Chi, Y. Meso-Structure Segmentation of Concrete CT Image Based on Mask and Regional Convolution Neural Network. *Mater. Des.* **2021**, *208*, 109919. [CrossRef]

30. Zhang, W.; Shi, D.; Shen, Z.; Zhang, J.; Zhao, S.; Gan, L.; Li, Q.; Chen, Y.; Tang, P. Influence of Chopped Basalt Fibers on the Fracture Performance of Concrete Subjected to Calcium Leaching. *Theor. Appl. Fract. Mech.* **2023**, *125*, 103934. [CrossRef]
31. Zhang, W.; Shi, D.; Shen, Z.; Wang, X.; Gan, L.; Shao, W.; Tang, P.; Zhang, H.; Yu, S. Effect of Calcium Leaching on the Fracture Properties of Concrete. *Constr. Build. Mater.* **2023**, *365*, 130018. [CrossRef]
32. Zhang, W.; Shi, D.; Shen, Z.; Shao, W.; Gan, L.; Yuan, Y.; Tang, P.; Zhao, S.; Chen, Y. Reduction of the Calcium Leaching Effect on the Physical and Mechanical Properties of Concrete by Adding Chopped Basalt Fibers. *Constr. Build. Mater.* **2023**, *365*, 130080. [CrossRef]
33. Roslin, A.; Marsh, M.; Piché, N.; Provencher, B.; Mitchell, T.R.; Onederra, I.A.; Leonardi, C.R. Processing of Micro-CT Images of Granodiorite Rock Samples Using Convolutional Neural Networks (CNN), Part I: Super-Resolution Enhancement Using a 3D CNN. *Miner. Eng.* **2022**, *188*, 107748. [CrossRef]
34. Dawson, H.L.; Dubrule, O.; John, C.M. Impact of Dataset Size and Convolutional Neural Network Architecture on Transfer Learning for Carbonate Rock Classification. *Comput. Geosci.* **2023**, *171*, 105284. [CrossRef]
35. Roslin, A.; Lebedev, M.; Mitchell, T.R.; Onederra, I.A.; Leonardi, C.R. Processing of Micro-CT Images of Granodiorite Rock Samples Using Convolutional Neural Networks (CNN). Part III: Enhancement of Scanco Micro-CT Images of Granodiorite Rocks Using a 3D Convolutional Neural Network Super-Resolution Algorithm. *Miner. Eng.* **2023**, *195*, 108028. [CrossRef]
36. Karimpouli, S.; Faraji, A.; Balcewicz, M.; Saenger, E.H. Computing Heterogeneous Core Sample Velocity Using Digital Rock Physics: A Multiscale Approach. *Comput. Geosci.* **2020**, *135*, 104378. [CrossRef]
37. Wang, Y.D.; Shabaninejad, M.; Armstrong, R.T.; Mostaghimi, P. Deep Neural Networks for Improving Physical Accuracy of 2D and 3D Multi-Mineral Segmentation of Rock Micro-CT Images. *Appl. Soft Comput.* **2021**, *104*, 107185. [CrossRef]
38. Dobry, R. Simplified Methods in Soil Dynamics. *Soil Dyn. Earthq. Eng.* **2014**, *61–62*, 246–268. [CrossRef]
39. Sutera, S.P.; Skalak, R. The History of Poiseuille's Law. *Annu. Rev. Fluid Mech.* **1993**, *25*, 1–20. [CrossRef]
40. Whitaker, S. Flow in Porous Media I: A Theoretical Derivation of Darcy's Law. *Transp. Porous Media* **1986**, *1*, 3–25. [CrossRef]
41. Ronneberger, O.; Fischer, P.; Brox, T. U-Net: Convolutional Networks for Biomedical Image Segmentation. In Proceedings of the Medical Image Computing and Computer-Assisted Intervention—MICCAI 2015, Munich, Germany, 5–9 October 2015; Navab, N., Hornegger, J., Wells, W.M., Frangi, A.F., Eds.; Springer International Publishing: Cham, Switzerland, 2015; pp. 234–241.
42. Ho, Y.; Wookey, S. The Real-World-Weight Cross-Entropy Loss Function: Modeling the Costs of Mislabeling. *IEEE Access* **2020**, *8*, 4806–4813. [CrossRef]
43. Petrou, M.M.P.; Petrou, C. *Image Processing: The Fundamentals*; John Wiley & Sons: New York, NY, USA, 2010.
44. Björck, Å. Least Squares Methods. In *Handbook of Numerical Analysis*; Elsevier: Amsterdam, The Netherlands, 1990; Volume 1, pp. 465–652, ISBN 1570-8659.
45. Benesty, J.; Chen, J.; Huang, Y.; Cohen, I. Pearson Correlation Coefficient. In *Noise Reduction in Speech Processing*; Cohen, I., Huang, Y., Chen, J., Benesty, J., Eds.; Springer: Berlin/Heidelberg, Germany, 2009; pp. 1–4. ISBN 978-3-642-00296-0.
46. Chan, F.H.Y.; Lam, F.K.; Zhu, H. Adaptive Thresholding by Variational Method. *IEEE Trans. Image Process.* **1998**, *7*, 468–473. [CrossRef] [PubMed]
47. Gostick, J.T. Versatile and Efficient Pore Network Extraction Method Using Marker-Based Watershed Segmentation. *Phys. Rev. E* **2017**, *96*, 23307. [CrossRef] [PubMed]
48. Kozeny, J. Unber Kapillare Leitung Des Wassers Im Boden. *Sitz. Akad. Wiss WIEN* **1927**, *136*, 271–306.
49. Xu, P.; Yu, B. Developing a New Form of Permeability and Kozeny–Carman Constant for Homogeneous Porous Media by Means of Fractal Geometry. *Adv. Water Resour.* **2008**, *31*, 74–81. [CrossRef]
50. Henderson, N.; Brêttas, J.C.; Sacco, W.F. A Three-Parameter Kozeny–Carman Generalized Equation for Fractal Porous Media. *Chem. Eng. Sci.* **2010**, *65*, 4432–4442. [CrossRef]
51. Javad Azarhoosh, M.; Koohmishi, M. Prediction of Hydraulic Conductivity of Porous Granular Media by Establishment of Random Forest Algorithm. *Constr. Build Mater.* **2023**, *366*, 130065. [CrossRef]
52. Terzaghi, K.; Peck, R.B.; Mesri, G. Soil Mechanics in Engineering Practice. Available online: https://www.wiley.com/en-us/Soil+Mechanics+in+Engineering+Practice%2C+3rd+Edition-p-9780471086581 (accessed on 8 June 2022).
53. Carman, P.C. Permeability of Saturated Sands, Soils and Clays. *J. Agric. Sci.* **1939**, *29*, 262–273. [CrossRef]
54. Hazen, A. Discussion of Dams on Sand Foundation by A.C. *Transp. ASAE* **1911**, *73*, 199–203.
55. Safari, M.; Gholami, R.; Jami, M.; Ananthan, M.A.; Rahimi, A.; Khur, W.S. Developing a Porosity-Permeability Relationship for Ellipsoidal Grains: A Correction Shape Factor for Kozeny-Carman's Equation. *J. Pet. Sci. Eng.* **2021**, *205*, 108896. [CrossRef]
56. Chapuis, R.P.; Aubertin, M. On the Use of the Kozeny-Carman Equation to Predict the Hydraulic Conductivity of Soils. *Can. Geotech. J.* **2003**, *40*, 616–628. [CrossRef]
57. Côté, J.; Fillion, M.-H.; Konrad, J.-M. Estimating Hydraulic and Thermal Conductivities of Crushed Granite Using Porosity and Equivalent Particle Size. *J. Geotech. Geoenvironmental Eng.* **2011**, *137*, 834–842. [CrossRef]
58. Belkhatir, M.; Schanz, T.; Arab, A. Effect of Fines Content and Void Ratio on the Saturated Hydraulic Conductivity and Undrained Shear Strength of Sand-Silt Mixtures. *Environ. Earth Sci.* **2013**, *70*, 2469–2479. [CrossRef]
59. Su, L.J.; Zhang, Y.J.; Wang, T.X. Investigation on Permeability of Sands with Different Particle Sizes. Available online: https://www.researchgate.net/publication/288721987_Investigation_on_permeability_of_sands_with_different_particle_sizes (accessed on 8 June 2022).

60. Xu, S.; Zhu, Y.; Cai, Y.; Sun, H.; Cao, H.; Shi, J. Predicting the Permeability Coefficient of Polydispersed Sand via Coupled CFD–DEM Simulations. *Comput. Geotech.* **2022**, *144*, 104634. [CrossRef]
61. Liu, Y.F.; Jeng, D.S. Pore Scale Study of the Influence of Particle Geometry on Soil Permeability. *Adv. Water Resour.* **2019**, *129*, 232–249. [CrossRef]
62. Garcia, X.; Akanji, L.T.; Blunt, M.J.; Matthai, S.K.; Latham, J.P. Numerical Study of the Effects of Particle Shape and Polydispersity on Permeability. *Phys. Rev. E Stat. Nonlin. Soft Matter. Phys.* **2009**, *80*, 021304. [CrossRef]
63. Regassa, B.; Xu, N.; Mei, G. An Equivalent Discontinuous Modeling Method of Jointed Rock Masses for DEM Simulation of Mining-Induced Rock Movements. *Int. J. Rock Mech. Min. Sci.* **2018**, *108*, 1–14. [CrossRef]
64. Dai, H.; Wang, Z.; Zhao, J.; Jia, X.; Liu, L.; Wang, J.; Abbasi, H.N.; Guo, Z.; Chen, Y.; Geng, H.; et al. Modeling and Optimizing of an Actual Municipal Sewage Plant: A Comparison of Diverse Multi-Objective Optimization Methods. *J. Environ. Manag.* **2023**, *328*, 116924. [CrossRef]
65. Dai, H.; Zhao, J.; Wang, Z.; Chen, C.; Liu, X.; Guo, Z.; Chen, Y.; Zhang, S.; Li, J.; Geng, H.; et al. Optimal Control of Sewage Treatment Process Using a Dynamic Multi-Objective Particle Swarm Optimization Based on Crowding Distance. *J. Environ. Chem. Eng.* **2023**, *11*, 109484. [CrossRef]
66. Zhao, J.; Dai, H.; Wang, Z.; Chen, C.; Cai, X.; Song, M.; Guo, Z.; Zhang, S.; Wang, X.; Geng, H. Self-Organizing Modeling and Control of Activated Sludge Process Based on Fuzzy Neural Network. *J. Water Process. Eng.* **2023**, *53*, 103641. [CrossRef]

water

MDPI

Article

Dynamic Forecasting and Operation Mechanism of Reservoir Considering Multi-Time Scales

Chengyu Han [1], Zhen Guo [2], Xiaomei Sun [2,*] and Yuquan Zhang [3,4]

1 School of Human Settlements and Civil Engineering, Xi'an Jiaotong University, Xi'an 710049, China; xbnlkjdx06@163.com
2 School of Water Conservancy and Hydropower, Xi'an University of Technology, Xi'an 710048, China; zhen_g@stu.xaut.edu.cn
3 State Grid Xi'an Electric Power Supply Company, Xi'an 710049, China; djyjyth@163.com
4 School of Management, Xi'an Jiaotong University, Xi'an 710049, China
* Correspondence: xiaomeisun1020@foxmail.com

Abstract: This paper proposes a feedback, rolling and adaptive operation decision-making mechanism for coupling and nesting of time scales. It is aimed at the change of time scale and the dynamics in the operation process, considering the relationship between operation period and multi-time scales. The key point is to integrate forecasting and operation in order to adapt to the multi-time scales dynamic change in the operation process. The operation process is divided into different time scales; forecasting and operation model method libraries are constructed, and the progressive updating and nesting mechanism are used to realize the process dynamic operation, according to the regulation period or operation period of the reservoir. Taking the Miyun Reservoir in Beijing, China as the research object, the operation mechanism is integrated into the operation process, and the complex forecasting operation and control mechanism are integrated, based on the integrated platform and using modern information technology. The forecasting and operation method uses classic different models, which can be selected based on different goals. The forecasting inflow is used as input, and the output is the water distribution plan, more importantly, the mechanism in the operation process is the key point. This is a rolling modification of the inflow process in the next stage, and the operation plan also changes accordingly. The feasibility, effectiveness, rationality and flexibility of the reservoir dynamic and adaptive operation are verified, so that the reservoir operation is dynamically changing and adapting to the changing demand. The proposed operation mechanism has scientific value and guiding significance to improve the reservoir operation theory, and it provides decision support for the actual reservoir operation and operation business.

Keywords: reservoir forecasting and operation; dynamic process; multi-time scales; integrated platform

Citation: Han, C.; Guo, Z.; Sun, X.; Zhang, Y. Dynamic Forecasting and Operation Mechanism of Reservoir Considering Multi-Time Scales. *Water* **2023**, *15*, 2472. https://doi.org/10.3390/w15132472

Academic Editor: Roohollah Noori

Received: 25 May 2023
Revised: 3 July 2023
Accepted: 4 July 2023
Published: 5 July 2023

1. Introduction

The problem of water scarcity in the 21st century is more and more serious, and there is a serious imbalance between supply and demand. It is related to the country's economy, social development and people's livelihood. As far as the water resources in China are concerned, the water resources are unevenly distributed in time and space, and contradictions in water use are prominent. In the case of serious water scarcity, the construction of water conservancy projects can redistribute water resources spatially, effectively alleviate the uneven distribution of water resources in time and space, and also alleviate the problem of regional water use [1–3]. Therefore, how to make the scientific and reasonable operation of the reservoir is the key for reservoir regulation, which directly affects the utilization of water resources. So, it is necessary to carry out research on the rational use of the reservoir operation.

The dynamic operation problem is a more common problem, which is more in line with the actual production needs. Many scholars pay attention to its research [4–6]. At present, it has become one of the hot spots in the field of operation [7,8]. In essence, the whole process of reservoir operation is the control of reservoir discharge process in a period of time. It is a dynamic process. In this process, the inflow of the reservoir is dynamic, and the factors describing the state of the reservoir (water level, storage capacity, etc.) are also dynamic Reservoir operation determines the control strategy or decision of the whole process of reservoir discharge in the whole dynamic process according to the objectives of the reservoir operation [9–11]. K.C. Abbaspour [12] applied an attractive procedure for obtaining model parameters in recent years to estimating hydraulic parameters in a lysimeter experiment. Stochastic dynamic programming (SDP) has been widely used in reservoir operation strategy considering runoff uncertainty [13–15]. Dariane [16] used an intelligent water drop algorithm to solve the reservoir optimal operation model. Ahmed [17] successfully used a genetic algorithm (GA) to solve the multi-objective reservoir optimal operation problem and achieved good optimization results. Moeini [18] used fuzzy dynamic programming to solve the optimal operation problem of cascade hydropower stations. Afshar [19] used a particle swarm optimization (PSO) algorithm in the reservoir-group operation. Sharma S. [20] updated the flood forecast in real time through in-depth research, and established a set of perfect real-time operation systems in the flood season. At present, the research on static operation is mature and has practical application. However, a lot of literature [7,21–25] also pointed out that due to the uncertainty of reservoir operation, even if the static planned operation obtains optimal results, it will become infeasible due to random disturbance in actual operation, that is, "the plan can't keep up with the change". Therefore, dynamic operation becomes more and more important.

At present, scholars at home and abroad have carried out a lot of theoretical research on reservoir operation and had many achievements. However, the decision making of the reservoir operation is an extremely complex process. The randomness of inflow runoff, the multiplicity of reservoir functions, the dynamic and real-time characteristics of the decision-making process, the limitations of mathematical models and solving technologies and the complexity of human factors make the reservoir operation problems present obvious unstructured or semi-structured characteristics. In order to combine the theoretical results with the actual application, a new reservoir forecasting and operation mechanism that can take the forecasting and operation as a whole is required in such a complicated setting. Therefore, considering the relationship between operation period and multi-time scales, this paper proposes a feedback, rolling and adaptive operation decision-making mechanism for coupling and nesting of time scales. The reservoir is dynamic during the operation process, including changes in incoming water, water volume, time scale, and water demand and so on. The operation plan is also dynamic, as long as there is a change in the real-time operation plan. Other time scale plans will also change due to the rolling adjustment of incoming water. The operation plans at different time scales are rolling nested, which is similar to a chain reaction between different layers.

2. Methodologies

2.1. Main Ideas

According to the existing forecasting mode and operation methods, aiming at the uncertainty and adaptability of dynamic changes in reservoir forecasting and operation, this paper studies the dynamic mechanism of reservoir forecasting and operation process. This mechanism reflects a double-layer adjustment. On the one hand, rolling adjustment refers to the change in water supply or operation plans at a certain moment, and subsequent predicted water supply will also change accordingly; nested adjustment refers to the nested adjustment of forecasting incoming water or operation plans between levels at different time scales. Based on the whole process of the reservoir operation cycle, the regulation is closely related to forecasting in the process. Only in each link of the operation process, mathematical models and methods are used to solve specific problems. The multi-time scales rolling nested mutual feedback mechanism is constructed in the process of forecasting and operation. On the visual integrated platform, the mechanism is realized, and a new paradigm of reservoir forecasting and operation is formed, which provides the scientific support for reservoir forecasting and operation under changing conditions.

In the process of reservoir operation in the operation period, the dynamic changes of factors are fully considered, and the forecasting and operation are closely related in the process. According to the rolling correction of forecast results (long-term, medium-term, short-term and real-time), the corresponding operation plan is also modified, and multi-time scales are nested in the operation process. The operation plan is used to guide the whole process of the operation cycle, and the real-time operation scheme is used to guide the implementation of operation. In the working process, the operation plan and the implementation operation scheme are fed back to each other (see in Figure 1).

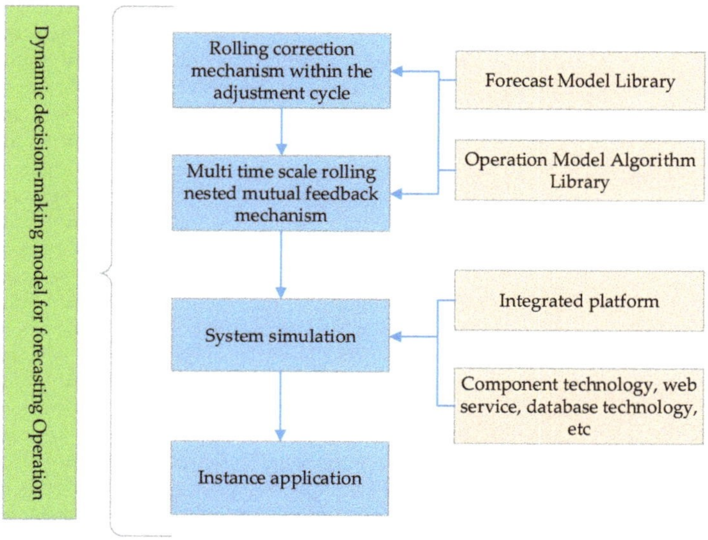

Figure 1. Main ideas.

During the adjustment cycle, if there is a change in incoming water at a certain moment, the forecasting of incoming water in the later period will also be reforecasted based on the latest actual incoming water. When there are new time changes, subsequent forecasting of incoming water will change and be repeated, which will lead to changes throughout the entire operation cycle. Forecasting of incoming water at other time scales will also be reforecasted based on the latest short-term results, and rolling nested reactions will occur. The operation plan will also be adjusted accordingly.

2.2. Rolling Correction Mechanism Mode

Reservoir forecasting and operation is an uninterrupted process, a continuous calculation process, and continuous feedback and a rolling correction process. It needs to be able to continuously provide a rolling forecast of the coming water, and the operation will also move with it.

It is assumed that the general form of the hydrological forecasting model is as follows:

$$y = f(x) \tag{1}$$

where x is the sample data, $f(x)$ is the forecasting model, and it can be any model of forecasting.

(1) The rolling process of forecast: suppose that the coming water at a certain time t_1 is used to forecast the coming water at the time t_2, and $\{x|x_0, x_1, \cdots, x_t\}$ is selected as the

sample datum of the forecast model, then when it is necessary to forecast the coming water at the time t_3, the sample data increase x_{t+1} and becomes $\{x|x_0, x_1, \cdots, x_t, x_{t+1}\}$. At this time, in order to forecast, a new sample must be used, which can be the same length as t_1. At this time, the sample value x_0 which is far away from time t_3 is removed, and the new sample is $\{x|x_1, \cdots, x_t, x_{t+1}\}$, thus forming a rolling forecast.

(2) For the rolling process with measured data, it is necessary to revise the previous forecasting process. The rolling correction principle of forecast in an operation period is shown in Figure 2. The yellow part represents the historical data, the purple part represents the measured data, the orange part represents the short-term forecast results, and the red part represents the whole operation cycle. Suppose that there is a measured value $x_{t'}$ at a time t_1, it needs to modify the forecasting at time t_2, then the sample changes from the original $\{x|x_0, x_1, \cdots, x_t\}$ to $\{x|x_0, x_1, \cdots, x_{t'}\}$. At this time, the measured value is added to the sample, and the new sample is used for subsequent forecasting, so as to modify the forecasting in the process of rolling. In this way, the measured value can improve the accuracy of forecasting.

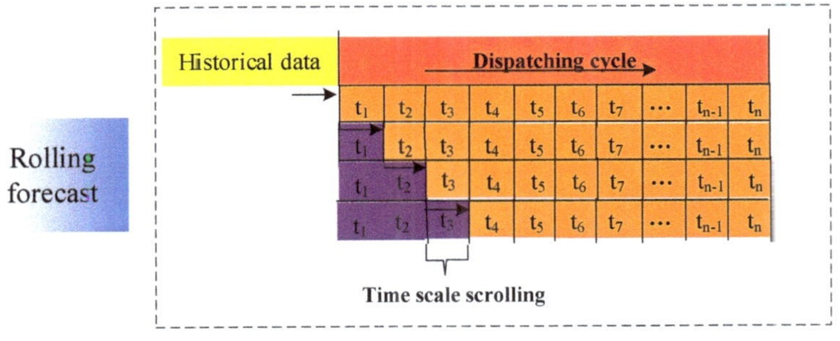

Figure 2. Principle of forecast rolling correction in operation cycle.

According to the rolling correction mechanism flow shown in Figure 3, suppose that the value of forecast at t_0 in the operation cycle is $\{Q|Q_{t1}, Q_{t2}, \cdots Q_{tn}\}$, when there is a measured value $\overline{Q}_1$ at t_1, the measured value $\overline{Q}_1$ needs to be added to the forecast samples, and the rolling correction forecast needs to be carried out again for the subsequent period. After rolling forecast, the forecast results of the reserved time in the cycle become $\{Q|\overline{Q}_1, Q'_{t2}, \cdots, Q'_{tn}\}$ At this time, the operation plan also changes with the change of the forecast result, and the operation plan has been performing a rolling correction, and the operation correction result is $\{Z|\overline{Z}_1, \overline{Z}_2, Z''_3, \cdots, Z''_n\}$. When there is a measured value Q_2 at t_2, the forecasting samples need to add the measured value $\overline{Q}_2$ to the forecasting samples, and the rolling correction forecast needs to be carried out again for the subsequent period. After rolling forecasting, the forecasting result of the reserved time in the cycle becomes $\{Q|\overline{Q}_1, \overline{Q}_2, Q'_{t3}, \cdots, Q'_{tn}\}$. At this time, the operation plan scheme also changes with the change of forecast, and the operation plan scheme has performed a rolling correction. The result of operation correction is $\{Z|\overline{Z}_1, \overline{Z}_2, Z''_3, \cdots, Z''_n\}$ and it has been rolling circularly. At this time, the measured value is added to the sample, and the new sample is used for subsequent forecasting, so as to roll correction forecasting in the process until the end of the operation cycle. With the continuous rolling correction of the forecast, the accuracy will also increase. The operation will change with the change of the incoming water forecast. The operation plan will be continuously modified to make the dynamic operation plan.

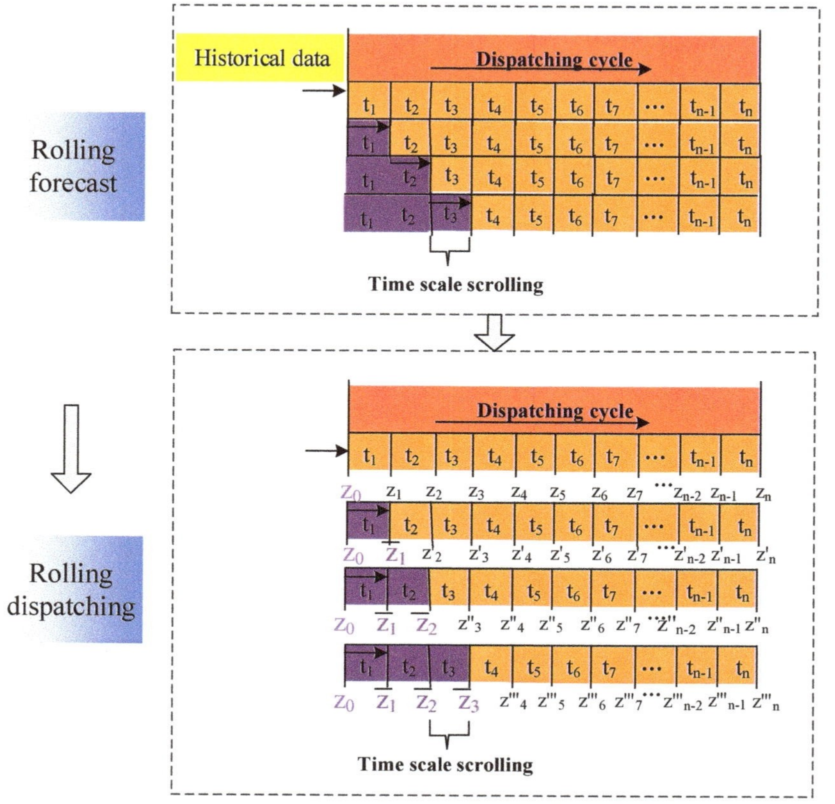

Figure 3. Rolling correction mechanism flow of forecast operation in operation cycle.

2.3. Multi-Scale Rolling Nested Mutual Feedback Mode

In this study, a mutual feedback mechanism between multi-time scales operation scheme and implementation operation scheme is established under the rolling correction mechanism to adapt to the change of conditions through process change. The working principle of the mutual feedback mechanism is shown in Figure 4. At the beginning of the operation cycle, the reservoir is operated according to the predetermined operation plan. The black arrows refer to the lengths of the time scales, and the dots refer to a long sequence of time. The yellow part represents historical data, the purple part represents measured data, the orange part represents short-term forecast results, and the red part represents the entire scheduling cycle.

(1) According to the actual inflow situation, if the real-time operation in the operation period is consistent with the operation plan, then at the end of an operation period, under the rolling nested forecasting mechanism, it is only necessary to judge whether the forecast sample data are increased. If it is increased, it will be forecasted again, and then the subsequent operation plan is formulated.

(2) The change in operation conditions leads to a deviation between the operation and the planned operation during the operation period. If an emergency occurs: an oil leakage water pollution incident occurs in the downstream of the reservoir, and the discharge flow needs to be increased for the purpose of diluting pollutants, or in management of floods, the currently implemented operation plan needs to respond to changes and make adjustments. After an operation period (as shown in t_1 in Figure 4), because the reservoir does not carry out the operation according to the original plan,

the follow-up operation of this period will be delayed. The operation plan should also be changed and reformulated. The actual boundary conditions (i.e., the current reservoir water level) are fed back to the upper short-term operation. Based on the rolling forecast results and the new boundary conditions, the operation model is modified, and the short-term operation plan after t_1 period (as shown in the yellow part of the figure, which is updated after t_1 period) is formulated again. The new operation plan is implemented for the next period (t_2 period) and the new operation plan is implemented in the whole period. According to this method, the operation period is recursively extended to medium-term operation and long-term operation until the end of the reservoir operation period.

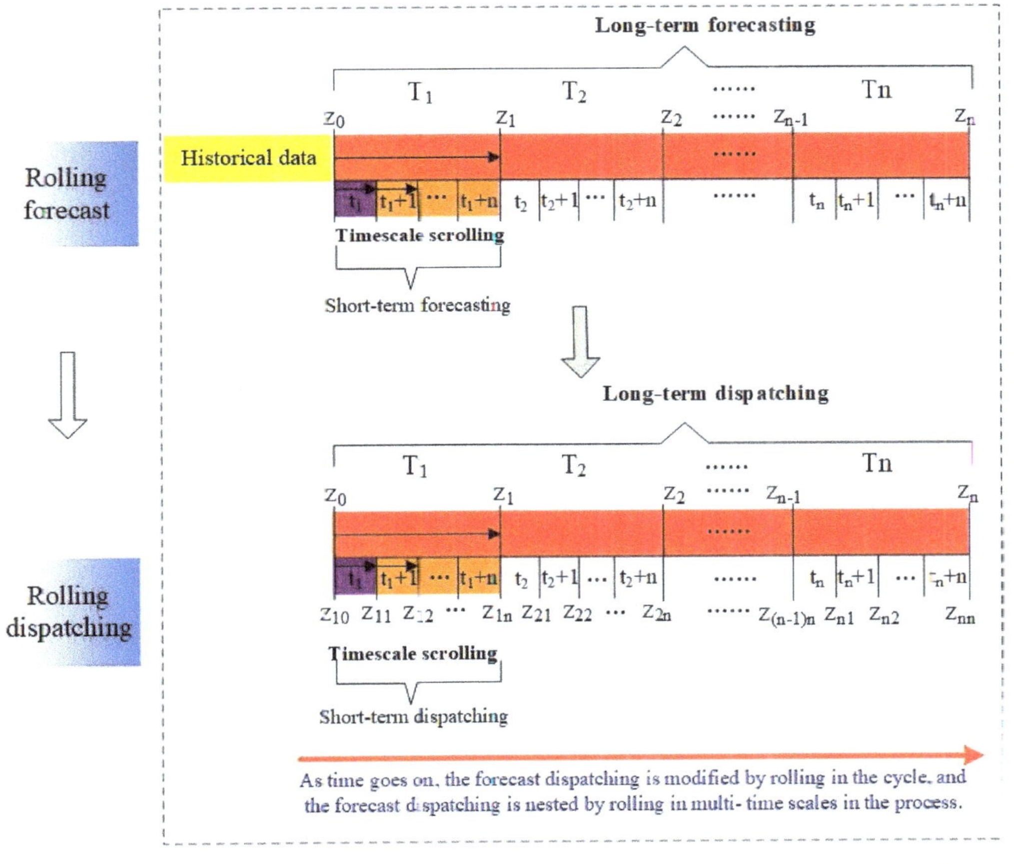

Figure 4. Schematic diagram of multi-time scale rolling nesting mechanism of forecast operation.

2.4. Forecasting and Operation Models

On the basis of existing reservoir forecasting and operation models, forecasting and operation model method libraries are constructed using information technology to serve different needs of forecasting and operation (see in Table 1), in order to make rapid calculation and analysis, such as long-term forecasting, mid-term forecasting, short-term forecasting, single objective operation, multi-objective operation, etc.

Table 1. Runoff forecast and operation models.

No.	Forecasting Models	Operation Models
00001	BP	DP
00002	GM(1,1) model	POA
00003	AR_P model	Particle swarm optimization
00004	SVR model	Genetic algorithm
00005	XinAnjiang model	NSGA-II
00006	Tank madel	MOEA/D
00007	NASH unit hydrograph model	MOEA/D-AWA
...	...	...

3. Implementation Means

3.1. Technical Support

(1) Technical basis: The integrated platform is a platform developed based on the National 863 project in China. The platform is designed and constructed according to the requirements of the water conservancy industry standard of the People's Republic of China "technical regulations for water conservancy information processing platform" (SL538-2011). The comprehensive integration platform can integrate Web information, XML information, report information, GIS information, statistical graphics, word documents, PPT and other information, and the user layer is mainly for users with login permissions to access the platform. The overall framework is shown in Figure 5.

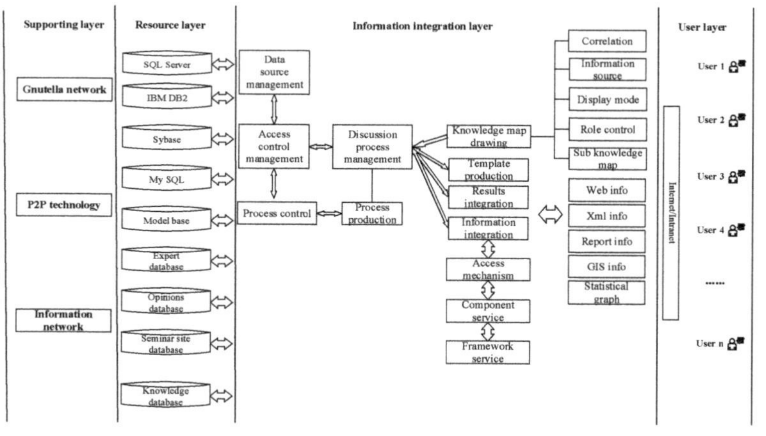

Figure 5. Framework design of water operation integrated platform.

(2) Technical means: Based on the comprehensive integration platform, this paper uses the key technologies such as knowledge graph technology [26], component technology, web service technology, database technology and integrated application technology to develop and realize the dynamic decision-making system of the forecast operation process.

3.2. Implementation Process

In this paper, the dynamic decision-making mechanism of reservoir forecasting and operation process is established. Based on the integrated platform, the system is constructed. The hierarchical structure based on the idea of reservoir forecasting and operation is shown in Figure 6. The business and components are coupled to realize the related business requirements.

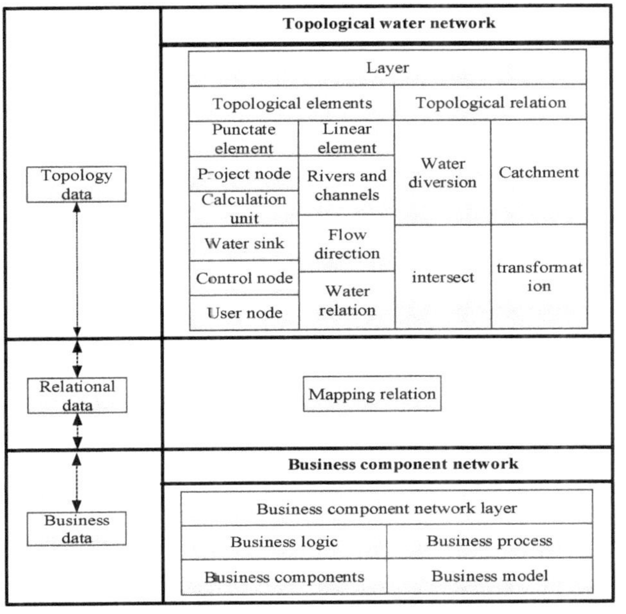

Figure 6. The hierarchical structure of topology and business implementation ideas.

The topological water network is composed of many nodes, and each node has a logical relationship and data flow relationship. The business of each node is supported by the components of the business component layer. Through the mapping relationship between the two, the components are customized to the nodes, and through the combined application of business components, the related business of reservoir forecasting and operation is realized.

4. Case Study

4.1. Study Area

Beijing, Tianjin and Hebei are the main platforms in China that show their strong vitality and competitiveness in the international economic system. At the same time, the region only uses 2.3% of the national land area to carry 8% of the population, and also contributes 10% of the national GDP by using less than 1% of the water. Beijing, Tianjin and Hebei are one of the regions in China and the world that are most affected by human activities and one of the regions that have the most difficulty in ensuring water safety, and with the largest water resources bearing capacity and the highest risk level [27]. Therefore, it is of great practical significance to study the water resources in this area.

The Miyun reservoir is located 13 km north of Miyun District, Beijing, in the hills of the Yanshan Mountains. It was built in September 1960. The area is 180 square kilometers, with a circumference of 200 km around Miyun Reservoir. The Miyun Reservoir is shaped like an equilateral triangle. The Miyun reservoir has a storage capacity of 4 billion cubic meters and an average water depth of 30 m, it serves as the largest and only source of drinking water supply in the capital city of Beijing. The characteristics of the Miyun reservoir are shown in Table 2. The Miyun Reservoir has two major inflow rivers, namely the Bai River and the Chao River. In the 30 years after the completion of the Miyun reservoir, it has generated huge benefits in various directions such as flood control, irrigation, urban water supply, power generation, fish farming and tourism.

Table 2. Miyun reservoir characters.

Miyun Reservoir Characters	Water Level (m)	Corresponding Water Surface Area (km^2)	Storage Capacity (10^8 m^3)
flood level	158.5 m	183.6	41.9
normal water level	157.5 m	179.33	40.08
flood limit water level	147.0 m	137.54	23.38
dead water level	126.0 m	46.154	4.37

For the accuracy of the forecasting, the historical data of incoming water were selected from 1990 to 2021 for the forecasting.

4.2. Models Application and System Simulation

The implementation of the rolling correction mechanism in the forecasting and operation process is shown in Figure 7. On the interface, you can see the incoming water forecasting and rolling forecasting nodes, which are supported by components in the component layer. You can see that the following blue line is also the starting condition of rolling forecasting. Based on the platform and supported by components, the rolling correction in the operation cycle can be realized on the visual flow chart. All arrows refer to the inflow and outflow of data flow. The dotted boxes refer to data nodes.

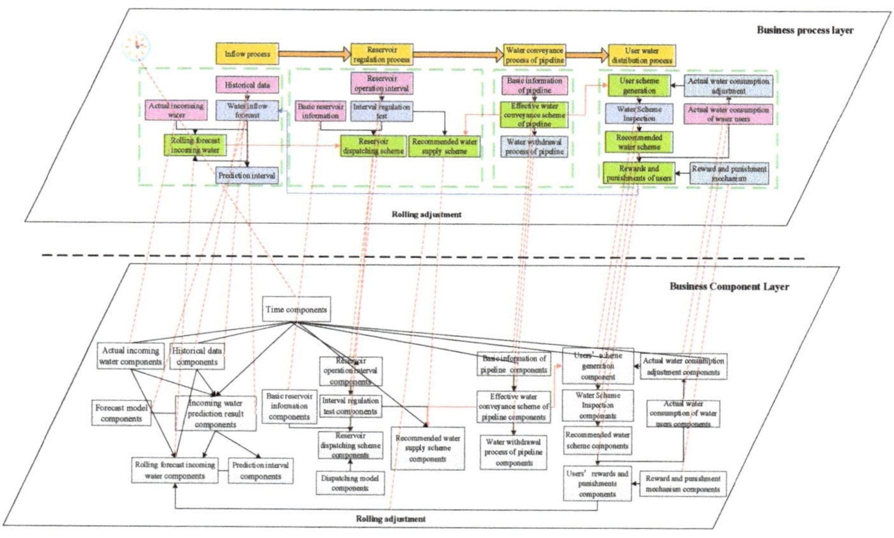

Figure 7. Implementation of rolling correction mechanism in forecast operation process.

4.3. Scenario Settings

The whole process of water adaptive regulation and control of the Miyun reservoir in China is based on the following process: water supply process $\rightarrow$ reservoir regulation process $\rightarrow$ pipeline water delivery process $\rightarrow$ user water distribution process. Through the rolling analysis of the "operation" and "distribution" process of each object, the dynamic adjustment of each link is realized, so as to achieve the process dynamic decision making of reservoir forecast and operation, and realize the global adaptive regulation.

The realization of the mutual feeding business between the multi-time scales operation scheme and the implementation operation scheme is shown in Figure 8. On the interface, we can see that the interface is composed of two parts: the water consumption plan and the operation scheme. On the business process layer, we can see that there are two-way arrows between the planned water consumption and the operation scheme, and there are nesting and mutual feeding relationships between them. In Figure 8, all arrows refer to the inflow and outflow of data flow, the red arrow refers to the outflow of results from the normal process of forecasting operation, and the blue arrow refers to the data flow for feedback adjustment operation.

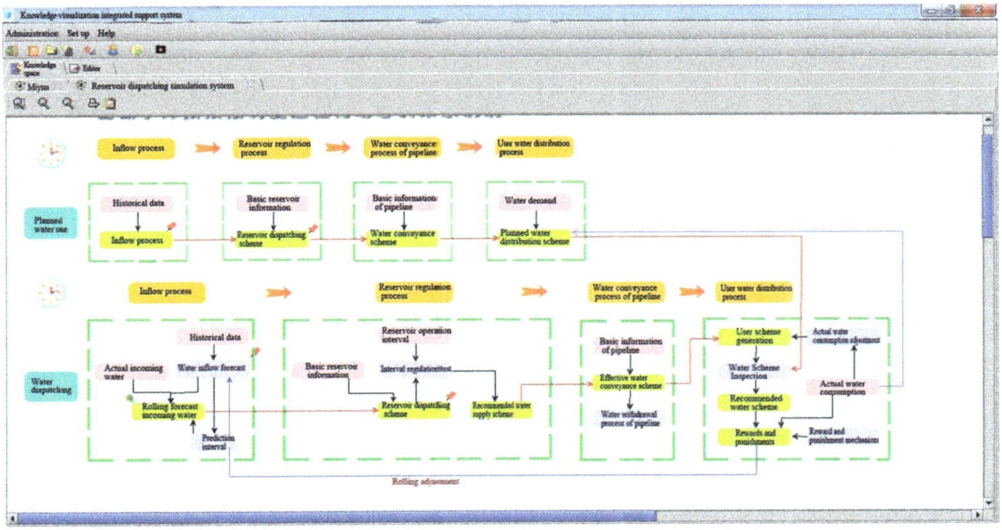

Figure 8. Adaptive control interface of Miyun Reservoir.

Based on the platform and supported by components, it realizes the rolling nesting mutual feed of multi-time scales in the process of the operation cycle on the visual flow chart.

4.3.1. Scenario 1: Rolling Forecast

The inflow process needs to be forecasted [28–30] according to the historical flow, and then the forecasted inflow process is compared with the measured inflow in the hydrological year. The rolling forecasting of the inflow process is realized by selecting the rolling forecasting time. If you choose the rolling forecast time as June 2015, the flow before June 2015 is the actual measured value, and the forecast value after that. Then the forecast is modified by rolling. The change of inflow will also lead to the change of the reservoir operation plan, so the operation is also in the state of rolling correction. Figure 9 is the rolling forecast of incoming water. In the figure, a comparison was made between the forecasted flow rate and the flow rate after rolling forecast adjustment. It can be seen that the flow rate after rolling adjustment is closer to the actual incoming water; therefore, the accuracy of the forecasting is higher.

4.3.2. Scenario 2: Water Distribution Plan

The operation scheme is based on the operation method library and operation target library established in advance. The operation method component and operation target component are customized on the corresponding nodes in the knowledge graph drawn on the platform. Combined with the actual operation requirements, the operation target and operation method are selected in the corresponding components to calculate the operation scheme. Reservoir regulation calculation is carried out according to its characteristic water level and storage capacity, as well as water level discharge curve and water level storage capacity curve. The pipeline calculates the water delivery capacity of the pipeline and generates the pipeline water delivery plan according to the water discharge process of the reservoir and the attributes of the pipeline itself. The user generates the user's planned water distribution scheme according to the user's water demand and the pipeline's water delivery scheme. The operation scheme and the user's planned water distribution scheme are shown in Figure 10. In the figure, a comparison was made between the user planned water allocation and the recommended water consumption after rolling forecast adjustment. It can be seen that the recommended water consumption after rolling adjustment is closer to the actual water use, thus reflecting the applicability of this mechanism.

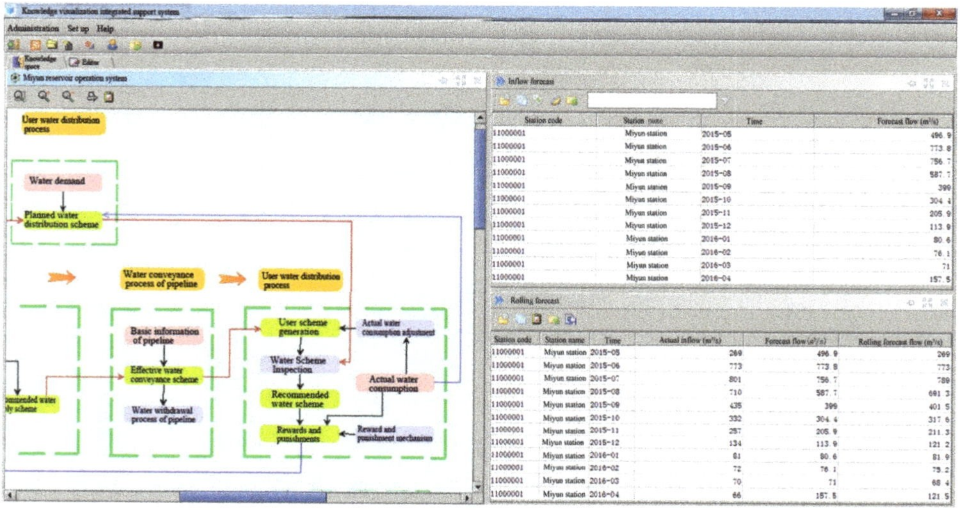

Figure 9. Rolling forecasting results of incoming water.

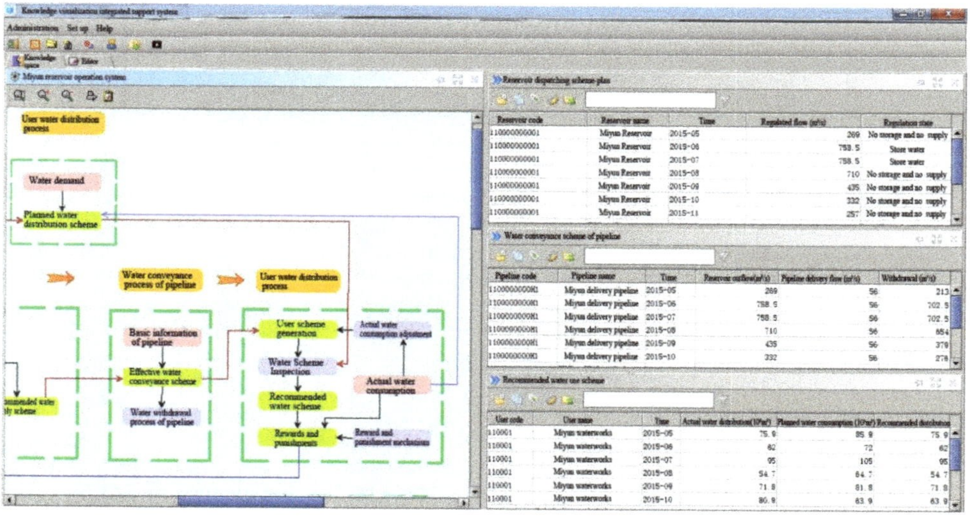

Figure 10. Water distribution plan of reservoir operation plan and water user plan.

4.3.3. Scenario 3: Water Use Plan Inspection

In the process of reservoir operation, it is necessary to conduct interval regulation inspection on the reservoir operation plan. If the discharge flow is unreasonable, the operation regulation can be reselected or the operation range can be recalibrated. If the result is reasonable, the recommended operation scheme can be generated. Figure 11 shows the recommended water use scheme generated. The actual water use plan can be verified by the user planned water distribution plan generated above, and the excess water amount is expressed by a red bar graph. After the inspection, a recommended water use plan can be generated. Finally, according to the designated step-by-step over-plan water price as a reward and punishment mechanism, users will be punished for exceeding the planned water volume. For users with penalties, they can adjust their next water use plan, and they need to adjust the incoming water on a rolling basis to achieve dynamic rolling control of the entire water volume allocation process.

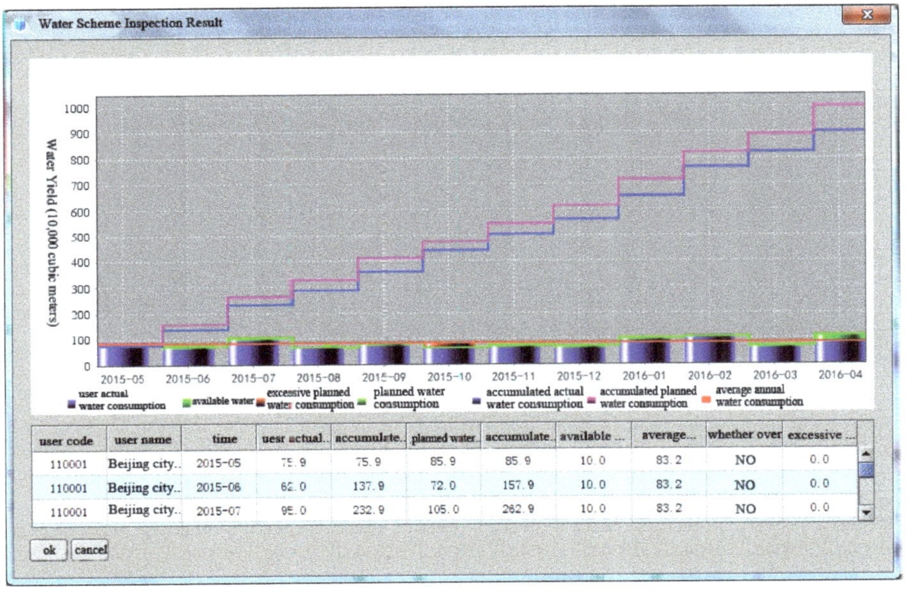

Figure 11. Test results of water use plan.

5. Results and Conclusions

The rolling correction mechanism can form a procedural operation method: in the whole reservoir operation cycle, the changing factors bring many unknowns to the operation, and also the actual operation results are not consistent with the planned scenario, and it is not clear how much the project benefits will play. In the process of operation, the time period and time scale have a great influence on the forecast operation.

The integration of reservoir forecasting, operation and decision making is realized on the platform, which can well adapt to the changes of the reservoir and achieve the purpose of dynamic operation. Through the construction of reservoir forecasting and operation process dynamic decision-making system, the rationality of the new mechanism is verified through the application of an example, which provides a new solution and means for reservoir operation business. In this paper, aiming at the time period change and the dynamic characteristics in the operation process, a process dynamic forecast operation mechanism of reservoirs is proposed, which forms a rolling correction mechanism for forecast operation in the operation period and a multi-scale rolling nested mutual feedback mechanism. Taking Miyun Reservoir in Beijing as an example, the reservoir process dynamic forecasting and operation system are carried out based on the integrated platform and computer technologies. Through the example application, the rationality of the new mechanism is verified, which provides a new solution and means for reservoir operation.

The dynamic operation of the reservoir itself is relatively complex. There are a lot of uncertainties in the operation of the reservoir. The theory and technology involved in the realization of the dynamic operation of the reservoir are relatively complex, it needs to be further solved and improved in the follow-up work. In the future, further research in this area is needed.

Author Contributions: Conceptualization, C.H. and Z.G.; methodology, C.H.; software, C.H., Y.Z. and Z.G.; validation, C.H. and X.S.; formal analysis, C.H. and Z.G.; investigation, C.H.; resources, X.S.; data curation, Z.G.; writing—original draft preparation, C.H.; writing—review and editing, X.S. and Y.Z.; visualization, C.H.; supervision, X.S.; project administration, X.S.; funding acquisition, X.S. All authors have read and agreed to the published version of the manuscript.

Funding: This research was funded by the National Natural Science Foundation of China (12272291, 11872295).

Data Availability Statement: The data presented in this study are available on request from the corresponding author.

Acknowledgments: We are grateful for the helpful and constructive review comments that helped to improve this work significantly. Therefore, we would also like to thank the reviewers.

Conflicts of Interest: The authors declare no conflict of interest.

References

1. Bozorg-Haddad, O.; Moravej, M.; Loaiciga, H.A. Application of the Water Cycle Algorithm to the Optimal Operation of Reservoir Systems. *J. Irrig. Drain. Eng. ASCE* **2015**, *141*, 04014064. [CrossRef]
2. Ming, B.; Chang, J.X.; Huang, Q.; Wang, Y.M.; Huang, S.Z. Optimal Operation of Multi-Reservoir System Based-On Cuckoo Search Algorithm. *Water Resour. Manag.* **2015**, *29*, 5671–5687. [CrossRef]
3. Noori, R.; Maghrebi, M.; Mirchi, A.; Tang, Q.; Bhattarai, R.; Sadegh, M.; Noury, M.; Torabi Haghighi, A.; Kløve, B.; Madani, K. Anthropogenic depletion of Iran's aquifers. *Proc. Natl. Acad. Sci. USA* **2020**, *118*, e2024221118. [CrossRef] [PubMed]
4. Zhang, W.; Liu, P.; Chen, X.; Wang, L.; Ai, X.; Feng, M.; Liu, D.; Liu, Y. Optimal Operation of Multi-reservoir Systems Considering Time-lags of Flood Routing. *Water Resour. Manag.* **2016**, *30*, 523–540. [CrossRef]
5. Basu, M. Artificial immune system for dynamic economic dispatch. *Electr. Power Energy Syst.* **2011**, *33*, 131–136. [CrossRef]
6. Xiaohua, X.; Jiangfeng, Z.; Elaiw, A. An application of model predictive control to the dynamic economic dispatch of power generation. *Control. Eng. Pract.* **2011**, *19*, 638–648. [CrossRef]
7. Ilich, N. WEB.BM—A Web Based River Basin Management Model with Multiple Time Step Optimization and 47 the SSARR Channel Routing Options. *Hydrol. Sci. J.* **2021**, *6*, 175–190.
8. Alsumait, J.S.; Qasem, M.; Sykulski, J.K. An improved Pattern Search based algorithm to solve the Dynamic Economic Dispatch problem with valve-point effect. *Energy Convers. Manag.* **2010**, *51*, 2062–2067. [CrossRef]
9. Shengping, Y.; Ruixia, L.; Binglin, Z. Simulation system for logistics in steelmaking process based on Flexsim. *J. Syst. Simul.* **2009**, *21*, 5145–5149. [CrossRef]
10. Guo, X.; Hu, T.; Zhang, T.; Lv, Y. Bilevel model for multi-reservoir operating policy in inter-basin water transfer-supply project. *J. Hydrol.* **2012**, *424–425*, 252–263. [CrossRef]
11. Wurbs, R.A. Reservoir-system simulation and optimization models. *J. Water Resour. Plan. Manag.* **1993**, *119*, 455–472. [CrossRef]
12. Abbaspour, K.C.; Schulin, R.; van Genuchten, M.T. Estimating unsaturated soil hydraulic parameters using ant colony optimization. *Adv. Water Resour.* **2001**, *24*, 827–933. [CrossRef]
13. Lei, X.H.; Tan, Q.F.; Wang, X.; Wang, H.; Wen, X.; Wang, C.; Zhang, J.W. Stochastic optimal operation of reservoirs based on copula functions. *J. Hydrol.* **2018**, *557*, 265–275. [CrossRef]
14. Celeste, A.B.; Billib, M. Evaluation of stochastic reservoir operation optimization models. *Adv. Water Resour.* **2009**, *32*, 1429–1443. [CrossRef]
15. Cervellera, C.; Chen, V.C.P.; Wen, A. Optimization of a large-scale water reservoir network by stochastic dynamic programming with efficient state space discretization. *Eur. J. Oper. Res.* **2006**, *171*, 1139–1151. [CrossRef]
16. Dariane, A.B.; Sarani, S. Application of Intelligent Water Drops Algorithm in Reservoir Operation. *Water Resour. Manag.* **2013**, *27*, 4827–4843. [CrossRef]
17. Ahmed, J.A.; Sarma, A.K. Genetic Algorithm for Optimal Operating Policy of a Multipurpose Reservoir. *Water Resour. Manag.* **2005**, *19*, 145–161. [CrossRef]
18. Moeini, R.; Afshar, A.; Afshar, M.H. Fuzzy rule-based model for hydropower reservoirs operation. *Int. J. Electr. Power Energy Syst.* **2011**, *33*, 171–178. [CrossRef]
19. Afshar, M.H. Large scale reservoir operation by constrained particle swarm optimization algorithms. *J. Hydroenviron. Res.* **2012**, *6*, 75–87. [CrossRef]

20. Sharma, S.; Siddique, R.; Reed, S.; Ahnert, P.; Mendoza, P.; Mejia, A. Relative effects of statistical preprocessing and postprocessing on a regional hydrological ensemble forecasting system. *Hydrol. Earth Syst. Sci.* **2018**, *22*, 1831–1849. [CrossRef]
21. Maggioni, V.; Massari, C. On the performance of satellite precipitation products in riverine flood modeling: A review. *J. Hydrol.* **2018**, *558*, 214–224. [CrossRef]
22. Radigan, W.J.; Carlson, A.K.; Fincel, M.J.; Graeb, B.D.S. Otolith chemistry as a fisheries management tool after flooding: The case of Missouri River gizzard shad. *River Res. Appl.* **2018**, *34*, 270–278. [CrossRef]
23. Peng, A.; Peng, Y.; Zhou, H.; Zhang, C. Multi-reservoir joint operating rule in inter-basin water transfer-supply project. *Sci. China Technol. Sci.* **2015**, *58*, 123–137. [CrossRef]
24. Bistline, W.G., Sr.; Banerjee, S.; Banerjee, A. RTSS: An interactive decision support system for solving real time scheduling problems considering customer and job priorities with schedule interruptions. *Comput. Oper. Res.* **1998**, *25*, 981–995. [CrossRef]
25. Noori, R.; Ansari, E.; Bhattarai, R.; Tang, Q.; Aradpour, S.; Maghrebi, M.; Haghighi, A.T.; Bengtsson, L.; Kløve, B. Complex dynamics of water quality mixing in a warm mono-mictic reservoir. *Sci. Total Environ.* **2021**, *777*, 146097. [CrossRef]
26. Liu, X.; Li, X.; Lei, Z. Summary of knowledge map. *J. Eng. Math.* **2000**, *17*, 33–40. (In Chinese)
27. Bao, C.; He, D.M. Spatiotemporal characteristics of water resources exploitation and policy implications in the Beijing-Tianjin-Hebei Urban Agglomeration. *Prog. Geogr.* **2017**, *36*, 58–67. (In Chinese) [CrossRef]
28. Huang, S.; Chang, J.; Huang, Q.; Chen, Y. Monthly streamflow prediction using modified EMD-based support vector machine. *J. Hydrol.* **2014**, *511*, 764–775. [CrossRef]
29. Lima, A.R.; Cannon, A.J.; Hsieh, W.W. Forecasting daily streamflow using online sequential extreme learning machines. *J. Hydrol.* **2016**, *537*, 431–443. [CrossRef]
30. Shiri, J.; Kisi, O. Short-term and long-term streamflow forecasting using a wavelet and neuro-fuzzy conjunction model. *J. Hydrol.* **2010**, *394*, 486–493. [CrossRef]

water

MDPI

Article

Intelligent Detection Method for Concrete Dam Surface Cracks Based on Two-Stage Transfer Learning

Jianyuan Li [1,2,3], Xiaochun Lu [1,2,*], Ping Zhang [1,2] and Qingquan Li [1,2]

1 College of Hydraulic and Environmental Engineering, China Three Gorges University, Yichang 443002, China; lijianyuan22@163.com (J.L.)
2 Hubei Key Laboratory of Construction and Management in Hydropower Engineering, China Three Gorges University, Yichang 443002, China
3 State Key Laboratory of Simulation and Regulation of Water Cycle in River Basin, China Institute of Water Resources and Hydropower Research, Beijing 100048, China
* Correspondence: luxiaochun1014@163.com; Tel.: +86-13986791594

Abstract: The timely identification and detection of surface cracks in concrete dams, an important public safety infrastructure, is of great significance in predicting engineering hazards and ensuring dam safety. Due to their low efficiency and accuracy, manual detection methods are gradually being replaced by computer vision techniques, and deep learning semantic segmentation methods have higher accuracy and robustness than traditional image methods. However, the lack of data images and insufficient detection performance remain challenges in concrete dam surface crack detection scenarios. Therefore, this paper proposes an intelligent detection method for concrete dam surface cracks based on two-stage transfer learning. First, relevant domain knowledge is transferred to the target domain using two-stage transfer learning, cross-domain and intradomain learning, allowing the model to be fully trained with a small dataset. Second, the segmentation capability is enhanced by using residual network 50 (ResNet50) as a UNet model feature extraction network to enhance crack feature information extraction. Finally, multilayer parallel residual attention (MPR) is integrated into its jump connection path to improve the focus on critical information for clearer fracture edge segmentation. The results show that the proposed method achieves optimal mIoU and mPA of 88.3% and 92.7%, respectively, among many advanced semantic segmentation models. Compared with the benchmark UNet model, the proposed method improves mIoU and mPA by 4.6% and 3.2%, respectively, reduces FLOPs by 36.7%, improves inference speed by 48.9%, verifies its better segmentation performance on dam face crack images with a low fine crack miss detection rate and clear crack edge segmentation, and achieves an accuracy of over 85.7% in crack area prediction. In summary, the proposed method has higher efficiency and accuracy in concrete dam face crack detection, with greater robustness, and can provide a better alternative or complementary approach to dam safety inspections than the benchmark UNet model.

Keywords: dam surface cracks; transfer learning; intelligent detection; small-scale datasets

check for updates

Citation: Li, J.; Lu, X.; Zhang, P.; Li, Q. Intelligent Detection Method for Concrete Dam Surface Cracks Based on Two-Stage Transfer Learning. *Water* **2023**, *15*, 2082. https://doi.org/10.3390/w15112082

Academic Editor: Jianjun Ni

Received: 27 April 2023
Revised: 20 May 2023
Accepted: 23 May 2023
Published: 31 May 2023

1. Introduction

Concrete dams are an important infrastructure for developing and utilising hydropower and water resources in human society, and they play an extensive role in promoting sustainable socioeconomic development and ensuring energy security. However, concrete dams are exposed to atmospheric and water environments for long periods and are subjected to external effects such as water erosion, temperature changes, dry and wet conditions, and freezing and thawing [1,2]. Thus, defects inevitably appear on their surfaces [3], with cracks being the most significant factor threatening dam safety and stability [4]. Cracks not only exist on dam surfaces but, if left untreated, will also extend to the inside of the dam, affecting the strength and service life of the dam and eventually causing safety accidents such as leakage and collapse [5]. Therefore, timely identification and detection of cracks on concrete dam surfaces are of great importance in predicting engineering hazards and ensuring dam safety. Due to the low efficiency and accuracy of manual detection and the weak adaptability of traditional image segmentation methods, an intelligent detection method for concrete dam surface cracks based on two-stage transfer learning is proposed, which can achieve accurate and intelligent dam crack segmentation from unmanned aerial vehicle (UAV) images.

To identify the safety hazards posed by cracks in concrete dams, early crack detection was mainly manual, such as inspectors taking a census of cracks through auxiliary tools such as binoculars [6] or by means of manual hanging baskets close to the dam surface [7]. However, manual detection is costly, susceptible to subjective influences, inefficient and inaccurate, and dangerous [8]. Recently, researchers have been investigating the role of dynamic sensor-based crack detection, and together with online singular spectrum analysis with real-time eigen perturbations can make a great contribution to crack monitoring [9]. Real-time single-sensor online damage monitoring technology based on eigen perturbation has been rapidly developed. This method can process data online and detect damage in structures in real time, improve the accuracy of assessing damage by generating new data through eigen perturbation, and monitor the deformation and direction of cracks in real time, enabling temporal and spatial advantages in crack identification [10,11]. However, information also needs to be collected before employing real-time single-sensor and singular spectrum analysis monitoring based on characteristic perturbations [12,13]. The location of cracks is found in time and information such as the specific shape and area of cracks is obtained to better judge the crack hazard level and repair. The greatest difference between dam surface cracks and other surface cracks is in the spatial distribution. Additionally, other cracks are mainly located on land, which is more convenient for using instruments and equipment directly for close-range exclusion. Information related to cracks in dams can be obtained efficiently and accurately by processing UAV images with computer vision technology [14,15].

In recent years, crack image segmentation methods based on computer vision have been researched and can be divided into traditional image segmentation methods and deep learning semantic segmentation methods [16,17]. Traditional image segmentation methods mainly use lower-order visual information such as a lower grey value of the crack's image pixels than the background [18], image binarization, image filters, and numerical images to achieve segmentation for crack recognition [19–21]. For example, Talab et al. [22] first changed an image to a greyscale image and then classified the background and foreground in the greyscale image using a suitable threshold to segment out the cracks. Zhou and Liu [23] used a threshold segmentation algorithm to extract concrete cracks, which was more effective in identifying concrete with no colour difference on the surface. Cho et al. [24] identified and classified crack candidate pixels, then performed filtering to remove noise, and finally searched, filled, and thresholded the crack region to detect cracks. Compared with manual direct detection, traditional image segmentation methods have a certain degree of safety and feasibility; however, when there is considerable noise in the background, the method is vulnerable to interference from external factors and segmentation results are not accurate [25]. Moreover, segmentation methods have more steps for crack detection and require humans to continuously adjust the parameters to adapt to the segmentation scene. Thus, with low detection efficiency [26] that cannot meet practical engineering needs in terms of accuracy and efficiency, an accurate and efficient intelligent crack segmentation method must be established.

Compared with traditional image segmentation methods based on low-order visual information, the higher-order visual information used in deep learning semantic segmentation methods has higher accuracy and robustness [27]. Semantic segmentation algorithms classify each image pixel by using higher-order visual information to fully consider the association between each pixel point [28], thus allowing for more efficient and accurate segmentation of image classes.

After analysing existing studies, semantic segmentation algorithms are feasible for crack detection. Jiang et al. [29] proposed a concrete crack image recognition method based on the UNet model to segment cracks. Then, they used the maximum inner tangent circle algorithm to calculate crack width. Jian et al. [30] used the DeepLabV3 model to experiment with the public dataset Crack500, and the test set IoU was 67%, which is less accurate. However, most studies are based on experimental studies of cracks in housing buildings, pavement, and bridges using open datasets [31–33], while there are fewer studies on safe methods for concrete dam crack detection. Tang et al. [34] proposed a visual crack width measurement method to improve the crack shape estimation efficiency by simplifying the redundant data in crack images according to an improved image refinement algorithm and width measurement scheme after crack image segmentation using the UNet model. Huang et al. [35] proposed an improved DeepLabV3+ network based on hydraulic concrete crack segmentation but could not solve the application scenario of fine crack segmentation. Chen et al. [36] proposed A_DCDNet for dam crack detection based on an FCNN, which possesses efficient detection efficiency, but the model is too coarse for crack edge segmentation. Wang et al. [37] constructed a dam crack dataset and used the benchmark model SegNet to achieve dam crack segmentation, but crack leakage occurred. Cheng et al. [38] proposed an improved U^{2}Net model for embankment crack detection with good detection accuracy, but the dataset is small and prone to underfitting and inadequate training if trained directly. Therefore, if the existing methods are directly applied to concrete dam surface crack detection, the following problems remain: (1) Concrete dam cracks are still different from cracks in the public dataset in distribution and pixel characteristics; thus, the public dataset is not suitable as a training dataset for concrete dam crack detection (see Figure 1). The cracks in the public dataset are more obvious, while the concrete dam cracks are more obscure, small, and low contrast, so the public crack dataset cannot be directly expanded into a dam face crack image, which will change the probability distribution of the training set data and reduce the recognition accuracy and edge segmentation precision. Additionally, the small sample size in the concrete dam crack dataset will lead to inadequate model training and missed detection. (2) Concrete dam images have few pixel points occupied by cracks, while the background takes up the vast majority of the image, resulting in the model's inadequate crack information extraction for cracks in the dam face features and weak segmentation of fine cracks. (3) Cracks in actual projects are more complex, and existing algorithms have low robustness and pixel accuracy for detecting cracks in concrete dams due to uneven lighting and the low contrast between cracks and the surrounding environment, making edge segmentation smooth.

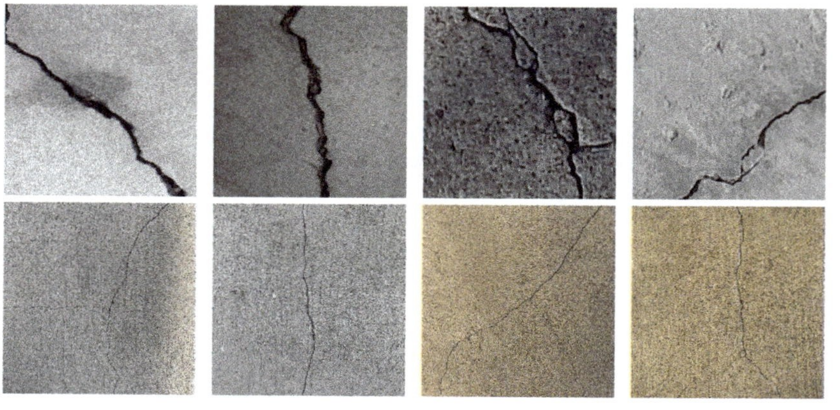

Figure 1. Comparison of crack features. The first row shows images of cracks in the open dataset [31,39], and the second row shows images of cracks in concrete dam surfaces.

To solve the above problems and improve concrete dam crack detection efficiency and accuracy, this paper explores an intelligent detection method for concrete dam surface cracks based on two-stage transfer learning. A two-stage transfer learning strategy is designed to fully train the model and improve accuracy by using the ResNet50 network as the backbone feature extraction network to fully extract the dam face crack feature information Multilayer parallel residual attention (MPR) is constructed and added to improve the model robustness and edge segmentation performance. We conducted experiments on a self-built dam surface crack dataset with UAVs collecting images at 3–10 m from the dam surface with crack widths between 3 and 20 mm to verify the correctness and feasibility of the proposed method. Finally, the experimental results show that the mIoU, mPA, and frames per second (FPS) of the proposed method in this paper are 88.3%, 92.7%, and 36.5, respectively, showing the best performance in dam surface crack segmentation.

The main contributions of this paper are as follows:

1. An intelligent detection method for concrete dam surface cracks based on two-stage transfer learning is proposed to share parameters and features through cross-domain and intradomain learning. The proposed approach can alleviate low segmentation accuracy due to the lack of concrete dam crack datasets that results in inadequate model training and overfitting and reduce the phenomenon of fine crack miss detection.
2. The ResNet50 network is used as the backbone feature extraction network to enhance crack feature information extraction by the UNet encoder and improve the ability to segment fine cracks.
3. The designed multilayer parallel residual attention (MPR) is integrated into its jump connection path to suppress the interference of extraneous regions on crack segmentation to improve pixel accuracy and make crack edge contours clearer.

2. Related Work

2.1. Deep Learning-Based Crack Detection

Deep learning currently has two popular directions: object detection methods and semantic segmentation methods. Usually, two main types of target detection exist: (1) two-stage detection [40] and (2) one-stage detection [41]. The two-stage detection approach first generates a fixed-size feature map for each candidate region using a RoI pooling layer and then obtains the result using bounding box regression, which is computationally intensive, although the accuracy is high. The one-stage detection method predicts the object class and bounding box in one step, as the name suggests, with high computational efficiency. Huang et al. [42] detected cracks in dams by improving YOLOX, and the results achieved high accuracy. Min et al. [43] proposed an improved YOLOv4 to achieve accurate bridge crack identification. The object detection method is accurate and efficient, enabling not only the identification of the type of detection target but also the localization of the target; however, it lacks the identification of specific contours in the detection target.

Deep learning semantic segmentation focuses on using convolutional neural networks to understand the real-world objects represented by each pixel in an image, that is, to segment the different objects in an image at the pixel level so that a specific outline of the target can be obtained. Since the specific shape and area of cracks are obtained to better determine the crack hazard level and to repair them, this paper uses semantic segmentation to conduct research on crack detection in concrete dams. In 2015, the fully convolutional network (FCN) was proposed [44] to reach a new milestone in image segmentation for deep learning. FCN uses a softmax function on top of a convolutional layer to classify each pixel to complete fine image segmentation. FCN cannot use global scene category information, which will cause feature loss. In the same year, Ronneberger et al. proposed the UNet model [45] based on the FCN network architecture, which modified the FCN encoder and decoder to obtain better performance on fewer training data, although the feature extraction capability is not optimal. Meanwhile, to solve the problem that FCN cannot utilise global scene information, Zhao et al. proposed the pyramid scene parsing network (PSPNet) [46] in 2017, which enables semantic segmentation models to fully access contextual relationships through the pyramid structure. In 2018, Chen et al. proposed a codec structure with null-separable convolution for image semantic segmentation (DeepLabV3+) [47], which achieves excellent segmentation results; however, it is computationally intensive and may lead to long training times. Subsequently, a number of excellent semantic segmentation models also emerged [48] and were applied to crack segmentation in various scenarios [49]. Liu et al. [50] used the DeepLabV3+ model for crack detection in ageing buildings and bridges to obtain information on crack changes in buildings. Rill-García et al. [51] trained an improved fully convolutional neural network (FCNN) on the public CrackForest dataset and achieved road crack detection but with low segmentation accuracy. Liu et al. [52] used Swin-UNet to perform experiments on a public fracture dataset, which can obtain fracture contours efficiently, but with an IoU of 70% and low accuracy. In summary, although semantic segmentation achieves better performance for crack detection than traditional image segmentation, targeted optimisation is needed for concrete dam surface crack detection where cracks are obscure and datasets are lacking.

2.2. Transfer Learning-Based Crack Detection

Deep learning algorithms require large datasets to support training, and insufficient datasets can seriously affect crack feature extraction and image detection. To address the lack of datasets limitation, the use of transfer learning is one of the most effective and practical solutions [53]. Transfer learning is the transfer of knowledge learned from one domain to another dataset domain; the knowledge is transferred from the source domain, and the knowledge is given to the target domain. In general, transfer learning can obtain relatively good test results even with small samples of labelled data [54]. For example, Fan et al. [3] proposed a transfer learning-based underwater dam crack image segmentation model that can accurately segment underwater dam crack images with a small dataset. Li et al. [55] improved crack detection accuracy by sharing model parameters in the source domain through a transfer learning approach. In theory, transfer learning can be performed between any domains, but if the pretrained dataset is more relevant to the task target dataset, then the better the pretrained model for knowledge transfer, the better the a priori knowledge acquisition for transfer learning will be [56]. Therefore, this paper proposes a two-stage transfer learning strategy to enable better transfer learning.

2.3. Crack Detection Backbone Network

Concrete dam surface cracks are more obscure and minute, and feature information extraction is difficult; therefore, it is necessary to make targeted changes to the model backbone feature extraction network to enhance the encoder information extraction capability. The dominant backbone network structures in current crack recognition are AlexNet [57], VGG [58], and ResNet [59]. Dung and Anh [60] replaced the backbone feature extraction network of the FCN model with a VGG network. The model achieved a comprehensive evaluation index, average accuracy of 90%, and can predict concrete cracks more accurately than the original model. Based on relevant research experience, network depth is crucial for visual recognition tasks; however, in practical experiments, it has been found that with the superposition of convolutional and pooling layers, the network learning effectiveness decreases and the error rate increases [61]. With the introduction of residual networks [59], it is possible to reach a deeper network level with a low classification error rate. Residual networks are widely used for this reason. Zhao et al. [62] addressed the low recognition accuracy problem of the UNet model by using ResNet18 as the backbone network to enhance the feature extraction capability of the network. Xu et al. [63] used the ResNet34 residual network as a model encoder to better extract crack detail information. Considering the stronger feature extraction capability of ResNet, this paper replaces the UNet backbone feature extraction network with ResNet50 to increase the depth of the feature extraction network and improve the segmentation accuracy of the model.

2.4. Crack Detection Based on Attention Mechanisms

In computer vision, attention mechanisms can mimic the human visual system, thus assigning different weights according to the importance of different content to improve attention to key information and enhance the network feature representation [64]. Yu et al. [65] addressed the low accuracy problem of existing deep learning road crack detection methods and improved the model's ability to detect cracks by embedding an attention mechanism in the UNet model. Given the application flexibility of the attention mechanism and the excellent performance that can be achieved in image processing and inspired by the residual module [59] and the atrous-spatial pyramid pooling (ASPP) module in DeepLabV3+ [47], an MPR attention mechanism was designed and integrated into the jump connection path of UNet to increase the model's attention to crack feature information, reduce the effect of noise on crack edge segmentation results, and improve the model's pixel segmentation accuracy.

3. Proposed Approach

3.1. Model Framework

The UNet model [45], as one of the most commonly used semantic segmentation models for deep learning, has the advantages of being lightweight and easy to deploy and can achieve good segmentation results even for small-scale dataset training; therefore, the UNet model is chosen for concrete dam surface crack detection in this paper. The UNet model can be divided into two parts: the encoder and the decoder. The encoder is used to extract features from the input image to gradually obtain higher-order semantic feature information in the image. The decoder is used to upsample and convolve the input feature map to gradually recover the image size and the number of feature map channels. Finally, the classification of pixels in the image is predicted. To enhance the information transfer between the encoder and decoder, the UNet model is designed with a jump connection so that the high-resolution shallow information output from the corresponding stage of the encoder is fed directly to the decoder, supplementing some of the feature information lost due to downsampling or convolution.

To further improve the segmentation ability and edge recognition accuracy of the UNet model for obscure and small cracks and to make the segmentation of concrete dam surface cracks more accurate, the UNet model is improved in this paper; the structure of the improved model is shown in Figure 2. The specific improvement measures are as follows: (1) A ResNet50 network is built as the feature extraction network of the UNet model encoder. The number of network layers and extraction capacity are deepened to fully obtain the crack feature information through the residual module so that the model can effectively learn the deep features of the dam face cracks and improve the crack segmentation model accuracy. (2) MPR attention added to the jump connection layer is designed to enhance the feature representation of the model by acquiring more semantic information, which not only suppresses feature responses in irrelevant regions and increases the importance of effective feature information channels but also allows the network to focus on crack feature information, complementing the loss of detail and enabling the model to segment the dam face crack images more accurately.

3.1.1. ResNet50 Backbone Network

Appropriately increasing the network depth can strengthen the feature extraction ability of the model and obtain more deep semantic information, but some researchers have found that as the structure of the network model deepens, training becomes increasingly difficult and encounters problems such as disappearing and exploding gradients. To effectively solve the degradation problem caused by deepening the network while strengthening crack feature information extraction by the UNet backbone network and improving crack detection accuracy, the ResNet50 backbone network, shown in Table 1, is proposed as the UNet encoder in this paper. Each residual module in the ResNet50 network consists of three concatenated convolutional layers of 1×1, 3×3, and 1×1, and a residual structure is added to retain some of the shallow information, which can prevent the model degradation problem due to gradient explosion. The residual module is divided into 2 types, as shown in Figure 3. If the number of input feature map channels matches the number of output channels, the residual module is A. If the number of input feature map channels does not match the number of output channels, the residual structure needs to perform a 1 × 1 convolution to adjust the number of channels, and the residual module is B.

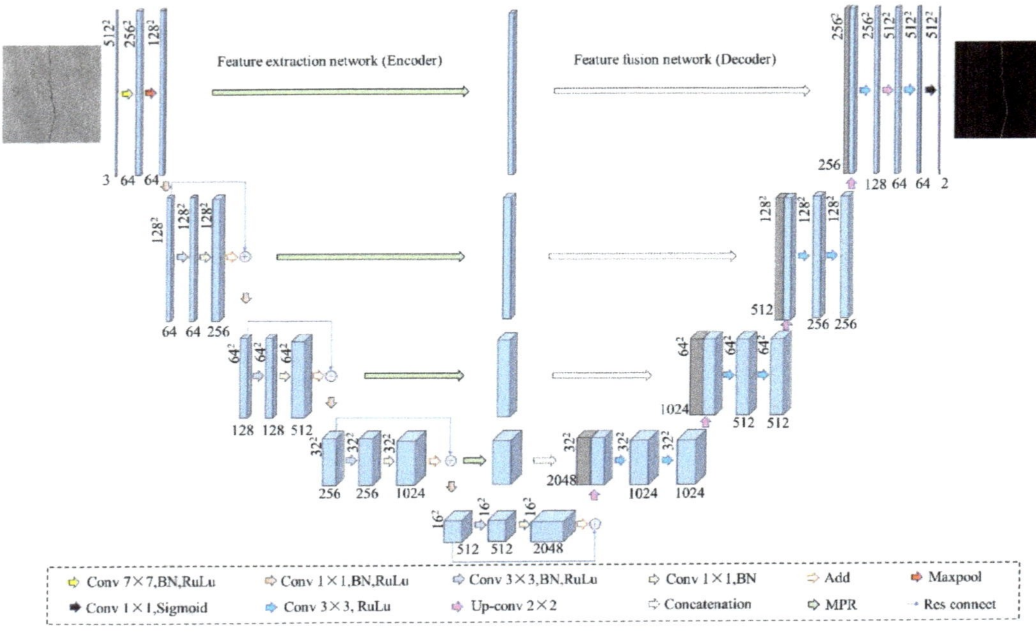

Figure 2. Model framework for this paper.

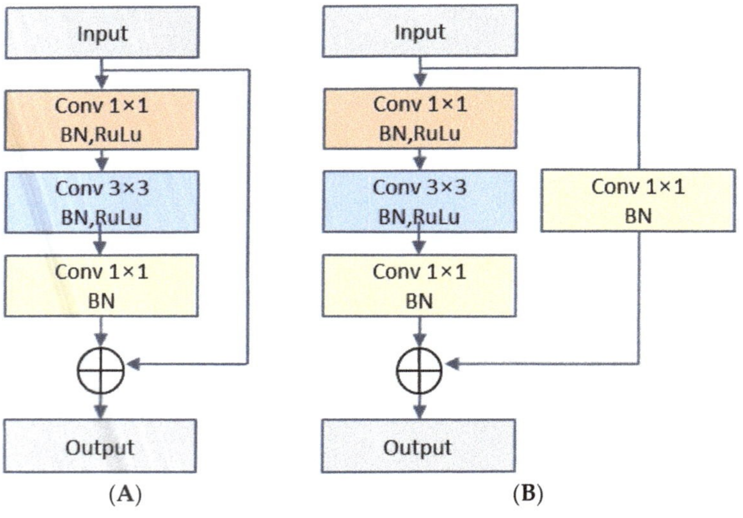

Figure 3. Residual module. (**A**) When the number of input channels matches the number of output channels, and (**B**) when the number of input channels does not match the number of output channels.

Table 1. ResNet50 network structure.

Layer Name	Operation	Output Size	Multiplier
Layer0	Input Conv7 × 7 Maxpool	$512^2 \times 3$ $256^2 \times 64$ $128^2 \times 64$	×1
Layer1	Conv1 × 1 Conv3 × 3 Conv1 × 1	$128^2 \times 64$ $128^2 \times 64$ $128^2 \times 256$	×3
Layer2	Conv1 × 1 Conv3 × 3 Conv1 × 1	$64^2 \times 128$ $64^2 \times 128$ $64^2 \times 512$	×4
Layer3	Conv1 × 1 Conv3 × 3 Conv1 × 1	$32^2 \times 256$ $32^2 \times 256$ $32^2 \times 1024$	×6
Layer4	Conv1 × 1 Conv3 × 3 Conv1 × 1	$16^2 \times 512$ $16^2 \times 512$ $16^2 \times 2048$	×3

3.1.2. Multilayer Parallel Residual Attention

MPR is designed to address the problem that cracks in the image provide little information and the pixel segmentation accuracy is low. MPR is added to the jump connection so that the decoder can obtain more important semantic information for learning and increase the edge segmentation accuracy. The MPR consists of 4 parts (Figure 4): the skip layer, the 1 × 1 convolution layer, the 3 × 3 convolution layer, and the atrous convolution layer with a dilation rate of 3. The 4 parts perform the convolution calculation in parallel, and the main codes are shown in Table 2. The skip layer skips the input feature map x and adds it directly to the output of $F_i(x)$. This step retains the feature information of the previous layer in the feature map y of the next layer to protect information integrity. The 1 × 1 convolution and 3 × 3 convolution layers allow the model to perform feature learning under multiscale convolution, extracting deep semantic information and increasing the nonlinear capability of the model. The cavity convolution layer uses a cavity convolution with a sampling rate of 3 to give the model a larger perceptual field for the same number of parameters and computational effort, which helps to improve the missing information problem for crack edge identification. Finally, the outputs of the four are summed to increase the amount of crack information in each feature map dimension, thus reducing the influence of irrelevant feature information on subsequent crack identification results.

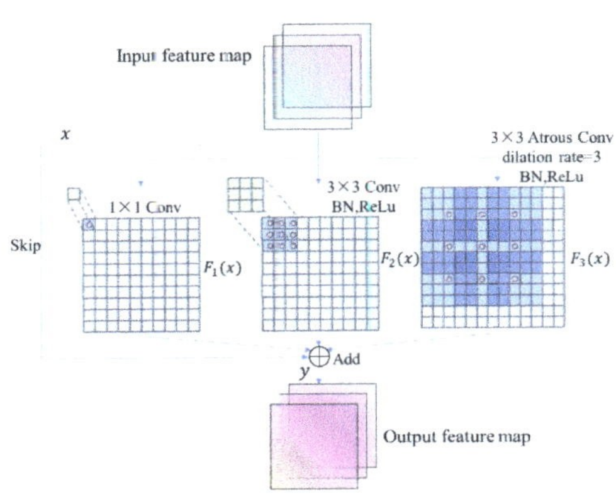

Figure 4. Multilayer parallel residual attention.

Table 2. MPR main codes.

```
class MPR (nn.Module):
    def __init__ (self, ch_in, ch_out):
        Super (FResidual, self).__init__ ()
        # first
        self.conv = nn.Sequential (
            nn.Conv2d (ch_in, ch_out, kernel_size = 3, stride = 1, padding = 1, bias = True),
            nn.BatchNorm2d (ch_out),
            nn.ReLU (inplace = True)
        )
        # second
        self.skip = nn.Sequential (
            nn.Conv2d (ch_in, ch_out, kernel_size = 1, stride = 1, padding = 0),
            nn.BatchNorm2d (ch_out),
        )
        # Third
        self.aspp_block1 = nn.Sequential (
            nn.Conv2d (ch_in, ch_out, 3, stride = 1, padding = 3, dilation = 3),
            nn.ReLU (inplace = True),
            nn.BatchNorm2d (ch_out),
        )
        # Fourth
        self.conv2 = nn.Conv2d (ch_in, ch_out, kernel_size = 1, stride = 1, padding = 0)
```

3.2. *Two-Stage Transfer Learning Strategy*

Semantic segmentation is a type of supervised learning that requires a large quantity of labelled training data, which is not easy to obtain on concrete dam surface cracks in complex environments. The labour cost of acquisition and data labelling is high, and the lack of datasets easily leads to inadequate model training and low detection accuracy. To solve the problem of insufficient training data for deep learning, transfer learning methods have been developed [66]. However, if the difference between two domains is particularly large, the results obtained by directly adopting the transfer learning method are often poor. Therefore, this paper designs a two-stage learning strategy for knowledge transfer to improve the low detection accuracy and poor modelling results on small sample datasets. The two-stage transfer learning training process is shown in Figure 5. (1) The first stage is cross-domain-based model knowledge transfer, i.e., sharing model parameters using the source domain to the target domain. The training results obtained from the PASCAL VOC 2012 dataset model in the source domain are used as the pretraining model for the target domain (DatasetA) in the first stage. The parameter information shared between the source and target domains is found so that DatasetA can be updated with parameters from the perspective of the algorithm and the model during the training process. This process avoids network learning from scratch and allows a better mentor model to be obtained in the second stage. (2) The second stage is based on intradomain feature knowledge transfer, i.e., mapping features from the source and target domains from the original feature space to the new feature space. The pretrained model of DatasetA after the first stage of transfer learning is used as the source domain in the second stage to learn the same information and knowledge structure from the related domain. Then, the same feature representation existing in the source domain as in the target domain is migrated to the target domain (DatasetB) through the transfer learning method. This ensures that the existing labelled data samples in the source domain can be better utilised for classification training in the new space and solves the poor accuracy problem caused by the sparse labelled dam face crack images.

Both phases were trained for 300 generations, but due to the similarity in data, tasks and models between the first and second phases, a freeze training step was added to the second phase transfer training to speed up the training efficiency and prevent the weights from being corrupted. This means that the backbone network was frozen for the first 150 training generations, only the decoding classifier was trained, and the whole network was unfrozen for the second 150 training and learning generations. The two-stage transfer learning strategy reduced the model training time and resulted in a highly accurate and robust model for intelligent concrete dam surface crack detection.

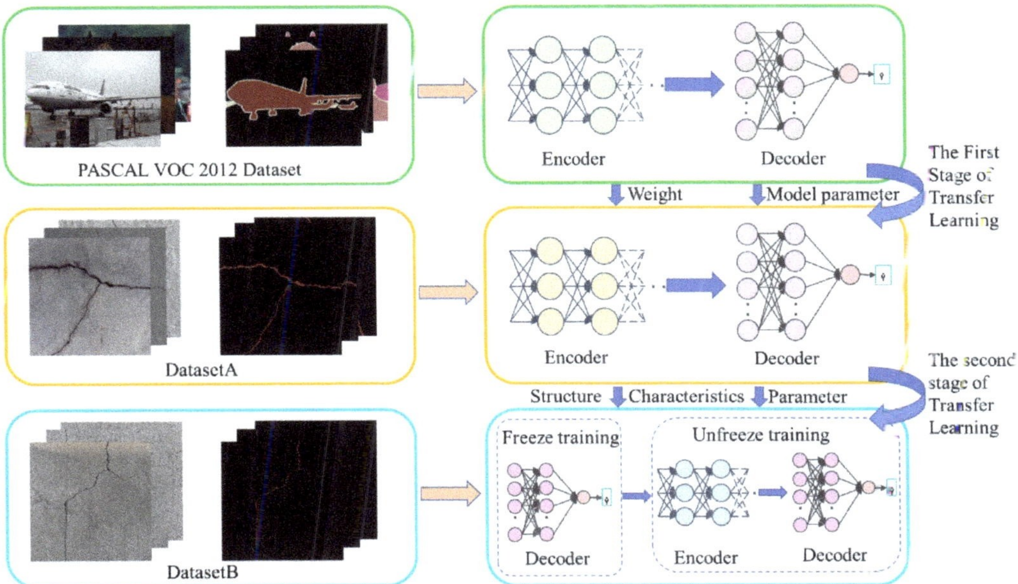

Figure 5. Two-stage transfer learning training process.

4. Experiments and Evaluation Indicators

4.1. Experimental Setup

The experiments in this paper were based on the Ubuntu 18.04 operating system, the CPU processor was an AMD EPYC7543 32-Core processor, the GPU used a GeForce RTX3090 graphics card with 24 GB of video memory, and the programming language was Python 3.8, and CUDA 11.1, the PyTorch 1.8.1 deep learning framework for network training was used. The specific model training parameters are shown in Table 3.

Table 3. Model training parameters.

Parameters	Configuration
Input image size	512 × 512
Batch_Size	8
Epoch	300
Optimiser	Adam
Momentum	0.9
Initial learning rate	0.0001
Initial learning rate decay type	CosineAnnealingLR
Loss function	Cross Entropy Loss

4.2. Datasets

The dataset used for the experiments in this paper has three parts: the PASCAL VOC 2012 dataset, the first stage cross-domain training crack dataset (DatasetA), and the second stage target domain concrete dam surface crack image dataset (DatasetB). The PASCAL

VOC 2012 dataset is a public dataset for world-class computer vision challenge image classification, detection, or semantic segmentation, consisting of 1464 images from the training set, 1449 images from the validation set, containing 20 categories and 1 background. DatasetA consists of 2 parts, partly derived from the publicly available datasets CrackForest [30], SDNET2018 [24], and Aft Original Crack DataSet Second [26], but the distribution and pixel characteristics of cracks on the concrete dam surfaces are very different from the open source dataset. Therefore, in the 3 publicly available datasets, only the images with fine and obscure cracks were selected, and those with similarity greater than 0.7 were excluded to improve the dataset quality; 280 images were finally obtained. Another part of DatasetA we acquired in actual concrete road cracks, and 700 small crack images were added to increase the realism of the dataset and make it closer to the actual engineering context. After the image enhancement algorithm retinex [67], adjustment of image luminosity, contrast, and spatial variation (random rotation, flip), DatasetA was expanded to 3152 images and used as the target domain dataset for cross-domain training in the first phase of the model. The selected open dataset images and road crack images are shown in Figure 6.

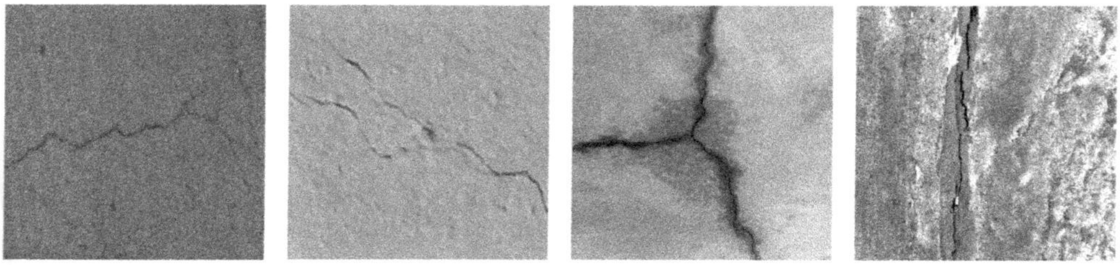

Figure 6. Images from DatasetA. From left to right, selected CrackForest, SDNET2018, Aft Original Crack Dataset. Second, a realistic collection of road cracks.

The second phase target domain DatasetB is our self-built concrete dam surface crack dataset, and the DJI Mavic 3 UAV was used to photograph several concrete dams in the upper Jinsha River in China to obtain concrete dam crack image data. The shooting distance was 3~10 m and focused on image diversity (different angles, background conditions, light intensity) with a total of 350 images collected, with crack width between 3 and 20 mm. To ensure that the dataset can adapt to different complex environments and to make the model more generalisable and robust, the images were annotated and then expanded using retinex enhancement, adjusting image luminosity, contrast, and spatial variation to obtain 1393 images.

The intelligent detection method for concrete dam surface cracks in this paper used supervised learning. To ensure the quality of the dataset and the accuracy of the target information, both the actual collected images of DatasetA and DatasetB were manually annotated at the pixel level using the LabelMe annotation tool. Finally, DatasetB was randomly divided into a training set and a validation set at a ratio of 8:2.

4.3. Evaluation Indicators

To objectively evaluate the performance of different models, typical evaluation metrics mean intersection over union (mIoU), mean pixel accuracy (mPA), parameters, and floating point operations (FLOPs) are introduced in this paper. Intersection of union (IoU) represents the overlap rate between the predicted mask and the actual mask. mIoU is the arithmetic mean of the IoU values for each category and assesses whether the overall image segmentation is precise. Pixel accuracy (PA) indicates the proportion of the number of correctly segmented pixels, and mPA is the arithmetic mean of each class of PA to assess the global accuracy of the model. Parameters measure the number of model parameters,

FLOPs measure the complexity of the model, i.e., the number of model computations, and FPS measures the speed of inference of the model.

$$mIoU = \frac{1}{k+1} \sum\nolimits_{i=0}^{k} \frac{p_{ii}}{\sum_{j=0}^{k} p_{ij} + \sum_{j=0}^{k} p_{ji} - p_{ii}} \tag{1}$$

$$mPA = \frac{1}{k+1} \sum\nolimits_{i=0}^{k} \frac{p_{ii}}{\sum_{j=0}^{k} p_{ij}} \tag{2}$$

where $k + 1$ denotes the category to be predicted plus a background, and p_{ij} denotes the number of pixels in category i that are predicted to be in category j. Therefore, p_{ii} is a positive sample, p_{ij} is a false negative sample, and p_{ji} is a false positive sample.

5. Results and Discussion

5.1. Ablation Experiments

To verify the effectiveness and advancement of the improvement strategies in this paper, eight sets of ablation experiments were designed to validate the model under the same dataset and experimental environment to verify the impact of different improvement strategies on the concrete dam surface crack extraction accuracy. In Table 4, it can be seen that the concrete dam surface crack segmentation performance improved under different strategies. After using ResNet50 as the backbone feature extraction network, mIoU and mPA improved by 1.2 and 1.3 percentage points, respectively, compared with the UNet model, the model computations were reduced by 64.1 percentage points, and the inference speed increased by 128.1 percentage points, indicating that the ResNet50 backbone network can obtain more crack feature information and segment the foreground crack region more accurately. Additionally, it can reduce the complexity of the UNet model and speed up the model inference capability. After the UNet model was embedded with MPR attention, mIoU and mPA improved by 1.1 and 1.6 percentage points, respectively, compared to the UNet model. The model's ability to extract important semantic information was enhanced, suppressing some of the useless background feature information, improving pixel segmentation accuracy, and making edge details more complete but sacrificing detection time. After using the two-stage transfer learning strategy, both mIoU and mPA substantially improved, proving that the method can effectively alleviate model overfitting and undertraining due to the lack of a dataset, reduce fragmentation of dam surface cracks, and reduce missed detections. In the overall analysis, the integrated capacity of the model was optimal after integrating the three improvements simultaneously. The mIoU and mPA of the proposed method were 88.3% and 92.7%, respectively, and were 4.6 and 3.2 percentage points higher than those of the UNet model. The computational volume required by this paper was 36.7 percentage points lower than that of the UNet model, and the inference speed was 48.9 percentage points higher than that of the baseline UNet model. This demonstrates that the method achieves better concrete dam surface crack segmentation performance. Although the increase in the number of method parameters in this paper led to an increase in the memory taken up by the model weights, the size of the memory taken up by the model had no direct effect on the concrete dam surface crack segmentation in this paper.

5.2. Comparative Trials of Different Attention Mechanisms

To reflect the advantages of MPR attention in this paper, Table 5 shows the experimental results of different attention mechanisms on model detection ability under the same experimental environment and dataset. The table shows that SE [68] attention had the least disturbance on the model inference speed but did not significantly improve model accuracy and precision. Compared to SE attention, the popular CBAM [69] attention mechanism achieved a more pronounced improvement to cleft IoU and PA, indicating that convolutional block attention module (CBAM) attention is more capable of extracting fine targets. The improvement in crack accuracy from adding CA [70] attention was negligible and had no significant effect on accuracy. There was a small decrease in accuracy with the addition of the lightweight ECA [71] model, with a 0.3% decrease in mIoU and a 0.7% decrease in mPA, indicating that lightweight efficient channel attention (ECA) is less suitable for fine crack detection and that attention is not always applicable to any scenario. Compared with other commonly used attention mechanisms, MPR attention, designed in this paper, is more effective in improving accuracy and precision and is more suitable for detecting concrete dam surface cracks. Although MPR will have some impact on the inference speed, the improvement in detection accuracy is undoubtedly more important for better ensuring dam safety and stability and is, therefore, more cost-effective.

Table 4. Comparison of the ablation experiment results.

Model	ResNet50	MPR	Transfer Learning	mIoU/%	mPA/%	Parameters/M	FLOPs/G	FPS
UNet				83.7	89.5	34.5	512.6	24.5
A	√			84.9	90.8	43.9	184.1	55.9
B		√		84.8	91.1	41.5	694.1	17.9
C			√	85.4	91.1	34.5	512.6	24.5
D	√	√		85.4	91.2	71.6	324.0	36.5
E		√	√	87.1	91.7	41.5	694.1	17.9
F	√		√	87.3	91.5	43.9	184.1	55.9
Our	√	√	√	88.3	92.7	71.6	324.0	36.5

Note:"√" indicates an improved strategy added to the original UNet model.

Table 5. Results of experiments with different added attention mechanisms (IoU and PA only for the crack class).

Model	IoU/%	mIoU/%	PA/%	mPA/%	FPS
UNet	68.0	83.7	79.2	89.5	24.5
UNet + SE	68.4	83.9	81.1	90.4	24.4
UNet + CBAM	68.9	84.2	81	90.4	19.1
UNet + CA	68.0	83.7	79.6	89.7	21.5
UNet + ECA	67.4	83.4	77.9	88.8	24.3
UNet + MPR	70.3	84.8	82.5	91.1	17.9

5.3. Training Trials with Different Transfer Learning Approaches

Table 6 shows the effect of different transfer learning approaches on improving the segmentation accuracy of the model. It is clear from Table 6 that the cross-domain-based transfer school approach achieved the lowest improvement in training accuracy for the target domain, with mIoU and mPA only improving by 0.6% and 0.3%, demonstrating that if the difference between the 2 domains is large, the pretrained model has little effect in improving the target domain. Compared with cross-domain transfer, the intradomain transfer learning approach achieved a more significant improvement in target domain accuracy, with the IoU and PA of the cracks improving by 1.9% and 1.5%, respectively, indicating that better target domain learning can be achieved through intradomain feature knowledge transfer. Compared to single-segment transfer learning, a two-stage transfer learning approach with cross-domain and in-domain learning was used in this paper for optimal results. The cross-domain training was first performed on the in-domain DatasetA through the out-of-domain PASCAL VOC 2012 dataset to avoid training from zero so that a better in-domain pretraining model was obtained. Then, the pretraining model was used as the tutor model for the target domain to achieve the best transfer learning effect. The final IoU and PA of the cracks improved by 5.6% and 2.9%, respectively, indicating that in transfer learning, in addition to the impact of the difference in data distribution between the source and target domains on the accuracy improvement, the merit of the mentor model also has an impact on transfer learning, which also verifies the effectiveness of the two-stage transfer learning strategy in this paper.

Table 6. Comparison results of different transfer learning approaches. (IoU and PA only for the crack class).

Transfer Learning Approach	Source Domain Datasets	Source Domain Model		Target Domain Model			
		mIoU	mPA	IoU	mIoU	PA	mPA
/	/	/	/	71.3	85.4	82.7	91.2
Cross-domain training	PASCAL VOC 2012	69.5	75.2	72.6	86.0	83.3	91.5
Intradomain training	DatasetA	88.7	93.5	73.2	86.4	84.2	91.9
Two-stage training	PASCAL VOC 2012 +DatasetA	90.2	94.4	76.9	88.3	85.6	92.7

5.4. Crack Segmentation Image Comparison

For a more intuitive analysis and to demonstrate that the proposed method has greater crack segmentation capability, the concrete dam surface cracks detection results are visualised in this section using the UNet model, UNet + ResNet50, UNet + ResNet50 + MPR, and the methods in this paper. The visualisation of the crack segmentation for the different models is shown in Figure 7. The selected dam face crack images were complex and contained obscure, microscopic cracks; therefore, segmentation was challenging. The UNet benchmark model in Figure 7 roughly extracted the dam face crack contours, but the extraction was incomplete and weak for fine cracks, leading to serious underidentification and noise phenomena, and the segmentation of crack edges was smoother and rougher, with low segmentation accuracy. With the addition of the ResNet50 backbone network to UNet, the model's ability to extract features from the image was enhanced, and the segmentation accuracy improved, allowing the outline of fine cracks to be roughly extracted. However, there were fractures and noise, and the edge details needed to be improved. UNet + ResNet50 then added MPR attention to reduce the interference of background regions on subsequent predictions, focus more on crack regions, and enhance fine crack extraction, thus improving the accuracy and precision of the model's crack edge shape segmentation with clearer contours and a better ability to retain details than the UNet model, but fragmented segmentation and missed detection were still present. After combining the three improved strategies, the proposed method achieved the highest degree of dam face crack segmentation completeness, which was closer to the manual annotation, and had a strong segmentation ability and robustness for cracks of different scales. These results show that the two-stage transfer learning strategy can compensate for the under-segmentation and overfitting phenomenon caused by the small quantity of data and can effectively improve the missed detection situation caused by thin cracks and few effective pixel points and process the edge details in a clearer, more detailed, and complete way. The combined subjective and objective results show that the proposed method exhibits the best performance in intelligent concrete dam surface crack segmentation and shows greater robustness in segmenting complex and obscure fine cracks.

5.5. Comparison of Typical Models

Table 7 shows the training evaluation results of this paper's model with the current typical semantic segmentation models UNet [28], PSPNet [29], DeepLabV3+ [30], and SegFormer [31] in the same experimental environment and with the same dataset. From Table 7, the DeeplabV3+ model had the lowest mIoU and mPA with only 71.6% and 75.0%, respectively, followed by PSPNet, with both models having lower segmentation values than expected. This indicates that using large networks such as DeepLabV3+ and other models with insufficient sample sizes can easily lead to overfitting. The better segmentation accuracy of the UNet and SegFormer models shows that these models have some stable segmentation ability even in small datasets and complex scenarios. Overall, the accuracy indexes of these methods were optimal, and the IoU and PA of cracks significantly improved compared with the UNet model before improvement, achieving 76.9% and 85.6%, respectively. This indicates that the segmentation accuracy of the improved method for concrete dam surface cracks is better than the other models and more suitable for concrete dam surface crack detection.

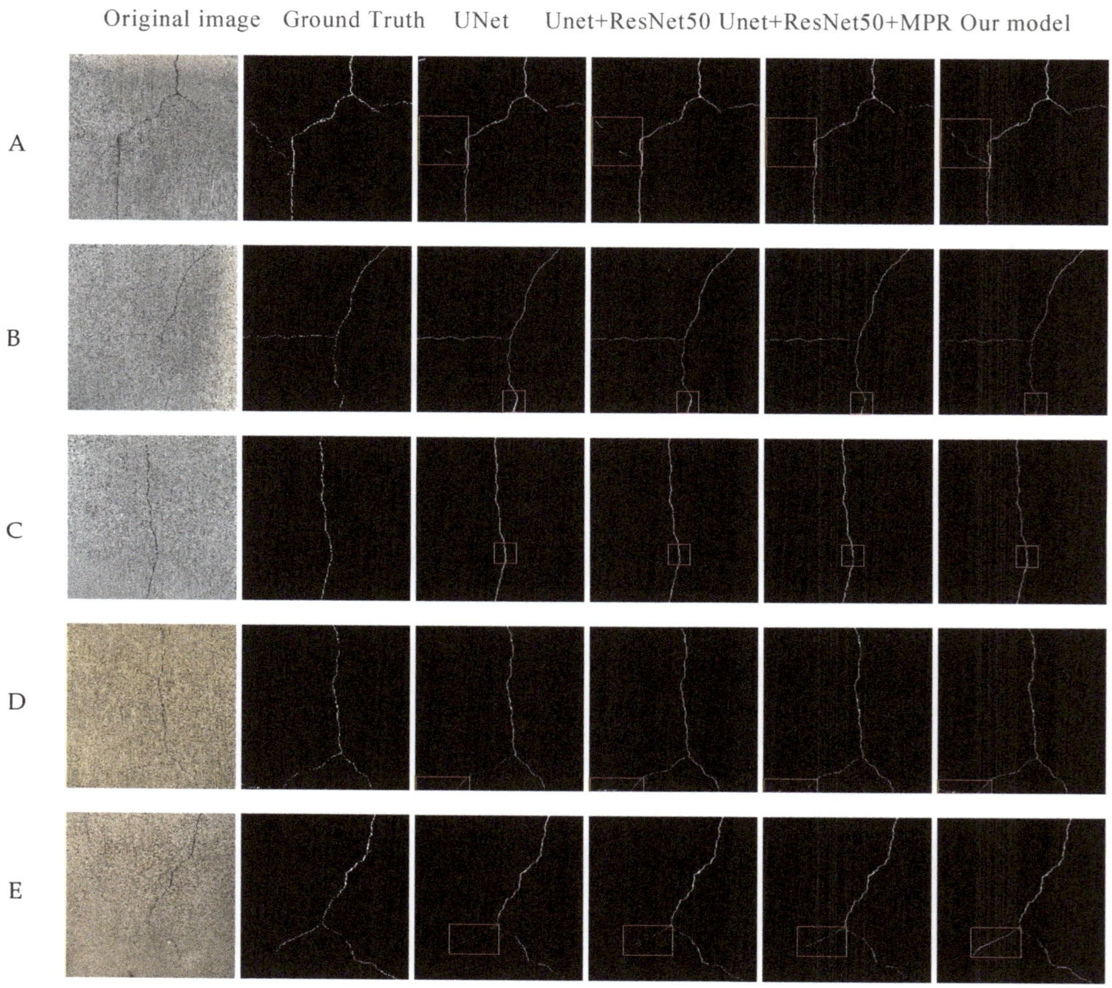

Figure 7. Crack segmentation results for different models. From left to right, original, ground truth, UNet, UNet + ResNet50, UNet + ResNet50 + MPR, our method. (**A–E**) denote the segmentation of different dam cracks in different models. The red box reflects a sharper contrast.

Table 7. Comparison of evaluation indicators of different models.

Model	IoU/%		mIoU/%	PA/%		mPA/%
	Crack	Background		Crack	Background	
UNet	68.0	99.4	83.7	79.2	99.8	89.5
PSPNet	52.3	99.2	75.7	62.9	99.7	81.3
DeepLabV3+	44.1	99.1	71.6	50.2	99.8	75.0
SegFormer	64.5	99.4	81.9	77.6	99.7	88.7
Our method	76.9	99.6	88.3	85.6	99.8	92.7

5.6. Discussion

5.6.1. The Effectiveness of Transfer Learning

Figure 8 shows the change curves of the ablation experimental metrics. It can be observed that the mIoU and mPA metric values of each experimental model increased with the increasing number of iterations, and the training loss decreased with the increasing number of iterations. tThe models that did not use the transfer learning strategy were trained for 260 iterations before the fluctuation in each metric gradually stabilised. Then, the mIoU, mPA, and training loss curves were parallel to the X-axis, indicating that the training could end. The improvement in mIoU and mPA metrics and the decrease in loss values were evident in the models trained with the transfer learning strategy. mIoU and mPA increased rapidly in the first 10 phases of the training process and converged smoothly in the latter phases of the training process, with the corresponding loss values dropping to much lower values. In summary, with the use of the transfer learning strategy, the segmentation accuracy of small datasets can be improved through knowledge and feature sharing, and high accuracy can be achieved in a short training time. Therefore, transfer learning can be applied not only to detection tasks with small datasets but also to scenarios with limited computational power for model training to achieve fast training results, reduce the number of training rounds, and save computational resources.

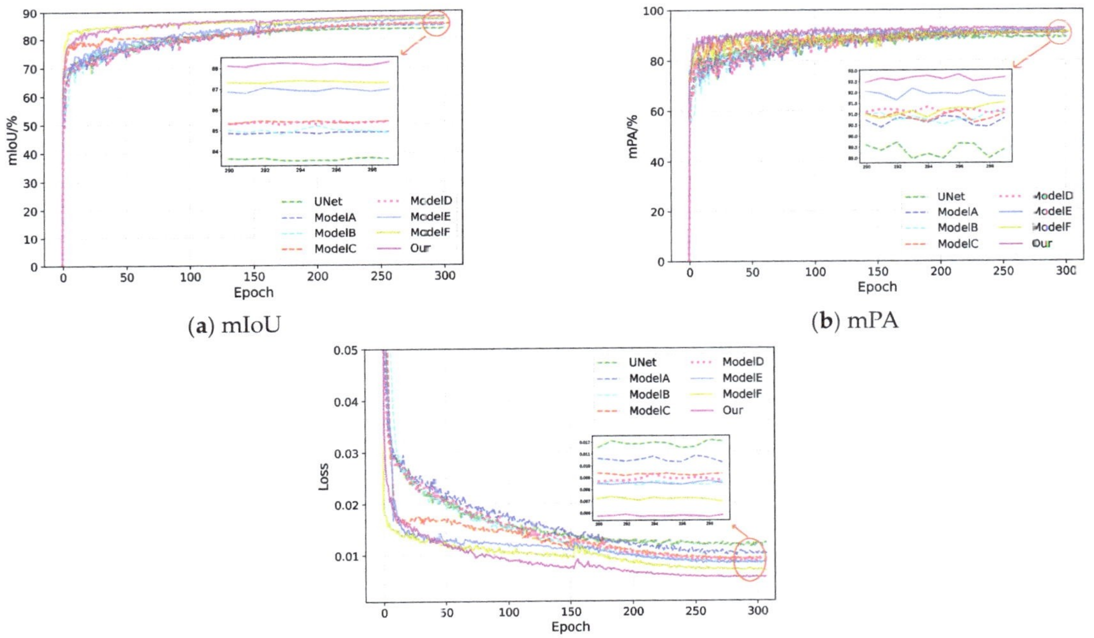

(a) mIoU (b) mPA (c) Loss

Figure 8. Variation curves of ablation experiment metrics. Figure 8 shows the metric values for each generation of weights obtained from training, from left to right, mIoU, mPA, and loss. The red circle in the figure shows the last 10 generations of the model metrics.

5.6.2. Crack Area Pixel Value Prediction

Tables 4, 6 and 7 and Figure 7 show that the proposed method performs best in the automatic concrete dam surface crack segmentation, with more accurate segmentation of complex and obscure small cracks, and can accurately and efficiently detect crack location and shape. This subsection, therefore, performs a pixel-level statistical analysis of the cracks obtained from the segmentation of this paper's method to predict the pixel value of the image crack area. Table 8 shows the statistical table of crack area prediction of the proposed method, and the selected images are consistent with those segmented in Figure 7. In Table 8, it can be seen that the crack area prediction accuracy by the proposed method reaches more than 85.7%, and the prediction accuracy of relatively obvious crack areas can reach more than 90%. The test results show that, in addition to the fast and accurate crack identification and segmentation, the proposed crack segmentation method can also accurately obtain crack area information. In summary, the UAV images can be fed into the proposed method to efficiently and accurately identify and segment concrete dam surface cracks, then the pixel values of the segmented crack areas can be derived, and the actual area values of concrete dam cracks can be obtained by converting the pixel points to the actual values [72]. The crack hazard level is judged based on the area obtained, and then the appropriate material is selected for remediation, enabling better targeted repair during dam health diagnosis and solving the low manual measurement efficiency and detection accuracy problems while avoiding material waste.

Table 8. Statistics on the predicted crack area of the model in this paper.

Picture		A	B	C	D	E
Pixel value	Ground True	78,556	57,768	38,576	45,377	33,107
	Our prediction	67,326	52,577	33,304	42,627	32,426
Precision/%		85.7	86.3	97.9	93.9	91.0

5.6.3. Practical Applications

Figure 9a,b gives the results of the dam crack image segmentation without annotation by the method in this paper. The method effectively detects the general outline and location of the cracks in the picture and can accurately segment different cracks, which proves the effectiveness and reliability of the method in this paper for detecting cracks in actual dams. Figure 9c,d shows the dam crack images with different environmental backgrounds from DatasetB, and we find that they can also obtain good recognition, probably because the migration learning process is used in pretraining the weights for crack images of similar environments, so crack segmentation in other environmental backgrounds also achieves good robustness.

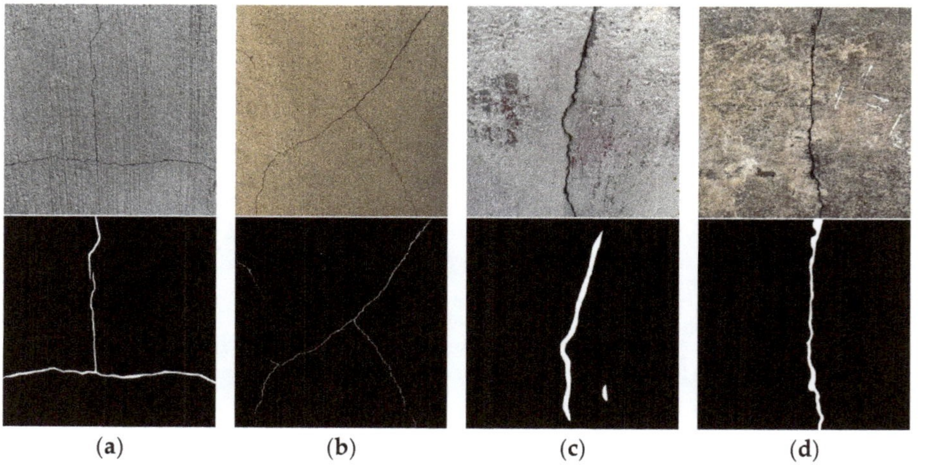

(a) (b) (c) (d)

Figure 9. The results of this paper's method for segmenting cracks in actual engineering dams. (**a**,**b**) are dam crack images without annotation training, (**c**,**d**) are dam crack images with different environmental backgrounds from DatasetB.

5.6.4. Future Work

Table 4 reflects the combined ability of the methods in this paper to reach optimality. The mIoU and inference speed were improved by ResNet50 as a feature extraction network, but the improvement in mPA was not significant. The mPA measures the proportion of pixels correctly segmented by the model, indicating that there is still room for improvement in the model's segmentation and edge recognition capabilities. The experiments found that the mPA improvement was more obvious after adding MPR to the UNet model. Therefore, the two models were combined, but there was still the fine crack missed detection problem, which was most likely caused by the small sample size of the dataset, making the model training insufficient. After the improved UNet model was trained using a two-stage transfer strategy, a significant reduction in model misses was observed in Figure 5, demonstrating that transfer learning is very effective in helping to train when using small datasets. The method ultimately has good inference speed while achieving the highest segmentation accuracy but also reflects the large number of parameters and the large space occupied by the method volume, resulting in weak model mobility, which is less suitable for detection in moving scenes. In the future, the model can be thinned by pruning the convolutional layers or replacing the model encoder with a lightweight feature extraction network to reduce the parameters and computations to make the model lightweight so that the model can also be deployed in mobile and embedded devices such as Jetson.

6. Conclusions

To improve concrete dam surface crack detection efficiency and accuracy, an intelligent concrete dam surface crack detection method based on two-stage transfer learning is proposed, which can achieve accurate and efficient UAV concrete dam surface crack image segmentation. Compared with other typical semantic segmentation models, the proposed method has the best overall capability and better performance in detecting cracks in concrete dams.

After training using two-stage transfer learning, the proposed method alleviates the inadequate training problem caused by small-scale dam crack datasets, reduces the fine crack miss detection phenomenon, and can achieve high accuracy in shorter training rounds. Additionally, using ResNet50 as a UNet model feature extraction network can fully extract crack feature information and improve the model segmentation capability. Finally, adding MPR to the jump connection path significantly improves pixel accuracy and makes crack edge segmentation more delicate and complete.

The experimental results show that the mIoU and mPA of the proposed method reach 88.3% and 92.7%, respectively, and the FPS is 36.5. The proposed method has the ability to segment cracks on concrete dam surfaces more efficiently and accurately and achieves an accuracy of over 85.7% for predicting the crack area, which can better detect cracks in dams safely

Author Contributions: J.L.: Conceptualization, methodology, software, writing—original draft, and writing—review and editing. X.L.: Data curation and investigation. P.Z.: supervision and conceptualization. Q.L.: Writing, methodology, review, and editing. All authors have read and agreed to the published version of the manuscript.

Funding: This research was funded by the National Natural Science Foundation of China (52009068), the Natural Science Foundation of Hubei Province (2022CFD168) and the Major Science and Technology Projects of the Ministry of Water Resources (SKS-2022102).

Institutional Review Board Statement: Not applicable.

Informed Consent Statement: Not applicable.

Data Availability Statement: Some of the datasets in this study are openly available from (https://github.com/cuilimeng/CrackForest-dataset, (accessed on 6 December 2022)), https://pan.baidu.com/s/1a5VCmzJunFbg_tyzgszOmQ?pwd=9kwl (accessed on 6 December 2022).

Acknowledgments: We would like to acknowledge the financial support of the National Natural Science Foundation of China (52009068), the Natural Science Foundation of Hubei Province (2022CFD168) and the Major Science and Technology Projects of the Ministry of Water Resources (SKS-2022102).

Conflicts of Interest: The authors declare no conflict of interest.

References

1. Ren, Q.; Li, M.; Shen, Y.; Zhang, Y.; Bai, S. A Pixel-Level morphological segmentation and feature quantification method for hydraulic concrete cracks. *J. Hydropower* **2021**, *40*, 234–246.
2. Salazar, F.; Vicente, D.J.; Irazábal, J.; De-Pouplana, I.; Mauro, J.S. A review on thermo-mechanical modelling of arch dams during construction and operation: Effect of the reference temperature on the stress field. *Arch. Comput. Methods Eng.* **2020**, *27*, 1681–1707. [CrossRef]
3. Fan, X.; Cao, P.; Shi, P.; Chen, X.; Zhou, X.; Gong, Q. An underwater dam crack image segmentation method based on multi-level adversarial transfer learning. *Neurocomputing* **2022**, *505*, 19–29. [CrossRef]
4. Huang, Y.; Zhang, X.; Xu, Z.; Huang, J.; Xu, X. Image stitching of underwater dam cracks based on connected domain a priori. *Chin. Body Vis. Image Anal.* **2020**, *25*, 408–418. [CrossRef]
5. Xu, B.; Xia, H. A review of research on concrete dam cracking morphology and its hazard analysis methods. *J. Water Resour. Water Eng.* **2016**, *27*, 162–168.
6. Deng, J. Census and treatment process of concrete dam cracks in a hydropower station in Lichuan, Hubei. *China Constr. Waterproofing* **2021**, *03*, 52–55. [CrossRef]
7. Huang, C.; Yang, X.; Xia, J. Inspection and treatment of water cracks on the upstream face of Danjiangkou initial project dam. *People's Yangtze River* **2015**, *46*, 45–48+74. [CrossRef]
8. Qu, C.; Wang, C. Concrete pavement crack detection based on attention mechanism and lightweight cavity convolution. *Comput. Sci.* **2023**, *50*, 231–236.
9. Wang, Y.; Gao, R.; Liu, W. Design of concrete crack displacement sensor based on optical fiber bending loss. *Sens. Microsyst.* **2023**, *42*, 87–90. [CrossRef]
10. Liu, Z.; Wang, Y.; Sun, F.; Jia, X.; Nie, F. An integration-assisted multi-objective optimization algorithm combining feature perturbation and allocation strategies. *Comput. Eng.* **2022**, *48*, 115–123. [CrossRef]
11. Bhowmik, B.; Krishnan, M.; Hazra, B.; Pakrashi, V. Real-time unified single-and multi-channel structural damage detection using recursive singular spectrum analysis. *Struct. Health Monit.* **2019**, *18*, 563–589.
12. Lin, Y.; Shen, Q. Building deformation monitoring based on singular spectrum analysis. *Mapp. Stand.* **2021**, *37*, 33–37.
13. Liu, H.; Hu, J.; Yuan, D.; Jun, C.; Xing, W. Research on seismic coherence attribute calculation method based on singular value spectrum analysis[C]//Professional Committee of Petroleum Physical Exploration of China Petroleum Society, Exploration Geophysics Committee of China Geophysical Society. In Proceedings of the 2022 China Petroleum Physical Exploration Annual Academic Conference, Haikou China, 27–29 September 2022.
14. Tian, G.; Liu, J.; Yang, W. A dual neural network for object detection in UAV images. *Neurocomputing* **2021**, *443*, 292–301. [CrossRef]
15. Wei, S.; Liu, Y.; Liu, J. Research on the application of concrete structure surface crack detection based on UAV and digital image method. *Spec. Struct.* **2020**, *37*, 107–111. [CrossRef]
16. Nguyen, A.; Gharehbaghi, V.; Le, N.T.; Sterling, L.; Chaudhry, U.I.; Crawford, S. ASR crack identification in bridges using deep learning and texture analysis. *Structures* **2023**, *50*, 494–507.
17. Zawad, M.; Shahriar, R.; Zawad, M.; Shahriar, F.; Rahman, M.; Priyom, S.N. A comparative review of image processing based crack detection techniques on civil engineering structures. *J. Soft Comput. Civ. Eng.* **2021**, *5*, 58–74.
18. SHanzaei, H.; Afshar, A.; Barazandeh, F. Automatic detection and classification of the ceramic tiles' surface defects. *Pattern Recognit.* **2017**, *66*, 174–189. [CrossRef]
19. Premachandra, C.; Waruna, H.; Premachandra, H.; Parape, C. Image based automatic road surface crack detection for achieving smooth driving on deformed roads. In Proceedings of the 2013 IEEE International Conference on Systems, Man, and Cybernetics, Manchester, UK, 13–16 October 2013; pp. 4018–4023.
20. Tang, Q.; Tan, Y.; Peng, L.; Cao, H. Research on the identification method of tunnel lining cracks based on digital image technology. *J. Railw. Sci. Eng.* **2019**, *16*, 3041–3049. [CrossRef]
21. Gupta, P.; Dixit, M. Image-based crack detection approaches: A comprehensive survey. *Multimed. Tools Appl.* **2022**, *81*, 40181–40229. [CrossRef]
22. Talab, A.M.A.; Huang, Z.; Xi, F.; Ming, L.H. Detection crack in image using Otsu method and multiple filtering in image processing techniques. *Optik* **2016**, *127*, 1030–1033. [CrossRef]
23. Zhou, Y.; Liu, T. Computer vision-based concrete crack identification. *J. Tongji Univ. (Nat. Sci.)* **2019**, *47*, 1277–1285.
24. Cho, H.; Yoon, H.; Jung, J. Image-based crack detection using crack width transform (CWT) algorithm. *IEEE Access* **2018**, *6*, 60100–60114. [CrossRef]
25. Yang, J.; Yang, D. A review of semantic segmentation based on deep learning. *Chang. Inf. Commun.* **2022**, *35*, 69–72. [CrossRef]

26. Zhang, S.W.; Bao, T.F.; Gao, X.H. A computer vision-based crack detection method for concrete dams. *Adv. Water Resour. Hydropower Sci. Technol.* **2021**, *41*, 83–88.
27. Li, L.; Zhang, X.; Lian, J.; Zhou, Y.; Zheng, W. Semantic SLAM algorithm combined with road structural features. *J. Harbin Inst. Technol.* **2021**, *53*, 175–183.
28. Zhang, X.; Yao, Q.; Zhao, J.; Jin, Z.; Feng, Y. A review of semantic segmentation methods for fully convolutional neural network images. *Comput. Eng. Appl.* **2022**, *58*, 45–57.
29. Jiang, T.; Tang, Y.; Lu, C.; Shen, G. Research on concrete crack image recognition method based on semantic segmentation. *Eng. Surv.* **2023**, *51*, 42–47.
30. Wang, J.; Yuan, H.; Rui, T.; Zhao, Q.; Yin, C. A semantic segmentation-based crack identification method for concrete structures. *Build. Struct.* **2022**, *52*, 923–929. [CrossRef]
31. Dorafshan, S.; Thomas, R.J.; Maguire, M. SDNET2018: An annotated image dataset for non-contact concrete crack detection using deep convolutional neural networks. *Data Br.* **2018**, *21*, 1664–1668. [CrossRef]
32. Yang, F.; Zhang, L.; Yu, S.; Prokhorov, D.; Mei, X.; Ling, H. Feature pyramid and hierarchical boosting network for pavement crack detection. *IEEE Trans. Intell. Transp. Syst* **2019**, *21*, 1525–1535. [CrossRef]
33. Li, L.; Ma, W.; Li, L.; Lu, C. Research on detection algorithm for bridge cracks based on deep learning. *Acta Autom. Sin* **2019**, *45*, 1727–1742.
34. Tang, Y.; Huang, Z.; Chen, Z.; Chen, M.; Zhou, H.; Zhang, H.; Sun, J. Novel visual crack width measurement based on backbone double-scale features for improved detection automation. *Eng. Struct.* **2023**, *274*, 115158.
35. Huang, S.; Bao, T.; Li, Y.; Niu, H. Semantic segmentation method of hydraulic concrete cracks based on improved DeeplabV3– network. *Adv. Hydropower Sci. Technol.* **2023**, *43*, 81–86.
36. Chen, B.; Zhang, H.; Li, Y.; Wang, S.; Zhou, H.; Lin, H. Quantify pixel-level detection of dam surface crack using deep learning. *Meas. Sci. Technol.* **2022**, *33*, 065402. [CrossRef]
37. Wang, Z.; Zhang, Q.; Fang, D.; Wang, X. Research on dam crack detection method based on deep learning. *Water Resour. Plan. Des.* **2022**, *1*, 90–94.
38. Cheng, H.; Li, Y.; Li, H.; Hu, H. Embankment crack detection in UAV images based on efficient channel attention U2Net. *Structures* **2023**, *50*, 430–443.
39. Cui, L.; Qi, Z.; Chen, Z.; Meng, F.; Shi, Y. Pavement distress detection using random decision forests. In *Data Science, Proceedings of the International Conference on Data Science, Sydney, Australia, 8–9 August 2015*; Springer: Cham, Switzerland, 2015; pp. 95–102.
40. Ren, S.; He, K.; Girshick, R.; Sun, Z. Faster R-CNN: Towards Real-Time Object Detection with Region Proposal Networks. In *IEEE Transactions on Pattern Analysis & Machine Intelligence*; IEEE: Piscataway, NJ, USA, 2017; Volume 39, pp. 1137–1149.
41. Redmon, J.; Divvala, S.; Girshick, R.; Farhadi, A. You Only Look Once: Unified, Real-Time Object Detection. In Proceedings of the 2016 IEEE Conference on Computer Vision and Pattern Recognition (CVPR), Las Vegas, NV, USA, 27–30 June 2016; pp. 779–788.
42. Huang, B.; Kang, F.; Tang, Y. Real-time detection method of concrete dam cracks based on target detection. *J. Tsinghua Univ. (Nat. Sci. Ed.)* **2023**, 1–9. [CrossRef]
43. Du, M.; Yang, G.; Zhang, H. Research on bridge crack detection method based on YOLOv4-EfficientNet B7. *J. Tianjin Urban Constr. Univ.* **2023**, *29*, 55–61. [CrossRef]
44. Long, J.; Shelhamer, E.; Darrell, T. Fully convolutional networks for semantic segmentation. In Proceedings of the IEEE Conference on Computer Vision and Pattern Recognition, Boston, MA, USA, 7–12 June 2015; pp. 3431–3440.
45. Ronneberger, O.; Fischer, P.; Brox, T. U-Net: Convolutional networks for biomedical image segmentation. In *Medical Image Computing and Computer-Assisted Intervention—MICCAI 2015*; Springer: Cham, Switzerland, 2015; pp. 234–241.
46. Zhao, H.; Shi, J.; Qi, X.; Wang, X.; Jia, J. Pyramid scene parsing network. In Proceedings of the IEEE Conference on Computer Vision and Pattern Recognition, Honolulu, HI, USA, 21–26 July 2017; pp. 2881–2890.
47. Chen, L.C.; Zhu, Y.; Papandreou, G.; Schroff, F.; Adam, H. Encoder-decoder with atrous separable convolution for semantic image segmentation. In Proceedings of the European Conference on Computer Vision (ECCV), Computer Vision Foundation, Munich, Germany, 8–14 September 2018; pp. 801–818.
48. Xie, E.; Wang, W.; Yu, Z.; Anandkumar, A.; Alvarez, J.M.; Luo, P. SegFormer: Simple and efficient design for semantic segmentation with transformers. *Adv. Neural Inf. Process. Syst.* **2021**, *34*, 12077–12090.
49. Yang, J.; Li, H.; Zou, J.; Jiang, S.; Li, R.; Liu, X. Concrete crack segmentation based on UAV-enabled edge computing. *Neurocomputing* **2022**, *485*, 233–241. [CrossRef]
50. Liu, Z.; Li, X.; Li, J.; Teng, S. A new approach to automatically calibrate and detect building cracks. *Buildings* **2022**, *12*, 1081. [CrossRef]
51. Rill-García, R.; Dokladalova, E.; Dokládal, P. Pixel-accurate road crack detection in presence of inaccurate annotations. *Neurocomputing* **2022**, *480*, 1–13. [CrossRef]
52. Liu, S. Research on concrete crack segmentation algorithm based on Swin-Unet. *Henan Sci. Technol.* **2022**, *41*, 13–17. [CrossRef]
53. Zhong, H.; Lv, Y.; Yuan, R.; Yang, D. Bearing fault diagnosis using transfer learning and self-attention ensemble lightweight convolutional neural network. *Neurocomputing* **2022**, *501*, 765–777. [CrossRef]
54. Zhang, H.; Feng, L.; Hao, Y.; Wang, Y. Dynastic recognition of ancient wall paintings based on attention mechanism and migration learning. *Comput. Appl.* **2023**, 1–9.

55. Li, Y.; Cheng, H.; Li, H.; Wang, J.; Hu, Q. UAV image based on improved U~2-net with migration learning for embankment crack detection. *Adv. Water Resour. Hydropower Sci. Technol.* **2022**, *42*, 52–59.
56. Yu, X.; Wang, J.; Hong, Q.Q.; Teku, R.; Wang, S.H.; Zhang, Y.D. Transfer learning for medical images analyses: A survey. *Neurocomputing* **2022**, *489*, 230–254. [CrossRef]
57. Krizhevsky, A.; Sutskever, I.; Hinton, G.E. Imagenet classification with deep convolutional neural networks. *Commun. ACM* **2017**, *60*, 84–90. [CrossRef]
58. Simonyan, K.; Zisserman, A. Very deep convolutional networks for large-scale image recognition. *arXiv* **2014**, arXiv:1409.1556.
59. He, K.; Zhang, X.; Ren, S.; Sun, J. Deep residual learning for image recognition. In Proceedings of the IEEE Conference on Computer Vision and Pattern Recognition, Las Vegas, NV, USA, 27–30 June 2016; pp. 770–778.
60. Dung, C.V.; Anh, L.D. Autonomous concrete crack detection using deep fully convolutional neural network. *Autom. Constr.* **2019**, *99*, 52–58. [CrossRef]
61. Kang, L.; Ye, P.; Li, Y.; Doermann, D. Convolutional neural networks for no-reference image quality assessment. In Proceedings of the IEEE Conference on Computer Vision and Pattern Recognition, Columbus, OH, USA, 23–28 June 2014; pp. 1733–1740.
62. Zhao, Y.; Hu, H.S.; Tang, M.; Chen, H. Research on crack detection method of concrete bridge based on improved coding-decoding network. *Guangzhou Archit.* **2022**, *50*, 1–7.
63. Xu, G.; Liao, C.; Chen, J.; Dong, B.; Zhou, Y. Extraction of concrete apparent crack information based on HU-ResNet. *Comput. Eng.* **2020**, *46*, 279–285. [CrossRef]
64. Zhao, X.B.; Wang, J.J. Bridge crack detection based on improved DeeplabV3+ and migration learning. *Comput. Eng. Appl.* **2022**, 1–10.
65. Yu, O.; Jing, P.; Zhang, W.; Xie, S.F.; Sli, Z.H.; Song, C. An improved U-Net model for road crack detection based on residuals and attention mechanism. *Comput. Eng.* **2022**, 1–14. [CrossRef]
66. Ge, W.; Yu, Y. Borrowing treasures from the wealthy: Deep transfer learning through selective joint fine-tuning. In Proceedings of the IEEE Conference on Computer Vision and Pattern Recognition, Hawaii, HI, USA, 21–26 July 2017; pp. 1086–1095.
67. Petro, A.B.; Sbert, C.; Morel, J.M. Multiscale retinex, Image Process. *Line* **2014**, 71–88. [CrossRef]
68. Hu, J.; Shen, L.; Sun, G. Squeeze-and-excitation networks. In Proceedings of the IEEE Conference on Computer Vision and Pattern Recognition, Salt Lake City, UT, USA, 18–23 June 2018; pp. 7132–7141.
69. Woo, S.; Park, J.; Lee, J.Y.; Kweon, I.S. Cbam: Convolutional block attention module. In Proceedings of the European Conference on Computer Vision (ECCV), Computer Vision Foundation, Munich, Germany, 8–14 September 2018; pp. 3–19.
70. Hou, Q.; Zhou, D.; Feng, J. Coordinate attention for efficient mobile network design. In Proceedings of the IEEE/CVF Conference on Computer Vision and Pattern Recognition, Nashville, TN, USA, 20–25 June 2021; pp. 13713–13722.
71. Wang, Q.; Wu, B.; Zhu, P.; Li, P.; Zuo, W.; Hu, Q. ECA-Net: Efficient channel attention for deep convolutional neural networks. In Proceedings of the IEEE/CVF Conference on Computer Vision and Pattern Recognition, Seattle, WA, USA, 13–19 June 2020; pp. 11534–11542.
72. Sun, X. *Design of Shellfish Size Detection System Based on Deep Learning*; Jiangnan University: Wuxi, China, 2022.

water

MDPI

Article

Comprehensive Evaluation Method for the Safety State of RCC Dams Based on Interval Number Theory

Xudong Chen [1,2], Peng Xu [3,*], Xinyi Liu [4] and Chen Su [5]

1 State Key Laboratory of Hydrology-Water Resources and Hydraulic Engineering, Nanjing Hydraulic Research Institute, Nanjing 210029, China
2 National Dam Safety Research Center, Wuhan 430019, China
3 Research Institute of Engineering Safety Science/Reservoir & Dam Center, PowerChina Guiyang Engineering Co., Ltd., Guiyang 550081, China
4 Information Management Department/Digital Engineering R&D Center and GIS Center, PowerChina Guiyang Engineering Co., Ltd., Guiyang 550081, China
5 Hohai University Design & Research Institute Co., Ltd., Nanjing 210098, China
* Correspondence: xup_gyy@powerchina.cn

Abstract: Roller Compacted Concrete (RCC) dams are critical infrastructure, playing an important role in economic and social development. However, dam failure can cause great losses. To mitigate hazards, studies of methods to deal with the uncertainty involved in the comprehensive evaluation process of the safety state of RCC dams are hot issues. Interval number theory is applied to quantify the uncertainty in this study. A comprehensive evaluation indicator system is explored, an approach to allocating the indicator weight rationally is proposed, and a comprehensive evaluation model is established. Comprehensive evaluation standards are developed. An RCC dam in China is used to illustrate the applicability of this comprehensive evaluation method based on interval number theory. The results indicate that the method and models proposed are suitable for comprehensively evaluating the safety state of RCC dams and can be used as a new procedure to monitor the safety of an RCC dam.

Keywords: RCC dams; safety state; comprehensive evaluation; interval number theory; uncertainty

Citation: Chen, X.; Xu, P.; Liu, X.; Su, C. Comprehensive Evaluation Method for the Safety State of RCC Dams Based on Interval Number Theory. *Water* **2023**, *15*, 2089. https://doi.org/10.3390/w15112089

Academic Editor: Paolo Mignosa

Received: 6 April 2023
Revised: 25 May 2023
Accepted: 29 May 2023
Published: 31 May 2023

1. Introduction

The concept of RCC dams was proposed by J.M. Raphael from the University of California, Berkeley, in 1970 [1]. RCC dams consist of concrete placed at a lower water-to-cement ratio compared with conventional concrete with the aid of compaction equipment and methodologies typically employed for earth-fill dam construction. This construction method permits a considerable reduction in the cost and construction time of dams compared not only with traditional concrete dams but also with earth-fill dams. To date, RCC dams have become the fastest-growing international dams. With the implementation of the strategy of Western power development and transportation from West to East, increasingly more RCC dams are under construction or planned to be constructed in China. The working behavior and safety state of RCC dams not only directly affect the expected benefits of the water conservation and hydropower projects but also the safety of the lives and property of people downstream. Therefore, a comprehensive evaluation of the safety state of RCC dams is of great significance for hazard prevention and mitigation.

The working behavior of RCC dams changes over time under the influence of external load and environmental changes, with specific reflections in deformation, seepage, instability, etc. Current research on dam safety evaluation can be classified into two categories: single indicator evaluation and comprehensive evaluation, including many indicators. Wang et al. [2] tried to find the leakage path using the tracer technique and evaluated the dam seepage state with entropy weight-set pair analysis. Chelidze et al. [3] established an effective remote monitoring system for dam deformation by connecting terminals, sensors, and a central processing unit to the monitoring sites through a global system for mobile communications or general packet radio service modems and obtained a real-time evaluation of the dam deformation state by comparing the monitoring values with the theoretical values. Li et al. [4] evaluated the dam crack state using a regression coefficient fluctuation method with a statistical model and catastrophe theory. Haralampos et al. [5] evaluated the dam strength state using laboratory and in situ tests. Chen [6] studied the dam seismic safety state evaluation methods, including the dam site-specific ground motion input, improved response analysis, dynamic model test verification, field experiment investigations, dynamic behavior of dam concrete, and seismic monitoring and observation. Li et al. [7] evaluated the dam instability state using the strength reserve factor method to simulate the gradual failure and possibly unstable modes of dams. Xi et al. [8] established a concrete dam deformation safety prediction model based on deep learning by using the open-source deep learning framework TensorFlow and the mature convolutional neural network technology in deep learning theory. Chen et al. [9] evaluated the method of combining radial basis neural network and kernel principal component analysis to monitor the deformation of super-high concrete dams. Wang et al. [10] proposed a dynamic matter-element-extension model which considered the correlation among indicators and established the seepage safety evaluation model of the dynamic concrete gravity dam. He et al. [11] introduced the cloud model into a fuzzy comprehensive evaluation model to solve the problems of fuzziness and randomness in the seepage security risk evaluation index. Yang et al. [12] applied the point-to-face variable comparison analysis mode to monitor the dam-shaped variables. Panvalkar et al. [13] applied borehole nuclear logging and tracer techniques to trace the suspected leakage zones and path through the body of the dam to evaluate the seepage situation of the dam.

The studies above provide useful tools for dam safety monitoring and evaluation. However, a single indicator evaluation can reflect only partial information regarding the dam's safety state. Comprehensive evaluation with many indicators is necessary for the overall monitoring of the dam's safety state. Mirzabozorga et al. [14] evaluated the dam safety state by combining multi-source in situ information and finite element calculation. Mata et al. [15] merged the physical quantities measured by the automated monitoring system of a dam, appropriately weighted, into a new single indicator belonging to any of two classes (normal behavior and development of a failure scenario). Lokke et al. [16] put forward a safety evaluation method for concrete gravity dams by building a dam-water-foundation system with response spectrum analysis. Chiganne et al. [17] demonstrated a new method for evaluating the flood overtopping failure scenario for embankment dams with concrete upstream slope protection considering geotechnical, hydraulic, and structural factors. Bretas et al. [18] presented a numerical model for the safety analysis of gravity dams based on the discrete element method and evaluated the sliding failure mechanism and the stress state. Li et al. [19] used the Delphi method combined with grey relational analysis to select the dominant factors and then selected back propagation neural network combined with the AdaBoost algorithm to build the dam safety evaluation model. Han et al. [20] proposed a safety evaluation model of earth-rock dams based on the combination method of the ideal point and cloud theory model, which can scientifically solve the uncertainty problem in the safety evaluation of earth-rock dams. Wen et al. [21] determined the comprehensive evaluation grade of dam safety behavior by using the fuzzy recognition model combining the subindex evaluation level and comprehensive weight. Shu et al. [22] proposed a dam safety evaluation model based on interval-valued intuitionistic fuzzy set and evidence theory for dam safety reliability evaluations, which aimed at the multi-sources, heterogeneity, and complexity of dam safety evaluation. Li et al. [23] used the modified moving average-cosine similarity method to quantify the lag effect of operation indicators and established a comprehensive safety evaluation index system on this basis. Sang et al. [24] proposed a new method combining an extended cloud model and an extended analytic hierarchy process to address the uncertainty problem in the evaluation of the overall safety trend of dams and the selection of safety trend warning indicators.

These studies on dam safety state evaluation provide powerful technical support for its operation. However, the construction interfaces of RCC dams make them different from other types of dams. The structural characteristics of the RCC dam make the construction layer easy to become a weak surface and the concentrated channel of seepage in the structure of the RCC dam. However, the current research on the characteristics of the RCC dam layer mainly focuses on the aspects of the layer strength and interlayer bonding qualities, while few pieces of research focus on the comprehensive evaluation of the characteristics of the RCC dam layer. The development of a comprehensive evaluation method for the safety state of RCC dams considering the specific structural characteristics of RCC dams and the uncertainty in the evaluation procedure is necessary. Extenics, rough set, fuzzy theory, gray system theory, attribute mathematics, and set pair analysis are normally used for dealing with uncertainty problems [25–32]. Each of the methods has its own advantages and disadvantages. Recently, the interval number theory [33] has shown advantages in dealing with the uncertainty problem. Therefore, the interval number theory is applied in this study to handle the uncertainty in the comprehensive evaluation of the RCC dam safety state considering the construction interfaces. Using interval numbers rather than natural numbers to describe the evaluation indicators and safety state is more reasonable and acceptable.

The rest of this paper is organized as follows. Based on interval number theory, the comprehensive evaluation indicator system is built correspondingly in Section 2, methods for calculating the interval number weights of these indicators are discussed in Section 3, and the comprehensive evaluation model is established in Section 4. Then, an RCC dam in China is utilized as a case study to verify the feasibility of this proposed method in Section 5. Finally, Section 6 is a summary that concludes this work.

2. Comprehensive Evaluation Indicators

2.1. Comprehensive Evaluation Indicator System

According to the principles of building the indicator system, i.e., scientific and operational, clearly prior and relatively independent, qualitative and quantitative, and reasonably hierarchical indicators were determined. As RCC dams are constructed in layers, the interface gap between adjacent layers is larger than in traditional concrete dams. This gap generates a certain pressure at the interface when compressing the cement concrete, creating an interface effect. The interface effect is mainly reflected in several aspects: stress concentration, interface exchange, and overall deformation. Therefore, in addition to the traditional dam evaluation items, the corresponding interface stresses, displacements, and other interface characteristics should be considered. We built the evaluation indicator system shown in Figure 1. The first level is the target level reflecting the safety state of the RCC dam. The second level is the criterion level, including the deformation, seepage, stress, and stability state of an RCC dam. The third level is the indicator level constituted by the displacement of the dam body, the seepage of the dam body, the stress of the dam heel, the stress of the dam toe, the stress of the dam abutment, the stress of the interfaces, and stability against slicing. The fourth level is the basic level, including horizontal displacement, vertical displacement, the ratio of the interface displacement to the total displacement, uplift pressure, quantity of leakage, ratio of the interface leakage to the total leakage, range of tensile stress, the normal vertical stress, the range of compressive stress, the maximum tensile stress, the maximum compressive stress, the maximum shear stress, stability against sliding of the dam foundation, stability against sliding of the interface, etc.

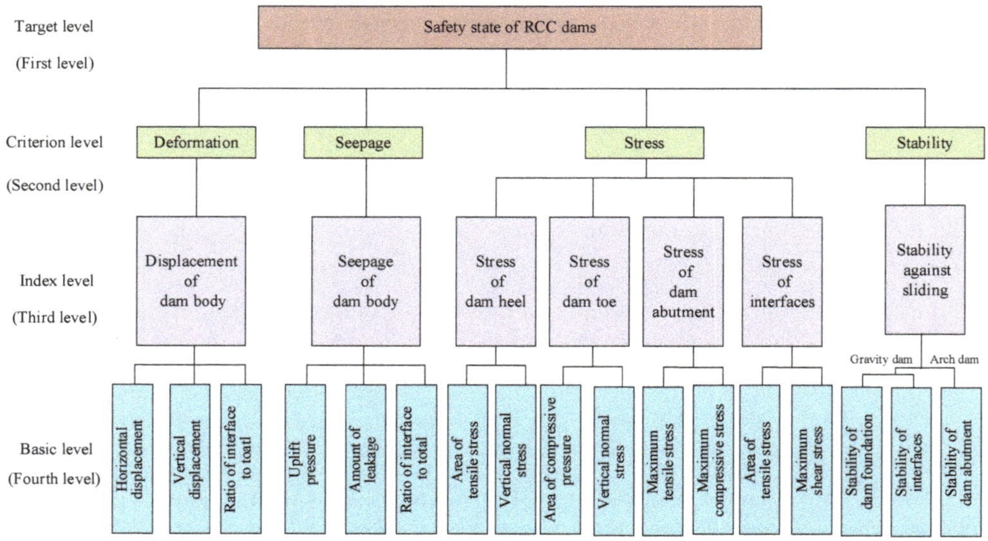

Figure 1. Comprehensive evaluation indicator system for the safety state of RCC dams.

2.2. *Grades of Evaluation Indicators*

Considering the guidelines on dam safety evaluation and the practical conditions of dam operation, we divide the evaluation indicators into five grades. Since some of the evaluation indicators are qualitative and others are quantitative, the respective grading methods are described.

2.2.1. Grades of Qualitative Evaluation Indicators

The qualitative indicators are usually graded with the experience of experts. To make the grades as accurate as possible, site management, design recheck, and construction review are investigated overall with the aid of professional regulations, relevant standards, and documents. The grading standards are shown in Table 1.

Table 1. Grading standards for qualitative evaluation indicators.

Grade	Site Management	Design Recheck	Construction Review
I	Structural integrity with no obvious damage and leakage	Strength and seepage meet the regulation requirement	Construction quality meets the design requirement fully
II	Local damage with a few cracks or little leakage	Strength and seepage mainly meet the regulation requirement	Construction quality mainly meets the design requirement
III	Obvious cracks or a few educts with hidden damage	Some strength and seepage do not meet the regulation requirement, but the stability is good	Some construction quality does not meet the design requirements
IV	Local deep cracks, serious leakage, obvious educts	Some strengths and seepage severely do not meet the regulation requirement, and no measures are taken	Construction quality cannot meet the design requirement
V	Large area with cracks, increased leakage	Extreme flood or earthquake, stability fails	–

2.2.2. Grading Standards for Quantitative Evaluation Indicators

The quantitative evaluation indicators are also divided into five grades: grades I, II, III, IV, and V. Grade I: the monitoring data change normally and do not exceed the allowed design value or the maximum historical value. Grade II: the monitoring data are close to or slightly exceed the allowed design value or the maximum historical value, and the monitoring data of the key positions in the dams are normal. Grade III: The monitoring data exceed the allowed design value or the maximum historical value, and the monitoring data for the key position in dams change normally. Grade IV: the monitoring data exceed the allowed design value or the maximum historical value, and the monitoring data for the key position in the dams obviously changes. Grade V: the monitoring data heavily exceed the allowed design value or the maximum historical value, and the monitoring data for the key position in the dams obviously change.

2.3. Interval Number Expressions of Evaluation Indicators

The interval number, denoted by $[A] = [a^L, a^U]$, where a^L is the lower boundary of an interval and a^U is the upper bound, represents a range that can describe the uncertainty, while a point value cannot. Thus, interval number theory is widely used in the uncertainty problem research field.

2.3.1. Interval Number Expressions of Qualitative Evaluation Indicators

If the interval $[0, 1]$ is used to represent a qualitative evaluation indicator, interval numbers for the five grades can be obtained by the equal interval partition method usually used in system engineering. The interval numbers are listed in Table 2.

Table 2. Interval numbers for five grades of qualitative evaluation indicators.

Grade	I	II	III	IV	V
Interval number	$[0.8, 1]$	$[0.6, 0.8)$	$[0.4, 0.6)$	$[0.2, 0.4)$	$[0, 0.2)$

2.3.2. Interval Number Expressions of Quantitative Evaluation Indicators

The main structural characteristic of the RCC dam is the existence of the layer, which is different from ordinary concrete dams. The layer influence zone is the weak zone of the RCC dam, and it is the key to influencing the strength, stability, and seepage of the RCC dam The deformation, seepage, stress, and stability of the layer are not only the main reflection of the characteristics of the layer but are also related to the safe operation of the RCC dam. Different from qualitative evaluation indicators, quantitative evaluation indicators are more complicated to grade with interval numbers. From the grading standards of the quantitative evaluation indicators described above, the interval number expressions of deformation, seepage, stress, and stability are studied.

(1) Interval number expressions of the deformation indicators

The deformation monitoring data are denoted by y, time by t, the monitoring values calculated with the deformation monitoring model by $\hat{y}$, the standard deviation of the deformation monitoring model by S, the time effect component of the deformation monitoring model by y_θ, the first level early warning value of deformation by y_{1n}, the second level early warning value of deformation by y_{2m}, and the third level early warning value of deformation by y_{3m}. Thus, the interval numbers of the five grades can be expressed using these variables. Grade I: $y \in [\hat{y} - 2S, \hat{y} + 2S]$ and $dy/dt < 0$. Grade II $y \in (\hat{y} + 2S, \hat{y} + 3S]$ or $y \in [\hat{y} - 3S, \hat{y} - 2S]$, $dy/dt < 0$, and $d^2y/dt^2 < 0$. Grade III: $y \in [y_{1m}, y_{2m}]$. Grade IV: $y \in [y_{2m}, y_{3m}]$. Grade V: $y = y_{3m}$.

The deformation feature of RCC dams is different from that of conventional concrete dams because of the interfaces existing in RCC dams. To display the interface influence, the ratio of the interface deformation to the total deformation is adopted. The interface deformation is denoted by y_i, and we express the grades as follows through reference to the available literature [34,35]. Grade I: $y_i/y \in [0, 0.1)$, Grade II: $y_i/y \in [0.1, 0.2)$, Grade III: $y_i/y \in [0.2, 0.3)$, Grade IV: $y_i/y \in [0.3, 0.4)$, and Grade V: $y_i/y \in [0.4, 1]$.

(2) Interval number expressions of the seepage indicators

The seepage monitoring data are denoted by q, time by t, the value calculated with the seepage monitoring model by $\hat{q}$, the standard deviation of the seepage monitoring model by S, the time effect component of the seepage monitoring model by q_θ, the designed value of seepage by q_D, and the maximum value of the seepage data series by q_m. Thus, Grade I: $q \in [\hat{q} - 2S, \hat{q} + 2S]$ and $dq_\theta/dt < 0$. Grade II: $q \in (\hat{q} + 2S, \hat{q} + 3S]$ or $q \in [\hat{q} - 3S, \hat{q} - 2S]$, $dq_\theta/dt > 0$, and $d^2q_\theta/dt^2 < 0$. Grade III: $q \in [0, q_D]$ and $d^2q_\theta/dt^2 = 0$. Grade IV: $q \approx [q_D]$ and $d^3q_\theta/dt^3 < 0$. Grade V: $q \in [q_D, q_m]$ and $d^3q_\theta/dt^3 = 0$.

Similarly, adapting the ratio of the interface seepage to the total seepage, we express the grades as follows. Grade I: $q_i/q \in [0, 0.1)$, Grade II: $q_i/q \in [0.1, 0.2)$, Grade III: $q_i/q \in [0.2, 0.3)$, Grade IV: $q_i/q \in [0.3, 0.4)$, and Grade V: $q_i/q \in [0.4, 1]$, where q_i is the interface seepage.

(3) Interval number expressions of the stress indicators

The stress monitoring data are denoted by σ, the designed value of stress by σ_D, the compressive stress, tensile stress or shear stress by σ_p, the maximum value of stress data series by σ_m, the stress in the control range by $\sigma(\Omega)$, and the maximum value of stress in the control range by $\sigma(\Omega)_m$. Assuming the compressive stress is positive, and the tensile stress is negative, the interval numbers of the five grades can be expressed as follows. Grade I: $\sigma \in [0, \sigma_D]$, Grade II: $\sigma \in (\sigma_D, \sigma_p]$, Grade III: $\sigma \in (\sigma_p, \sigma_m]$, Grade IV: $\sigma(\Omega) \in (\sigma_D, \sigma_p]$, and Grade V: $\sigma(\Omega) \in (\sigma_p, \sigma(\Omega)_m]$.

(4) Interval number expression of the stability indicators

According to the strength reserve coefficient method, the standard of stability safety coefficient against sliding is $K_e \geq [K_e]$, where K_e is the safety coefficient in a quasi-elastic state and $[K_e]$ is the safety coefficient allowed. Considering the water pressure overload, environmental degradation, and material parameter variation, $K_e/[K_e]$ is adopted to describe the grades of stability safety. Grade I: $K_e/[K_e] \in [1, 2]$, Grade II: $K_e/[K_e] \in [0.9, 1)$, Grade III: $K_e/[K_e] \in [0.8, 0.9)$, Grade IV: $K_e/[K_e] \in [0.7, 0.8)$, and Grade V: $K_e/[K_e] \in [0.6, 0.7)$.

In addition, the stability of interfaces in the RCC dams usually plays an important part in RCC dam safety. Using the cohesion and friction coefficient to describe the stability, we express the grades as follows. Grade I: $C_i/C \in [0.8, 1]$, Grade II: $C_i/C \in [0.7, 0.8)$, Grade III: $C_i/C \in [0.6, 0.7)$, Grade IV: $C_i/C \in [0.5, 0.6)$, and Grade V: $C_i/C \in [0, 0.5)$, where C_i is the interface cohesion and C is the RCC cohesion. Grade I: $f_i/f \in [0.95, 1]$, Grade II: $f_i/f \in [0.9, 0.95)$, Grade III: $f_i/f \in [0.8, 0.9)$, Grade IV: $f_i/f \in [0.7, 0.8)$, and Grade V: $f_i/f \in [0, 0.7)$, where f_i is the interface friction coefficient and f is the RCC friction coefficient.

2.4. Normalization Method of the Interval Number Indicators

From the interval number expressions of the evaluation indicators above, we can see that some indicators have no units, and others have different units. Normalization is necessary to make all kinds of indicators comparable. There are two main methods for the normalization of interval numbers: the "interval-point" method and the "interval-interval" method [36]. The "interval-interval" method is adopted in this study, and the general expression is

$$r_{ij} = y_{ij} / \max_{1 \leq j \leq m}(y_{ij}) \tag{1}$$

or

$$r_{ij} = y_{ij} / \min_{1 \leq j \leq m}(y_{ij}) \tag{2}$$

where $\max_{1 \leq j \leq m}(y_{ij}) = [\max_{1 \leq j \leq m}({y_{ij}}^L), \max_{1 \leq j \leq m}({y_{ij}}^U)]$ is the maximum interval, $\min_{1 \leq j \leq m}(y_{ij}) = [\min_{1 \leq j \leq m}({y_{ij}}^L), \min_{1 \leq j \leq m}({y_{ij}}^U)]$ is the minimum interval, L means the lower boundary of an interval, and U means the upper bound. The maximum interval and the minimum interval are obtained by interval sequencing based on the possibility degree theory. Let $[A_1] = [{a_1}^L, {a_1}^U]$ and $[A_2] = [{a_2}^L, {a_2}^U]$. The relationship between $[A_1]$ and $[A_2]$ can be

$$\text{Isolation}: \qquad {a_1}^U < {a_2}^L \text{ or } {a_2}^U < {a_1}^L \tag{3}$$

$$\text{Overlap} \quad {a_1}^L \le {a_2}^L < {a_1}^U \le {a_2}^U \text{or } {a_2}^L \le {a_1}^L < {a_2}^U \le {a_1}^U \tag{4}$$

$$\text{or Containment}: {a_1}^L \le {a_2}^L < {a_2}^U \le {a_1}^U \text{or } {a_2}^L \le {a_1}^L < {a_1}^U \le {a_2}^U \tag{5}$$

shown visually in Figure 2.

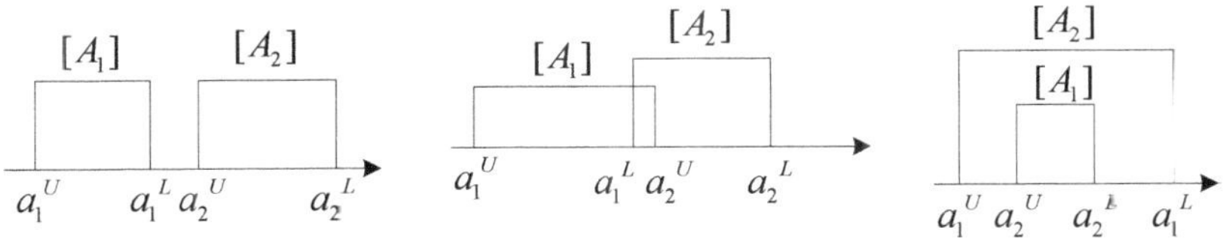

(a) Isolation (b) overlap (c) Containment

Figure 2. Three cases of the relationship between intervals.

Let $l(A_i) = {a_i}^U - {a_i}^L$ and $l(A_j) = {a_j}^U - {a_j}^L$; the possibility degree of $[A_i] \ge [A_j]$ is

$$p([A_i] \ge [A_j]) = \max\{1 - \max(\frac{{a_j}^U - {a_i}^L}{l(A_i) + l(A_j)}), 0\} \tag{6}$$

simply denoted by $p_{ij}(i, j = 1, 2, \cdots, n)$, and the possibility degree matrix is

$$P = \begin{bmatrix} p_{11} & p_{12} & \cdots & p_{1n} \\ p_{21} & p_{22} & \cdots & p_{21} \\ \cdots & \cdots & \cdots & \cdots \\ p_{n1} & p_{n2} & \cdots & p_{nn} \end{bmatrix} \tag{7}$$

Since $0 \le p_{ij} \le 1$, $p_{ij} + p_{ji} = 1$, $i, j = 1, 2, \cdots, n$, the matrix P is a fuzzy complementary judgment matrix with a sequencing vector $\omega = (\omega_1, \omega_2, \cdots, \omega_n)^T$, where

$$\omega_i = \frac{\sum\limits_{j=1}^{n} p_{ij} + \frac{n}{2} - 1}{n(n-1)}, \; i = 1, 2, \cdots, n \tag{8}$$

Thus, the interval can be sequenced by ω_i, and the interval number indicators can be normalized.

3. Interval Number Weights of the Comprehensive Evaluation Indicators

There are three main classes of methods for determining weights: subjective weighting methods, including expert evaluation, the Delphi method, etc., and objective weighting methods, including principal component analysis, factor analysis, the analytical hierarchy process, the entropy method, etc. The rationality of index weight directly affects the accuracy of comprehensive evaluation. The determination of weight is mainly divided into subjective methods and objective methods. Using one kind of method alone has a certain one-sidedness. The subjective method assigns weight according to experts' subjective emphasis on indicators, while the objective method determines weight according to data information. Combining subjective and objective methods can not only give consideration to expert experience but also reduce subjective arbitrariness so as to achieve the unity of subjective and objective weights for indicators.

Combined methods with subjective and objective weighting are more applicable. In this study, the weights of the comprehensive evaluation indicators are calculated with a combined method using expert evaluation and the improved analytical hierarchy process. In addition to the interval number weight, this combined method is used for the first time to reflect the fuzziness and uncertainty in the comprehensive evaluation.

3.1. Interval Number Weights Obtained by Expert Evaluation

Let n experts evaluate p indicators. The evaluation result of indicator j by expert k is denoted by $[w_{kj}] = [w_{kj}{}^L,\ w_{kj}{}^U]\ (1 \le k \le n,\ 1 \le j \le p)$, shown in Table 3.

Table 3. Data structure of interval number weights obtained by expert evaluation.

Indicator	1	2	...	p
Expert 1	$[w_{11}{}^L,\ w_{11}{}^U]$	$[w_{12}{}^L,\ w_{12}{}^U]$	...	$[w_{1p}{}^L,\ w_{1p}{}^U]$
Expert 2	$[w_{21}{}^L,\ w_{21}{}^U]$	$[w_{22}{}^L,\ w_{22}{}^U]$	...	$[w_{2p}{}^L,\ w_{2p}{}^U]$
⋮	⋮	⋮	⋮	⋮
Expert n	$[w_{n1}{}^L,\ w_{n1}{}^U]$	$[w_{n2}{}^L,\ w_{n2}{}^U]$	...	$[w_{np}{}^L,\ w_{np}{}^U]$
Expert group	$[w_{N1}{}^L,\ w_{N1}{}^U]$	$[w_{N2}{}^L,\ w_{N2}{}^U]$	...	$[w_{Np}{}^L,\ w_{Np}{}^U]$

In Table 3, $[w_{Nj}] = [w_{Nj}{}^L,\ w_{Nj}{}^U]$ means the interval number weight of indicator j given by the expert group. Since different experts have different experiences, different weights should be assigned to different experts. Thus, $[w_{Nj}]$ is amended as

$$[w_{Nj}] = \sum_{k=1}^{n} \rho_k \times [w_{kj}] \tag{9}$$

where ρ_k is the weight to show the degree of importance of expert k.

3.2. Interval Number Weights Obtained by the Improved Analytical Hierarchy Process

The analytical hierarchy process (AHP) is a method to determine the indicator weight by comparing the degree of importance of each indicator. The process can be divided into three steps: compare the degree of importance of each indicator, form the comparison matrix; check the consistency of the comparison matrix; calculate the weight of the indicator. AHP can be improved by making the comparison matrix consistent using an optimal transfer matrix, named the improved AHP, which can be used to obtain the weight of the indicator more quickly and conveniently. The application steps are as follows.

(1) On the basis of the degree of importance of the indicators, form the comparison matrix B using the reciprocal 1~9 scaling method, shown in Table 4, where b_{ij} means the comparative result of indicator i with indicator j.

Table 4. Reciprocal scale method of 1~9.

Scale	1	3	5	7	9	2, 4, 6, 8
Description degree	equally	slightly	obviously	seriously	extremely	in between

(2) Calculate the transfer matrix C

$$C = (\ln b_{ij})_{p \times p} \tag{10}$$

(3) Calculate the optimal transfer matrix D

$$D = (d_{ij})_{p \times p} \tag{11}$$

where $d_{ij} = \frac{1}{p}\sum_{k=1}^{p}(\ln b_{ik} - \ln b_{jk})$.

(4) Calculate the weight of indicator j

$$\overline{w_i} = \sqrt[p]{\prod_{i=1}^{p} e^{d_{ij}}} / \sum_{j=1}^{p} \sqrt[p]{\prod_{i=1}^{p} e^{d_{ij}}},\ j = 1,\ 2,\ \cdots,\ p \tag{12}$$

where $\overline{w_j}$ is the weight of indicator j, d_{ij} is the difference measure between the eigenvectors of the function, and p is the length of the eigenvectors.

3.3. Integrated Interval Number Weights Obtained by Combined Methods

Combining the weights obtained by expert evaluation with those by improved AHP, integrated weights can be obtained.

$$[w_j] = q_1[w_{Nj}] + (1 - q_1)[\overline{w_j}] \tag{13}$$

where $0 < q_1 < 1$.

Therefore, the integrated interval number weights of indicators can be expressed as

$$[\vec{W}] = ([w_1],\ [w_2],\ \cdots,\ [w_p]) \tag{14}$$

where $[w_i]$ means the integrated interval number weight of indicator i.

The weights are constrained by $\sum w_i = 1,\ 0 \le w_i \le 1$, where $i = 1,\ 2,\ 3,\ 4$ and w_i refers to the weight of deformation, seepage, stress, and stability.

4. Comprehensive Evaluation Models with Interval Number Indicators and Weights

The comprehensive evaluation interval number $[R]$ can be calculated with the interval number indicators $[\vec{A}]$ and weights $[\vec{W}]$, expressed as

$$[R] = [\vec{A}]\cdot[\vec{W}]^T \tag{15}$$

where $[\vec{A}] = [[a_1,\ a_2],\ [a_3,\ a_4],\ [a_5,\ a_6],\ [a_7,\ a_8]]$, with $[a_1,\ a_2]$, $[a_3,\ a_4]$, $[a_5,\ a_6]$ and $[a_7,\ a_8]$ as the interval number indicators of deformation, seepage, stress, and stability, respectively, $[\vec{W}] = [[w_1,\ w_2],\ [w_3,\ w_4],\ [w_5,\ w_6],\ [w_7,\ w_8]]$, expressing the interval number weights of those indicators, and $[R] = [R_1,\ R_2]$. R_1 and R_2 can be obtained by operation of $[\vec{A}]$ and $[\vec{W}]$ through an optimization algorithm. However, in Section 2.3.2, there is more than one interval number of deformation, seepage, stress, and stability. We can use one interval number such as $[a_1,\ a_2]$ to express the indicator of deformation by combining the indicators of the same category by expert evaluation. This method can also be used to process indicators of seepage, stress, and stability.

Once the comprehensive evaluation interval number $[R]$ is calculated, to obtain the evaluation result of the RCC dam safety state, reasonable evaluation standards should be set. According to enforcement rules for the hydropower dam safety inspections and some statistics, the evaluation grades are set as five: I: Normal, II: Almost Normal, III: Mildly Abnormal, IV: Abnormal, and V: Dangerous, shown in Table 5.

Table 5. Evaluation standards of the safety state for RCC dams.

Safety State	Normal	Almost Normal	Mildly Abnormal	Abnormal	Dangerous
Grade	I	II	III	IV	V
Standard	[0.8, 1]	[0.6, 0.8)	[0.4, 0.6)	[0.2, 0.4)	[0, 0.2)

In summary, the procedure of establishing the comprehensive evaluation model based on interval number theory can be shown in Figure 3.

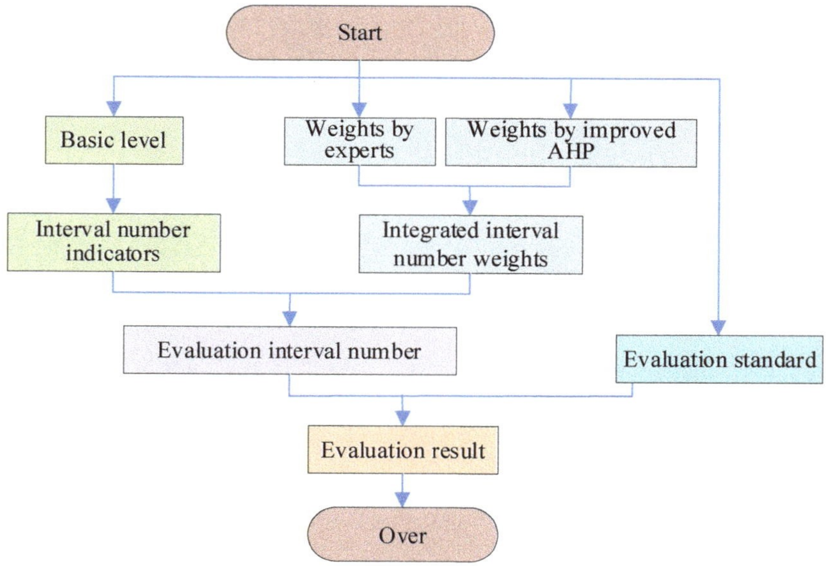

Figure 3. Comprehensive evaluation procedure of the safety state for RCC dams.

5. Case Study

To verify the validity and feasibility of this proposed method, an RCC Dam (Figure 4) located in the Guangxi Zhuang Autonomous Region on the south coast of China was taken as a case study. This RCC Dam is the most important engineering structure of the Red River cascade development project used for power generation, flood prevention, and shipping, with a dam height of 216.5 m.

Figure 4. An RCC dam.

The orthographic perpendicular lines and inverted perpendicular lines are placed in the dam body to monitor the dam deformation. Osmometers monitor dam osmotic pressure; stress and strain gauge groups monitor dam stress. Dam stability is obtained by stability analysis using the finite element method. Taking the typical block $11^{\#}$, for example, its monitoring system (see Figure 5) consists of osmometers UP_{11}-1 and UP_{11}-3; inverted perpendicular line IP_{11}, orthographic perpendicular lines PL_{11}-3, PL_{11}-2, and PL_{11}-1; groups of stress and strain gauges $S^5{}_{11}$-1, $S^5{}_{11}$-2, $S^5{}_{11}$-3, and $S^5{}_{11}$-4.

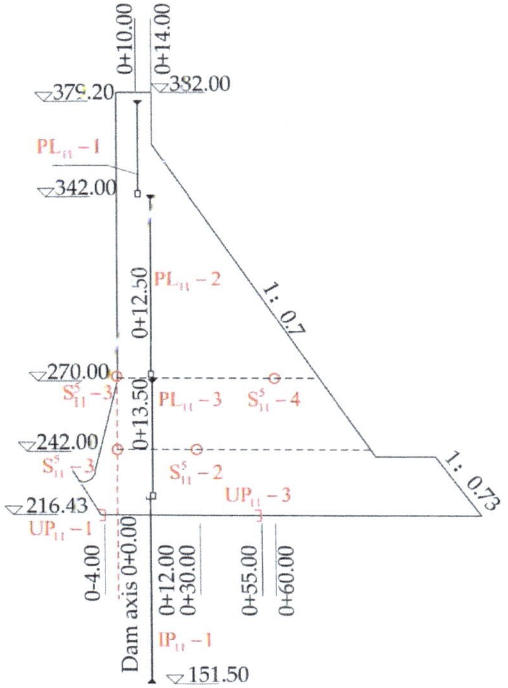

Figure 5. Monitoring instruments of block 11#.

The monitoring data series from 10 October 2006 to 28 August 2009 is studied. First, the interval number expressions and grades of the indicators are determined using the above procedure. These parameters are listed in Table 6.

Table 6. Interval number expressions and grades of the indicators.

Item	Deformation		Seepage		Stress	Stability	
Basic level	$\lvert y-\hat{y}\rvert$	y_i/y	$q/[q_D]$	q_i/q	$\sigma/[\sigma]$	C_i/C	f_i/f
Interval number	[0, 2S]	[0.04, 0.06]	[0, 1]	[0.13, 0.16]	[0, 0.25]	[0.6, 0.96]	[0.83, 0.97]
Grade	I	I	I	II	I	II~I	III~I

Combine the indicators of the same category and normalize them. The interval number indicators finally are $[\vec{A}]$ = [[0.86, 0.96], [0.73, 0.9], [0.9, 1], [0.53, 0.92]]. The weights $[\vec{W1}]$ calculated by expert evaluation are [[0.1, 0.3], [0.3, 0.4], [0.15, 0.2], [0.3, 0.4]], and the weights $[\vec{W2}]$ calculated by improved AHP are [[0.11, 0.13], [0.23, 0.29], [0.05, 0.07], [0.50, 0.61]]. If q_1 in Equation (13) is assumed to be 0.4, the integrated interval number weights $[\vec{W}]$ are [[0.11, 0.20], [0.26, 0.33], [0.09, 0.12], [0.42, 0.53]]. Finally, the comprehensive evaluation interval number $[R]$ is calculated by the optimization algorithm as $[R]$ = [0.654, 0.932]. Comparison of this result with the comprehensive evaluation standards shows that the comprehensive safety state of dam block 11# in this RCC dam is between "almost normal" and "normal"; in other words, the comprehensive safety grade is II~I.

Table 6 shows that for single indicator safety evaluation of dam block 11#, the deformation and stress state grades are both I, the seepage state is II~I, and the stability state is III~I. The reason the safety grades of seepage and stability are relatively low is that the interfaces in dam block 11# weaken its ability for anti-leaking and anti-sliding. The comprehensive evaluation result provides overall information for decision-makers.

Although we try to make the evaluation of the safety state for RCC dams comprehensive and objective, there is some subjectivity. For example, q_1 in Equation (13) is supposed to be 0.4, which is individually experiential. In addition, the evaluation safety state of dam block 11# is based on the site monitoring data and the back analysis result with the finite element method that we can obtain, so more attention is required when performing similar work. As many as possible of the indicators in Figure 1 are recommended for inclusion, with the purpose of providing the most reliable evaluation result. Finally, the research on easy-to-use indicators needs further testing and improvement by validation so that dam managers can use this procedure conveniently to evaluate and monitor the RCC dam safety state.

6. Conclusions

RCC dams have become the fastest-growing international dams, and the dam height has been increasing. Therefore, safety state monitoring and evaluation of RCC dams have been a major concern. However, RCC dams with a particular structure characteristic of interfaces behave differently from traditional concrete dams. In addition, the RCC dam structure is uncertain, non-conservative, and non-linear and consists of an anisotropic material. Therefore, the safety state evaluation of RCC dams is a hard job with much uncertainty, especially the comprehensive safety state evaluation. In this study, we developed a comprehensive evaluation method for the safety state of RCC dams based on interval number theory.

We built a comprehensive evaluation indicator system, developed the interval number expressions and weights of these indicators, established a comprehensive evaluation model, and set evaluation standards. In addition, we verified the performance of the proposed method with an RCC dam in China. The case study shows that this method is efficient and applicable.

The comprehensive evaluation of the safety state of RCC dams involves many indicators and levels. Uncertainty intersection is inevitable. The analytical techniques and methods call for more in-depth research.

Author Contributions: X.C.: conceptualization, formal analysis, writing—reviewing and editing. P.X.: conceptualization, software, writing—original draft. X.L.: conceptualization, methodology, writing—reviewing. C.S.: data curation, methodology. All authors have read and agreed to the published version of the manuscript.

Funding: This research was supported by the Belt and Road Special Foundation of the State Key Laboratory of Hydrology-Water Resources and Hydraulic Engineering (Funding number: Grant No. U2021nkms06; Funder: Xudong Chen), Fund of National Dam Safety Research Center (Funding number: CX2022B05; Funder: Xudong Chen), and the Fund of Research on Key Technologies of spatio-temporal information and collaborative design in engineering (Funding number: YJZD2020-01; Funder: Peng Xu).

Data Availability Statement: Not applicable.

Conflicts of Interest: The authors declare that they have no known competing financial interests or personal relationships that could have appeared to influence the work reported in this paper.

References

1. Raphael, J.M. *The Optimum Gravity Dam, Rapid Construction of Concrete Dams*; American Society of Civil Engineers: New York, NY, USA, 1970.
2. Wang, T.; Chen, J.S.; Wang, T.; Wang, S. Entropy weight-set pair analysis based on tracer techniques for dam leakage investigation. *Nat. Hazards* **2015**, *76*, 747–767. [CrossRef]
3. Chelidze, T.; Matcharashvili, T.; Abashidze, V.; Kalabegishvili, M.; Zhukova, N. Real time monitoring for analysis of dam stability: Potential of nonlinear elasticity and nonlinear dynamics approaches. *Front. Struct. Civ. Eng.* **2013**, *7*, 188–205. [CrossRef]

4. Li, Z.; Gu, C.; Wang, Z.; Wu, Z. On-line diagnosis method of crack behavior abnormality in concrete dams based on fluctuation of sequential parameter estimates. *Sci. China Technol. Sci.* **2015**, *58*, 415–424. [CrossRef]
5. Haralampos, G.; Christos, O. Determination of factors affecting compressive strength of lean RCC mixtures: The experience of Filiatrinos Dam. *Geotech. Geol. Eng.* **2014**, *32*, 1317–1327.
6. Chen, H.Q. Seismic safety of high concrete dams. *Earthq. Eng. Eng. Vib.* **2014**, *13*, 1–16. [CrossRef]
7. Li, B.; Zhang, Z.; Liu, Y.; Yang, S. Evaluation Standard for Safety Coefficient of Roller Compacted Concrete Dam Based on Finite Element Method. *Math. Probl. Eng.* **2014**, *2014*, 601418. [CrossRef]
8. Xi, W.; Yang, J.; Song, J.; Qu, X. Deep learning model of concrete dam deformation prediction based on CNN. *IOP Conf. Ser. Earth Environ. Sci.* **2020**, *580*, 012042. [CrossRef]
9. Chen, S.; Gu, C.; Lin, C.; Zhao, E.; Song, J. Safety monitoring model of a Super-high concrete dam by using RBF neural network coupled with kernel principal component analysis. *Math. Probl. Eng.* **2018**, *2018*, 1712653. [CrossRef]
10. Wang, X.; Yu, H.; Lv, P.; Wang, C.; Zhang, J.; Yu, J. Seepage safety assessment of concrete gravity dam based on matter-element extension model and FDA. *Energies* **2019**, *12*, 502. [CrossRef]
11. He, Y.; Zhao, M.; Wang, K.; Liu, P. Fuzzy comprehensive evaluation of earth rockfill dam seepage security risk based on cloud model. *Water Resour. Power* **2018**, *36*, 83–86.
12. Yang, C.; Yang, J.; Li, Y.; Han, L.; Chen, S. 3D laser scanning point cloud analysis method for dam deformation monitoring. *Bull. Surv. Mapp.* **2021**, *10*, 54–59.
13. Panvalkar, G.A.; Chunade, A.D. Tracing dam seepage using nuclear logging and tracer techniques—A case study. *ISH J. Hydraul. Eng.* **2018**, *24*, 311–316. [CrossRef]
14. Mirzabozorg, H.; Hariri-Ardebili, M.; Heshmati, M.; Seyed-Kolbadi, S. Structural safety evaluation of Karun III Dam and calibration of its finite element model using instrumentation and site observation. *Case Stud. Struct. Eng.* **2014**, *1*, 6–12. [CrossRef]
15. Mata, J.; Leitão, N.S.; de Castro, A.T.; da Costa, J.S. Construction of decision rules for early detection of a developing concrete arch dam failure scenario, a discriminant approach. *Comput. Struct.* **2014**, *142*, 45–53. [CrossRef]
16. Lokke, A.; Chopra, A.K. Response spectrum analysis of concrete gravity dams including dam-water-foundation interaction. *J. Struct. Eng.* **2015**, *141*, 1–9. [CrossRef]
17. Chiganne, F.; Marche, C.; Mahdi, T.F. Evaluation of the overflow failure scenario and hydrograph of an embankment dam with a concrete upstream slope protection. *Nat. Hazards* **2014**, *71*, 21–39. [CrossRef]
18. Bretas, E.M.; Lemos, J.V.; Lourenco, P.B. A DEM based tool for the safety analysis of masonry gravity dams. *Eng. Struct.* **2014**, *59*, 248–260. [CrossRef]
19. Li, H.; Ouyang, J.; Li, F.; Xie, X. Study on safety evaluation model of small and medium-sized earth-rock dam based on BP-AdaBoost algorithm. *IOP Conf. Ser. Mater. Sci. Eng.* **2019**, *490*, 032024. [CrossRef]
20. Han, L.; Liu, M.; Zhang, H.; Yao, L.; Ge, W. Safety assessment model of Earth-rock dam based on ideal point-cloud theory. *J. Eng Sci. Technol. Rev.* **2019**, *12*, 38–50.
21. Wen, L.; Yang, Y.; Li, Y.; Liu, Y.; Zhou, H. Comprehensive evaluation method for the concrete-face rockfill dams behavior based on the fuzzy recognition model. *J. Perform. Constr. Facil.* **2022**, *36*, 04022021. [CrossRef]
22. Shu, X.; Bao, T.; Li, Y.; Zhang, K.; Wu, B. Dam Safety Evaluation Based on Interval-Valued Intuitionistic Fuzzy Sets and Evidence Theory. *Sensors* **2020**, *20*, 2648. [CrossRef] [PubMed]
23. Li, M.; Si, W.; Ren, Q.; Song, L.; Liu, H. An integrated method for evaluating and predicting long-term operation safety of concrete dams considering lag effect. *Eng. Comput.* **2020**, *37*, 2505–2519. [CrossRef]
24. Sang, L.; Wang, J.; Sui, J.; Dziedzic, M. A new approach for dam safety assessment using the extended cloud model. *Water Resour. Manag.* **2022**, *36*, 5785–5798. [CrossRef]
25. Su, H.; Hu, J.; Yang, M.; Wen, Z. Assessment and prediction for service life of water resources and hydropower engineering. *Nat. Hazards* **2015**, *75*, 3005–3019. [CrossRef]
26. Zheng, G.; Jing, Y.; Huang, H.; Zhang, X.; Gao, Y. Application of life cycle assessment (LCA) and extenics theory for building energy conservation assessment. *Energy* **2009**, *34*, 1870–1879. [CrossRef]
27. Pawlak, Z. *Rough Sets: Theoretical Aspects of Reasoning about Data*; Springer Science & Business Media: Dordrecht, The Netherland, 1991.
28. Greco, S.; Matarazzo, B.; Slowinski, R. Rough sets theory for multicriteria decision analysis. *Eur. J. Oper. Res.* **2001**, *129*, 1–47. [CrossRef]
29. Deng, J.L. *Courses of Grey System Theory*; Huazhong University of Science and Technology Press: Wuhan, China, 1990.
30. Harris, J. *Fuzzy Logic Applications in Engineering Science*; Springer Science & Business Media: Dordrecht, The Netherland, 2005.
31. Miller, R.L.; Brickman, P.; Bolen, D. Attribution versus persuasion as a means for modifying behavior. *J. Personal. Soc. Psychol.* **1975**, *31*, 430–470. [CrossRef]
32. Zhao, K.Q. Set pair analysis and its preliminary application. *Explor. Nat.* **1994**, *13*, 67–72.
33. Sevastianov, P. Numerical methods for interval and fuzzy number comparison based on the probabilistic approach and dempster-shafer theory. *Inf. Sci.* **2007**, *177*, 4645–4661. [CrossRef]
34. Chen, L. Gradual Change Characteristics and Safety Monitoring Models of RCC Dams. Ph.D. Dissertation, University of Hohai, Nanjing, China, 2006.

35. Li, B. Gradual Change Laws of Mechanical Parameters and Comprehensive Analysis Methods of Interfaces for RCC Dams. Ph.D. Dissertation, University of Hohai, Nanjing, China, 2010.
36. Yu, X.H.; Xu, Z.S.; Chen, Q. A method based on preference degrees for handling hybrid multiple attribute decision making problems. *Expert Syst. Appl.* **2011**, *38*, 3147–3154. [CrossRef]

water

MDPI

Article

Analysis of the Hydromechanical Properties of Compact Sandstone and Engineering Application

Peng Tang †, Wenbing Zhang †, Haoyu Wang, Jiaxin Zhou, Yabin Dang and Zhiming Chao *

College of Ocean Science and Engineering, Shanghai Maritime University, Shanghai 200135, China; pengtang@shmtu.edu.cn (P.T.); zhangwb@shmtu.edu.cn (W.Z.); 202230410002@stu.shmtu.edu.cn (J.Z.); ybdang2022@163.com (Y.D.)
* Correspondence: zmchao@shmtu.edu.cn; Tel.: +86-199-6559-5212
† These authors contributed equally to this work.

Abstract: The paper proposes a method to simulate the mechanical behavior of compact rock considering hydromechanics by combining physical experiments and numerical analysis. The effectiveness of the constructed method is validated by the comparison between the numerical and physical results of triaxial shear experiments on sandstone in seepage conditions. Based on the validated method, the stability of underground water-sealed oil and gas storage caverns in surrounding compact sandstone during excavation is analyzed. The main findings are as follows: The intrinsic permeability of compact sandstone has a power function relationship with the porosity; the combination of the porous media elastic model and the modified Drucker–Prager plasticity model can preciously represent the mechanical properties of compact sandstone; the proposed method can accurately replicate the hydromechanical response of compact sandstone in seepage conditions; the effects of hydromechanical effects have significant impacts on the stability of surround compact sandstone during the excavation of underground water sealed oil and gas storage caverns, which causes the obvious increase in stress, deformation and plastic deformation zones of the surrounding compact sandstone and remarkable decrease in the stability safety factor.

Keywords: compact rock; gas permeability; underground storage caverns

Citation: Tang, P.; Zhang, W.; Wang, H.; Zhou, J.; Dang, Y.; Chao, Z. Analysis of the Hydromechanical Properties of Compact Sandstone and Engineering Application. *Water* **2023**, *15*, 2011. https://doi.org/10.3390/w15112011

Academic Editors: Jie Yang, Chunhui Ma and Lin Cheng

Received: 19 March 2023
Revised: 10 May 2023
Accepted: 17 May 2023
Published: 25 May 2023

1. Introduction

Compact rock is a main media in many engineering applications including underground water-sealed oil and gas storage caverns, coal mining, oil fields, and nucleolar waste storage caverns [1]. For engineering applications in compact surrounding rock, their stability is mainly controlled by the mechanical properties of the compact surrounding rock [2]. In practical engineering facilities, compact rock is inevitably under the effects of seepage, which has a significant influence on the mechanical properties of compact rock, causing detrimental impacts on the stability of relevant engineering facilities [3]. Many scholars have conducted research on the impact of fluid flow conditions on the mechanical properties of compact sandstone and found that fluid flow to some extent affects the mechanical properties of compact sandstone [4,5]. Thus, it is vital to have a deep understanding of the hydromechanical response of compact rock.

Conducting physical experiments is an important method to investigate the mechanical characteristics of compact rock, which has been adopted by many researchers [6–10]. However, the existing research mainly focuses on the mechanical characteristics of compact rock without considering the effects of seepage, and the investigation of the hydromechanical characteristics of compact rock is seldom. This is because compact rock has extra-low porosity and permeability, and fluid is difficult to permeate through it, causing the high difficulty and cost to measure the hydromechanical properties of compact rock in laboratory tests. Additionally, due to the limitation of experimental apparatus dimension, conducting laboratory tests cannot explore the hydromechanical response of large-scale compact rock.

Numerical techniques can simulate the mechanical properties of large-scale compact rock to provide guidance for the reasonable design of practical engineering applications [11–15]. However, currently, numerical research about the hydromechanical response of compact rock is rare. This is because the relationship between intrinsic permeability and porosity for compact rock is extremely difficult to determine, which is a necessary parameter to establish the numerical model that can consider hydromechanics [16–21]. Therefore, it is necessary to establish a method that can determine the relationship between intrinsic permeability and porosity for compact rock, which can provide the possibility for the numerical investigation of the hydromechanical response of large-scale compact rock.

Due to the properties of low permeability and porosity, underground compact rock is the optimum storage media for oil and gas, with the merits of high leakproofness and safety, funding and land-saving, etc. [22–24]. In recent years, many underground oil and gas storage caverns have been constructed in geological formations composed of compact rock around the world [25–27]. In petroleum engineering, the compact rock around the underground oil and gas storage caverns is usually in the saturated status via carrying out artificial water injection to further rise the leakproofness of caverns, aiming for preventing the leakage of oil and gas [28–31]. For the water-sealed underground oil and gas storage caverns, when experiencing engineering activities such as excavation of caverns, etc., the coupling effect between seepage and stress fields of the saturated surrounding compact is inevitable to occur, which has a significant influence on the stability of underground storage caverns [32–35]. Thus, correct analysis of the stability for underground water-sealed oil and gas storage caverns during engineering activities with considering hydromechanics has remarkable meaning for maintaining the safety of engineering projects.

In this paper, a method is presented for analyzing the hydromechanical properties of compact sandstone by considering the relationship between the intrinsic permeability and porosity of these rocks through a combined approach of experiments and numerical simulations. Based on the presented solution, numerical triaxial shear tests on compact sandstone in seepage conditions are conducted. Through the comparison between the results of numerical and physical triaxial shear experiments, the effectiveness of the established method to replicate the hydromechanical response of compact sandstone is validated. By using the validated method, the stability of the large-scale surrounding compact sandstone during the excavation of underground water-sealed oil and gas storage caverns is analyzed, with the effects of hydromechanics (permeability, porosity, etc.) on the stability of surrounding compact sandstone being studied. The research outcomes can provide guidance for the implementation of the excavation of underground water-sealed oil and gas storage caverns in compact sandstone.

2. Numerical Modelling

2.1. Seepage Theory

For this paper, Forchheimer seepage law is adopted to carry out the seepage analysis of compact sandstone, as shown in Equation (1) [36,37]:

$$k = \frac{k_{liq}}{\left(1+\theta\sqrt{Vw \cdot Vw}\right)} k_s \tag{1}$$

where, k represents the permeability coefficient (m^2); k_{liq} represents the intrinsic permeability (m^2) (The function of porosity); k_s represents the saturation permeability (m^2); θ represents the velocity coefficient (m^{-1}) (The function of porosity); Vw represents velocity of fluid (m/s).

According to the mass conservation equation, the fluid flow continual equation is gained, as shown in Equation (2) [1]:

$$-\left(\frac{\partial v}{\partial x}+\frac{\partial v}{\partial y}+\frac{\partial v}{\partial z}\right) = S\frac{\partial h}{\partial t_1} \tag{2}$$

where, t_1 represents time (t); v represents specific discharge (m/s); S represents phreatic storativity velocity (m^{-1}); h represents water pressure head (m).

It is assumed that the seepage is steady, namely, S equals zero. Thus, the continual equation for steady seepage is gained, as presented in Equation (3).

$$\frac{\partial v}{\partial x}+\frac{\partial v}{\partial y}+\frac{\partial v}{\partial z}=0 \tag{3}$$

After that, the stable seepage equation is deduced, as shown in Equation (4).

$$\frac{\partial}{\partial x}\left(k_x\frac{\partial h}{\partial x}\right)+\frac{\partial}{\partial y}\left(k_y\frac{\partial h}{\partial y}\right)+\frac{\partial}{\partial z}\left(k_z\frac{\partial h}{\partial z}\right)=0 \tag{4}$$

where, k_x, k_y, k_z represent the permeability coefficient for materials (m^2) along the x, y and z axis direction (m), respectively.

The permeability of materials is assumed to be isotropic and homogeneous; the Laplace Equation can be transformed into Equation (5) [38].

$$\frac{\partial^2 h}{\partial x^2}+\frac{\partial^2 h}{\partial y^2}+\frac{\partial^2 h}{\partial z^2}=0 \tag{5}$$

The fluid flow occurs mainly in a horizontal plane, and the flow in the third dimension (i.e., the vertical direction) is negligible or can be assumed to be constant [39]. When the seepage is two-dimensional, Equation (5) can be presented as in Equation (6)

$$\frac{\partial^2 h}{\partial x^2}+\frac{\partial^2 h}{\partial y^2}=0 \tag{6}$$

Finally, Equation (6) is transformed as in Equation (7).

$$\frac{\partial}{\partial x}\left(k_x\frac{\partial h}{\partial x}\right)+\frac{\partial}{\partial y}\left(k_y\frac{\partial h}{\partial y}\right)=0 \tag{7}$$

2.2. Constitutive Modelling

Linearization simplifies the mathematical expressions required for the analysis and makes it easier to solve the governing equations for the system. However, it is important to note that the linearized model is only valid within the specified range of property values, and outside this range, the model will not accurately represent the behavior of the material [40]. In geotechnical engineering, the modified Drucker–Prager model is widely applied to represent the mechanical properties of geotechnical materials. However, the modified Drucker–Prager model has a main deficiency that it cannot replicate the non-linear mechanical properties of materials during the elastic stage [41]. Hence, this paper combines the porous media elasticity model and the modified Drucker–Prager model to represent the constitutive relationship for compact sandstone.

2.2.1. Porous Media Elastic Model

The porous media elastic model is adopted to simulate the non-linear mechanical properties of compact sandstone during the elastic stage in this paper [42]. In this model, the change of elastic volume (the volume of a material that is able to recover its original shape after deformation under elastic conditions) is proportioned with the logarithm of confining pressure:

$$\frac{\kappa}{(1+e_0)}ln\left(\frac{p_0+p_t^{el}}{p+p_t^{el}}\right)=J^{el}-1 \tag{8}$$

where, κ represents logarithmic bulk modulus (MPa), which is an intrinsic property of a medium that represents its volumetric stiffness; e_0 represents initial porosity; p_0 represents initial confining pressure (MPa); J^{el} represents the elastic portion of the porosity; p_t^{el} represents elastic tensile strength (MPa); p represents current confining pressure (MPa), being defined in Equation (9).

$$p = -\frac{1}{3}trace\sigma = -\frac{1}{3}(\sigma_{11} + \sigma_{22} + \sigma_{33}) \tag{9}$$

where σ represents stress (MPa); $\sigma_{11}, \sigma_{22}, \sigma_{33}$ represents the principle stress (MPa).

2.2.2. Modified Drucker–Prager Plasticity Model

Compared with the traditional Drucker–Prager plasticity model, the modified Drucker–Prager model has an additive cap yield surface [43], which can reflect the yielding of materials under isotropic compression and control the stress-dilatancy [44]. The failure surface F_S of the model can be written as in Equation (10) [45]

$$F_s = t - p\tan\beta - d = 0 \tag{10}$$

where β, d represents friction angle (°) and cohesion (MPa), respectively; t represents deviatoric stress (MPa), as shown in Equation (11).

$$t = \frac{1}{2}q[1 + \frac{1}{K} - (1 - \frac{1}{K})(\frac{r}{q})^3] \tag{11}$$

where K represents the material parameter.

2.3. *Hydromechanics Theory*

2.3.1. The Effects of Seepage Fields on Stress Fields

It is assumed that seepage fields are varying the seepage body force for compact rock to impact stress fields. In two-dimensional seepage conditions (assume that the fluid is incompressible and the material is anisotropic), the seepage body force for compact rock is proportional to the gradient of the water head, being expressed in Equation (12) [46]:

$$\begin{Bmatrix} f_x \\ f_y \end{Bmatrix} = \begin{Bmatrix} -r_w\frac{\partial h}{\partial y} \\ -r_w\frac{\partial h}{\partial y} \end{Bmatrix} = \begin{Bmatrix} r_w & J_x \\ r_w & J_y \end{Bmatrix} f = \sqrt{f_x + f_y} \tag{12}$$

where f represents seepage body force (N/m^3); r_w represents the unit weight for fluid (N/m^3); f_x, f_y represents the component force for seepage body force along the x and y direction, respectively (N/m^3); J_x, J_y represents the water head gradient along the x and y direction, respectively (m^{-1}).

The finite element method involves dividing the physical domain into many small geometric shapes, calculating the external loads on each element by numerical integration using equivalent nodal forces, and converting them into equivalent forces at each node [47]. By utilizing Equation (13), the seepage body force is transformed to equivalent nodal force in numerical calculation.

$$\{F_s\} = \int_{\Omega} [N]^T \begin{Bmatrix} f_x \\ f_y \end{Bmatrix} dxdy \tag{13}$$

where $\{F_s\}$ represents the seepage equivalent nodal force (N); $[N]$ is the function of element shape.

$$\{\Delta F_s\} = \int_{\Omega} [N]^T \begin{Bmatrix} f_x \\ f_y \end{Bmatrix} dxdy \tag{14}$$

where $\{\Delta F_s\}$ represents the increment of seepage equivalent nodal force (N).

Therefore, the mathematical model for stress fields that can consider the impacts of the seepage field is obtained, being expressed in Equation (15).

$$\begin{cases} \left[k_{liq}\right]\{\Delta\delta\} = \{\Delta F\} + \{\Delta F_s\}, (x,y) \in \Omega \\ \{\delta\} = \{\delta_0\}, (x,y) \in S_u \\ \sigma_{ij}n_j = T, (x,y) \in S_a \end{cases} \tag{15}$$

where $\{\Delta F\}$ is the increment of nodal load (N); δ represents displacement (m); $\Delta\delta$ represents the increment of displacement (m); S_u is the known displacement boundary condition; S_a represents the known stress boundary condition.

2.3.2. The Impact of Stress Fields on Seepage Fields

Stress fields are affecting the porosity of compact rock to impact the seepage field. This is because the porosity of compact rock is controlled via stress fields, and the porosity can determine the permeability of compact rock to influence the seepage field. The mathematical model of a two-dimensional steady seepage field, which can consider the impacts of stress fields, is presented in Equation (16).

$$\begin{cases} \frac{\partial}{\partial x}\left[k_{liq}(\phi)\right]\frac{\partial h}{\partial x} + \frac{\partial}{\partial y}\left[k_{liq}(\phi)\right]\frac{\partial H}{\partial y} = 0, (x,y) \in \Omega \\ H(x,y) = H_1(x,y), (x,y) \in \Gamma_1 \\ k_{liq}(\phi)\frac{\partial H}{\partial n_2} = q_1(x,y), (x,y) \in \Gamma_2 \\ H(x,y) = y, k_{liq}(\phi)\frac{\partial H}{\partial n_3} = 0, (x,y) \in \Gamma_3 \end{cases} \tag{16}$$

where, ϕ represents porosity; q_1 represents fluid flow (m^2/s); H represents the hydraulic head; Γ_1 represents the boundary conditions with known water head (m); Γ_2 represents the boundary conditions with the known fluid flow (m^2); Γ_3 represents the mixed boundary conditions, that is to say, $h + \frac{\partial h}{\partial t_1}\alpha_1 = \beta_1$, α_1, β_1 represents constant values related to the mixed boundary conditions.

As shown in Equation (16), the core of ascertaining the influence of stress fields on seepage fields is to specify the relationship between porosity and intrinsic permeability for compact sandstone. As long as the relationship between the permeability and porosity is ascertained, the mathematical model for describing the hydromechanical properties of compact sandstone can be specified and the corresponding numerical analysis of the hydromechanical response of compact sandstone can be conducted.

3. Physical Experiments

3.1. Gas Permeability Experiment

3.1.1. Preparation of Gas Permeability Experiment

To ascertain the relationship between the porosity and intrinsic permeability of compact sandstone, gas permeability experiments on compact sandstone were carried out. The compact sandstone sample used in the experiment was obtained from Taoyuan Hydropower Station, which is located in Taoyuan County, Changde City, Hunan Province, China, along the main stream of the Yuanshui River. The sample is a typical tight and homogeneous sandstone with low porosity and permeability, exhibiting strong mechanical properties, and the region is situated in the Nan Yue Mountain area with mountainous and hilly terrain, within the undulating area of Jiangnan Hills. The sandstone was prepared as a cylindrical sample with a diameter of 50 mm × height of 100 mm The physical parameters of the dried sample are shown in Table 1.

Table 1. The parameters for samples.

Number	Diameter (mm)	Height (mm)	Mass (g)	Initial Density ($g \cdot cm^3$)	Initial Porosity (%)
V-1	49.99	99.09	246.16	2.59	7.19
V-2	50.02	99.90	246.19	2.39	6.99
V-3	50.04	99.90	249.09	2.50	6.99

The permeability of compact sandstone was measured by the permeability test system. The experimental system consists of a rock core pressure chamber, confining pressure regulation equipment, seepage pressure regulation equipment, upstream and downstream gas pressure controlling panels, a pressure monitor and a gas pressure gauge with high precision. Due to the extra-low permeability of the compact rock, when adopting fluid as the seepage media, it requires a long duration to penetrate through the compact rock sample, which may cause experimental errors due to the variation in the external environment during the long-term tests. Thus, the tests adopted Argon gas as seepage media, and the permeability for compact sandstone was calculated by using the gas flow method [48]. Additionally, according to the experimental system, the porosity of compact sandstone under different confining pressures was evaluated by using the volume expansion method. The specific experiment scheme is listed in Table 2.

Table 2. Experiment scheme.

Number	Measured Properties	Experiment Method	Gas Seepage Pressure/MPa	Confining Pressure/MPa
S-1, S-2, S-3	Gas permeability	Gas flow method	0.1, 0.2, 0.4, 0.6, 0.8, 1.0, 1.2, 1.4	3, 5, 10, 15, 25, 35
	Porosity	Volume expansion method	×	

3.1.2. Gas Slippage Effect

Utilizing gas as the seepage media to evaluate the permeability of compact rock may induce the Gas Slippage Effect. It means that in the seepage processes of gas in low-porosity porous media, due to the collisions between gas molecules and the pore wall, the gas molecule in pore walls is in motion states to generate the non-Darcy flow (non-linear flows), causing the measured gas permeability to be greater than the intrinsic permeability [49]. The phenomenon was discovered by Klinkenberg in 1941 through experiments [50]. Some researchers proposed that the relationship between the gas-measured permeability and intrinsic permeability can be represented by Equation (17) [51].

$$k = k_{liq}\left(1 + \frac{b}{p_c} - \frac{a}{{p_c}^2}\right) \tag{17}$$

where b represents the slippage factor, a represents the secondary slippage factor, and p_c represents seepage pressure (MPa).

As shown in Equation (17), as long as the gas permeability for compact sandstone in diverse seepage pressure can be obtained, by combining Equation (17), the intrinsic permeability of compact sandstone can be gained. A detailed analysis is as follows.

3.1.3. The Analysis for Gas Permeability Experimental Results

According to the gas permeability experiment results, the relationship between the gas permeability of compact sandstone and seepage pressure is drawn in Figure 1. Then, Equation (17) was used to fit the relationship curves between gas permeability and seepage pressure to attain the intrinsic permeability of compact sandstone under different confining pressures. The fitting degree is listed in Table 3.

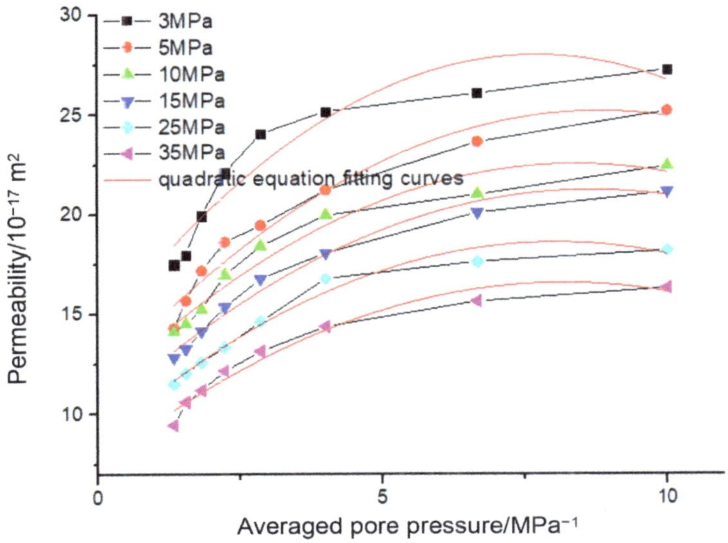

Figure 1. The relationships between the gas permeability for compact sandstone and seepage pressure.

Table 3. The fitting degree R^2 of Equation (2).

Confining Pressure (MPa)	3	5	10	15	25	35
Fitting Degree (%)	96	94	95	99	96	95

According to Table 3, the fitting results of Equation (17) show that the fitting degrees under each confining pressure are all higher than 94%, which means that Equation (17) can describe the Gas Slippage Effect of compact sandstone accurately and the intrinsic permeability of compact sandstone under different confining pressures that is attained based on Equation (17) is reliable, as shown in Table 4.

Table 4. Intrinsic permeability of compact sandstone under different confining pressures.

Confining Pressure (MPa)	3	5	10	15	25	35
Intrinsic Permeability (10^{-17} m^2)	15.04	12.95	12.17	9.30	8.90	7.60

According to Figure 2, the intrinsic permeability of compact sandstone reduces with the rise in confining pressure. Chao [6] established a power model to replicate the relationship between the intrinsic permeability of compact rock and confining pressure. Thus, in the paper, the power function model is used to describe the relationship between the intrinsic permeability of compact sandstone and confining pressure quantitatively, as shown in Equation (18), and the fitting curve is depicted in Figure 2. The fitting degree and parameter are shown in Table 5.

$$k_{liq} = d_1 e^{-h_1 p_c} \quad (18)$$

where d_1 and h_1 are constants that represent the parameters of the power function model used to describe the relationship between intrinsic permeability and confining pressure for compact sandstone.

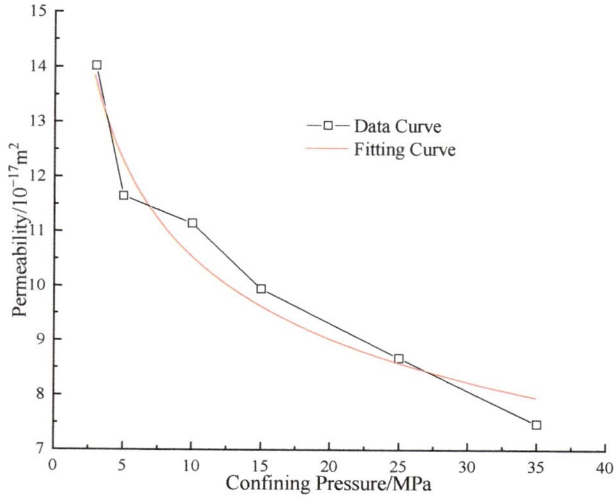

Figure 2. Intrinsic permeability for compact sandstone under diverse confining pressures.

Table 5. Fitting degree and parameters of the power function model.

Fitting Degree/R^2	d_1	h_1
96.93	13.68	0.02

According to Table 5, the fitting degree is 96.93%. It means that the power function model can replicate the relationship between the intrinsic permeability of compact sandstone and the confining pressure well.

According to the experimental results, the relationship between the porosity of compact sandstone and confining pressure is drawn in Figure 3.

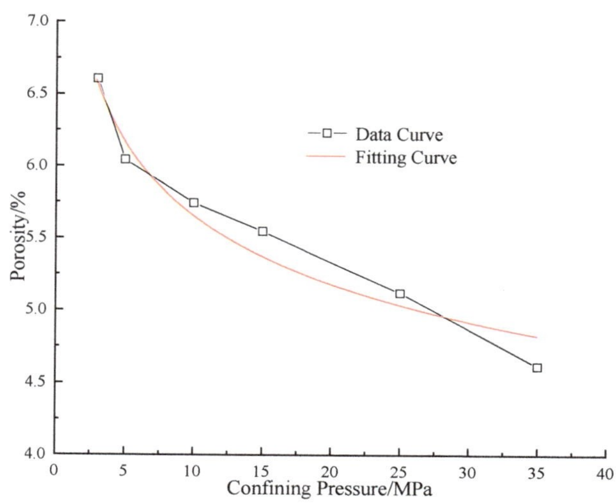

Figure 3. The relationship curve between the porosity for compact sandstone and confining pressure.

According to Figure 3, the porosity of compact sandstone reduces with the rise in confining pressure. Wang proposed that the porosity of compact rock has an exponential relationship with confining pressure [52]. Thus, an exponential function model is adopted to describe the relationship between the porosity of compact sandstone and confining pressure quantitatively, depicted in Equation (19). The fitting degree and parameter is listed in Table 6. The fitting curve is shown in Figure 3.

$$\phi = \phi_0 e^{-ip_c} \tag{19}$$

where, ϕ_0 represents the initial porosity of compact rock; i represents a constant.

Table 6. Fitting degree and parameters of the exponential function model.

Fitting Degree/R^2	i
97	13.68

Based on Table 6, the fitting degree is 97%, which indicates that the exponential function model is able to describe the relationship between the porosity of compact rock and confining pressure preciously. By the combination of Equations (18) and (19), the relationship between intrinsic permeability and porosity for compact sandstone is obtained, as shown in Equation (20).

$$p_c = \frac{ln\frac{d_1}{k_{liq}}}{h_1} = \frac{ln\frac{\phi_0}{\phi}}{i} \tag{20}$$

After the simplification of Equation (20), the relationship between the intrinsic permeability and porosity of compact sandstone is expressed in Equation (21).

$$\frac{k_{liq}}{d_1} = (\frac{\phi}{\phi_0})^{\frac{h_1}{i}} \tag{21}$$

According to Equation (21), the intrinsic permeability and porosity of compact sandstone is a power function relationship. Based on Tables 5 and 6, the value for parameters d_1, h_1, i and ϕ_0 is substituted into Equation (21), as shown in Equation (22).

$$k_{liq} = 0.313\phi^2 \tag{22}$$

Thus, according to the obtained relationship between the porosity and intrinsic permeability of compact sandstone, Equation (22) is substituted into Equation (16) to determine the impact of stress fields on seepage fields. The mathematical model for a two-dimensional stable seepage field that can consider the impact of stress fields is presented in Equation (23).

$$\begin{cases} \frac{\partial}{\partial x}\left[0.313\phi^2\right]\frac{\partial H}{\partial x} + \frac{\partial}{\partial y}\left[0.313\phi^2\right]\frac{\partial H}{\partial y} = 0, (x,y) \in \Omega \\ H(x,y) = H_1(x,y), (x,y) \in \Gamma_1 \\ 0.313\phi^2\frac{\partial H}{\partial n_2} = q(x,y), (x,y) \in \Gamma_2 \\ H(x,y) = y, 0.313\phi^2\frac{\partial H}{\partial n_3} = 0, (x,y) \in \Gamma_3 \end{cases} \tag{23}$$

where H represents the hydraulic head (water pressure) in the seepage field (m), Γ_2 represents the second normal stress difference, and Γ_3 represents the third normal stress difference.

The mathematical model for describing the hydromechanical characteristics of compact sandstone is then obtained by combining the mathematical models of stress fields and seepage fields, as shown in Equation (24).

$$\begin{cases} [K]\{\Delta\delta\} = \{\Delta F\} + \{\Delta F_s\} \\ \left[0.313\phi^2\right]\{H\} + \{f\} = 0 \\ k = 0.313\phi^2 \end{cases} \tag{24}$$

where, $[K]$ represents the stiffness matrix (N/m^2); $\{\Delta\delta\}$ represents the incremental displacement vector (m); $\{\Delta F\}$ represents the incremental force vector (N); $\{\Delta F_s\}$ represents the incremental force vector due to the stress fields represents (N); ϕ the ratio of the volume of void space to the total volume of the rock (%); $\{f\}$ represents the water head distribution function for seepage fields (m/s^2); k represents the permeability of the rock (m^2).

3.2. Triaxial Shear Experiment

3.2.1. Numerical Tests

Based on the proposed method for describing the hydromechanical properties of compact sandstone, Finite Element Method (FEM) numerical triaxial shear experiments on compact sandstone were carried out. In finite element analysis, meshing is the process of dividing the solution domain into small discrete elements. The use of meshing can improve the computational accuracy and efficiency of the model by dividing the large computation domain into small elements to approximate complex geometries, making numerical calculations and simulations more feasible [46]. The stress-strain relationship of compact sandstone is replicated by combining the porous media elastic model and the modified Drucker–Prager plasticity model. The numerical model parameters are listed in Table 7. The numerical compact sandstone has a diameter of 50 mm × length 100 mm. On the top and bottom end, there is a rigid plate fixing on it, respectively. During the numerical shear stage, the displacement of the bottom rigid plate is fixed, and the upper rigid plate moves downwards vertically at a certain velocity to simulate the real shearing process in laboratory triaxial tests. Meanwhile, the central nodal point of the model is used as the deformation and stress monitoring point. The experimental scheme is listed in Table 8. The obtained experimental outcomes are depicted in Figure 4.

Table 7. The parameters for the numerical model.

Bulk Modulus/MPa	Poisson's Ratio	Cohesion/MPa	Friction Angle/°	Cap Eccentricity	Initial Yield Surfaces Position	Transition Surface Position	Flow Stress Ratio	Initial Porosity /%	Initial Permeability/m^2	Density/ Kg/m^3	Logarithmic Bulk Modulus
5333	0.25	7.87	53.82	2	0.1	0.02	0.79	6.6	1.4×10^{-16}	2460	0.0133

Table 8. Triaxial shear test scheme.

Test Material	Confining Pressure/MPa	Seepage Pressure/MPa	Shearing Rate/MPa/min
Compact sandstone	4, 6, 8	1	0.75

3.2.2. Physical Tests

To verify the applicability of the established method, laboratory triaxial shear tests on compact sandstone were implemented to verify the correctness of the obtained hydromechanical response of the compact sandstone in the numerical triaxial shear tests. The physical triaxial shear tests were performed on a static triaxial shear system. This system consists of a pressure chamber, loading equipment, a voltage stabilizer, fluid pressure and water pressure systems, and a data acquisition computer system. The largest sample dimension for the triaxial pressure cell is 50 mm × 100 mm. The largest confining pressure of the cell is 60 MPa. The range of the imposed axial pressure is 0–100 MPa. In the laboratory tests, all the experimental parameters and schemes are consistent with those in the numerical simulation. The axial strain, radial strain and volumetric strain for the rock specimen were recorded in the shearing process.

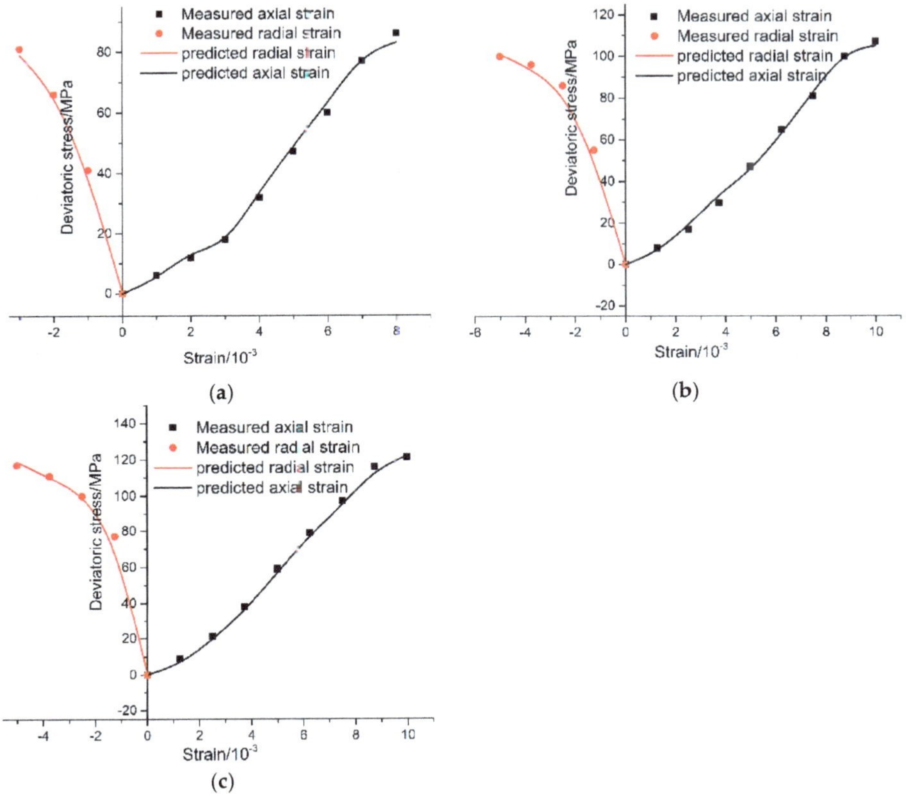

Figure 4. The comparison of the numerical and physical triaxial experiment results: (**a**) Confining pressure 4 MPa; (**b**) Confining pressure 6 MPa; (**c**) Confining pressure 8 MPa.

3.2.3. Analysis of Experimental Results

According to the experimental results, the relationship curves between stress and strain of compact sandstone under diverse confining pressure in seepage conditions are obtained, and the laboratory experiment results are compared with the results of numerical simulation, as shown in Figure 4.

Under diverse confining pressure, the relationship between strain and stress of compact sandstone in seepage conditions obtained by the numerical simulation is generally consistent with that gained in the physical experiments. It indicates that the proposed solution can describe the hydromechanical response of compact sandstone correctly, which lays the foundation for the numerical analysis of the hydromechanical properties of large-scale compact sandstone. It also proves that the combination of the porous media elastic model and the modified Drucker–Prager model plasticity model is able to represent the mechanical properties of compact sandstone accurately.

4. Engineering Application

4.1. Numerical Modelling

Mathematical models such as Darcy's law and more complex models such as dual-porosity models or flow models are often used to accurately describe the water flow behavior in compact sandstone, which involves a lot of calculations and simulations. Oil and gas storage caverns are often constructed underground below the groundwater level, with the water curtain system being installed above the roofs of the caverns to improve the sealing performance and avoid the leakage of oil [53]. This results in the underground oil and gas storage caverns being surrounded with water during the construction and operation period, and the seepage effects of water are inevitable to induce the hydromechanical response of surrounding compact rock and influence the stability of storage caverns. Once the instability of underground storage caverns occurs, the leakage of oil can cause huge contamination in the surrounding environment and endanger the health of the public. Thus, by using the proposed method, the stability of the surrounding compact sandstone during the excavation of caverns in seepage conditions was numerically analyzed. The numerical simulation outcomes can provide suggestions for the safe excavation and operation of underground water-sealed oil storage caverns in surrounding compact sandstone.

A FEM numerical model for the surrounding compact sandstone of underground water-sealed oil storage caverns was built by adopting the established approach, considering the effect of gravity. Four storage caverns were excavated during the numerical simulation, numbered from 1 to 4. The dimension of storage caverns is the same, with a width of 20 m and a height of 30 m. The anisotropic properties of compact sandstone are similar at different scales, and the permeomechanical properties are close. Previous studies have shown that the choice of model scale has exceeded the volume of its representative elementary unit [54,55]. The geometry, layout and number of the four storage caverns are shown in Figure 5. A water curtain tunnel is formed at 30 m above the caverns in the model, which determines the height of the groundwater level. The mechanical characteristics of the simulated surrounding compact sandstone are replicated by combining the porous media elastic model and modified Drucker–Prager plasticity model, with the model parameters listed in Table 7.

Figure 5. The schematic diagrams for the water-sealed storage cavern (unit: m).

The FEM numerical model is partitioned as 20,009 elements and 30,009 nodes, as shown in Figure 6. To guarantee computational precision, finer meshes are used around the caverns. To simulate the actual situation of underground water-sealed oil and gas storage caverns, the horizontal displacement for the left and right boundary and the vertical displacement for the bottom boundary of the model is fixed. Meanwhile, the bottom, left and right boundary of the model is set as impermeable, and the top boundary for the model is set as permeable, with the pore pressure being 0 kPa. The seepage is induced by the groundwater, and the groundwater level is located at a height of 30 m above the roof of caverns. After the excavation of the caverns, the pore water pressure for the boundary of the caverns and the bottom boundary for the water curtain tunnel is set as 0 kPa. Additionally, around the boundaries of each cavern, eight monitoring points are set to record the deformation and stress of the surrounding sandstone during the excavation, numbering from A to F, as shown in Figure 7. The stability of underground water-sealed oil storage caverns in non-seepage conditions during the excavation stage was also analyzed, which did not consider the influence of groundwater seepage. For other model parameters of the numerical models in non-seepage conditions, they are the same as those of the model in seepage conditions. Additionally, the strength reduction method was applied to acquire the stability safety factor of underground oil storage caverns in seepage conditions and in non-seepage conditions, respectively, for analyzing the overall stability of surrounding sandstone quantitatively. In this model, the instability condition of the strength reduction method is the mis-convergence of the established numerical models or the formation of consecutive plastic zones in surrounding rock.

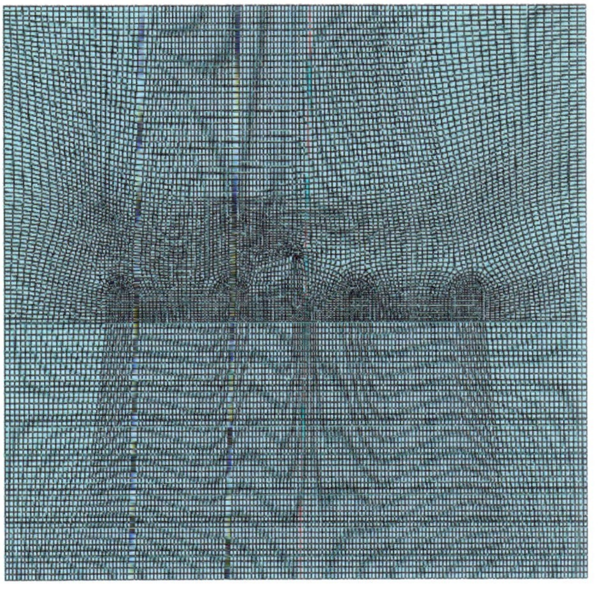

Figure 6. The meshes of the numerical model.

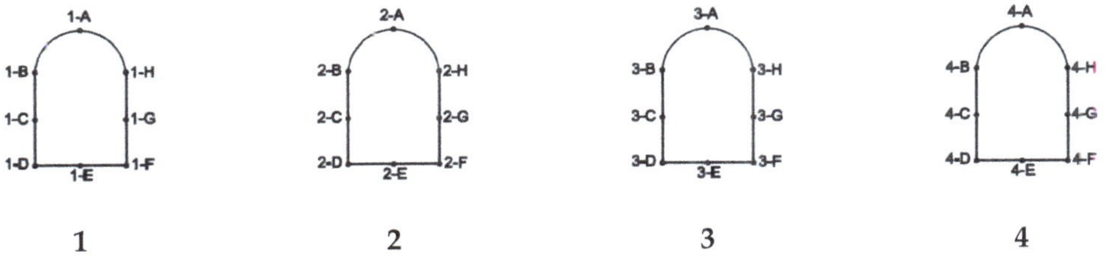

Figure 7. The monitoring points around caverns.

4.2. The Analysis of Numerical Simulation Results

4.2.1. Deformation Analysis

Based on the outcomes of numerical simulation, the deformation distribution and deformation vector diagrams of the surrounding compact sandstone of underground oil storage caverns in seepage conditions and non-seepage conditions are presented in Figure 8, respectively.

According to Figure 8, in seepage conditions, large deformation mainly occurs around the boundary of caverns during excavation. The vertical deformation in the top and bottom boundary of caverns occurs while the horizontal deformation in the side walls takes place. Additionally, there is obvious deformation in the area above the storage caverns, and the deformation extended to the top of the model. In comparison, in non-seepage conditions, the deformation mainly locates at the top and bottom boundary of the caverns, which is in the vertical direction. There is no obvious deformation in the side walls of the caverns and the area above the caverns. This indicates that the hydromechanical effect significantly increases the area of large deformation in surrounding compact sandstone. In practical underground water-sealed storage caverns, attention and corresponding measures should be adopted to control the deformation in the boundaries of caverns and the area above the caverns.

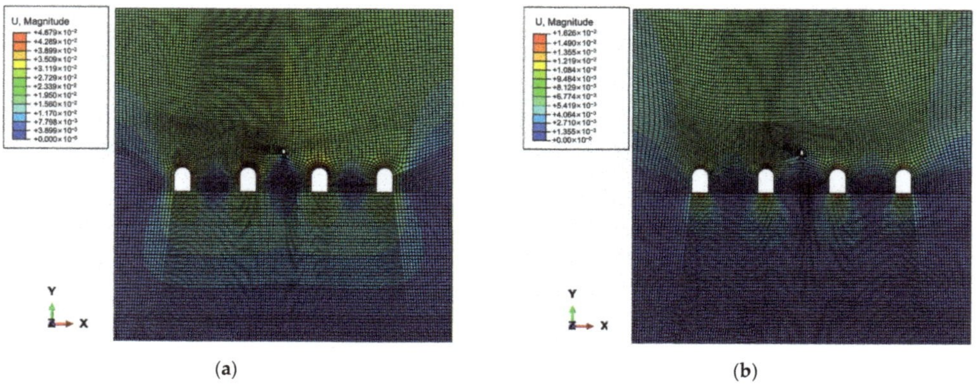

Figure 8. The deformation diagram: (a) The deformation distribution diagram in seepage conditions; (b) The deformation distribution diagram in non-seepage conditions.

To analyze the deformation of surrounding rock quantitatively, the average deformation recorded by different monitoring points in each cavern under seepage and non-seepage conditions is drawn in Figure 9. In Figure 9, 1 to 4 means the number of caverns.

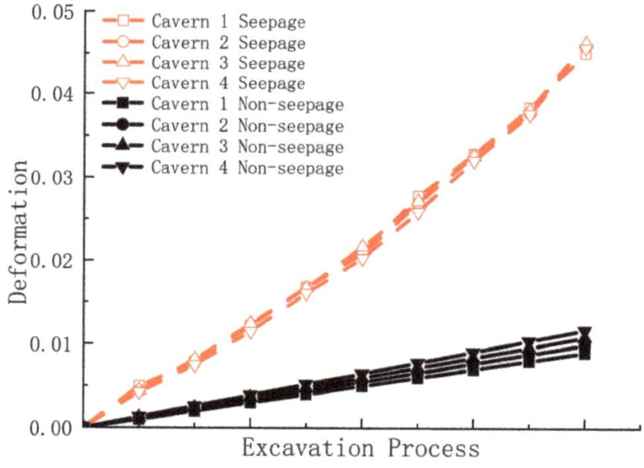

Figure 9. Average deformation recorded by different monitoring points in caverns.

As shown in Figure 9, the average deformation recorded by the different monitoring points around the boundary of caverns in seepage conditions and non-seepage conditions all rises during the excavation, indicating the excavation has a notable influence on the deformation of surrounding sandstone. Additionally, in general, the average deformation in seepage conditions is almost two or three times larger than that in non-seepage conditions after excavation. It means that hydromechanical effects raise the deformation magnitude of surrounding compact sandstone remarkably during excavation.

4.2.2. Stress Analysis

According to the outcomes of numerical simulation, the distribution and magnitude of Mises equivalent stress (Mises equivalent stress is a measure used to describe the strength and stress state of a material, also known as von Mises stress or Mises stress [56–60]) for the surrounding compact sandstone in seepage conditions and non-seepage conditions are shown in Figure 10.

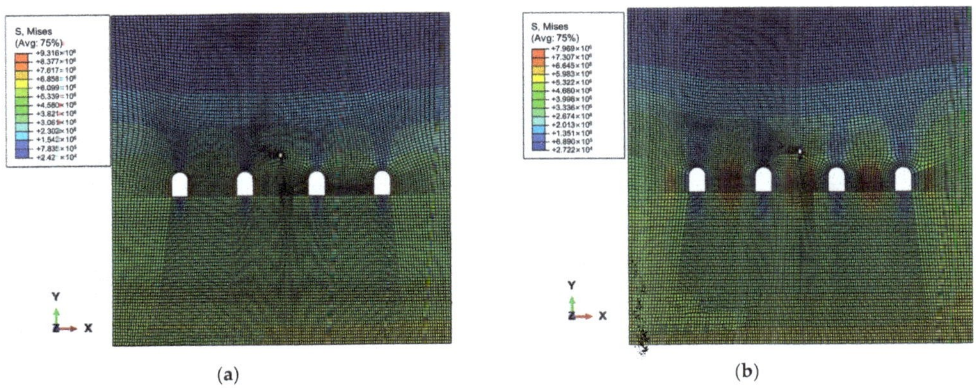

Figure 10. The Mises equivalent stress diagram: (a) Seepage conditions; (b) Non-seepage conditions

As shown in Figure 10, in seepage conditions, the stress in the boundaries of caverns is larger than in other areas of the surrounding compact sandstone during the excavation. Among the boundaries, the stress in the side walls of caverns is higher than the value in the top and bottom boundaries. In comparison, in non-seepage conditions, the maximum stress is located in the middle area between caverns, which is higher than the stress around the boundaries of the caverns. Overall, the maximum stress of the surrounding compact sandstone in seepage conditions is higher than that in non-seepage conditions, and at the same location, the stress in seepage conditions is higher than that in non-seepage conditions. This demonstrates that hydromechanical effects can increase the stress of the surrounding compact sandstone during excavation.

To analyze the stress of surrounding rock quantitatively, the average Mises equivalent stress recorded via using diverse monitoring points around caverns in seepage and non-seepage conditions are drawn in Figure 11.

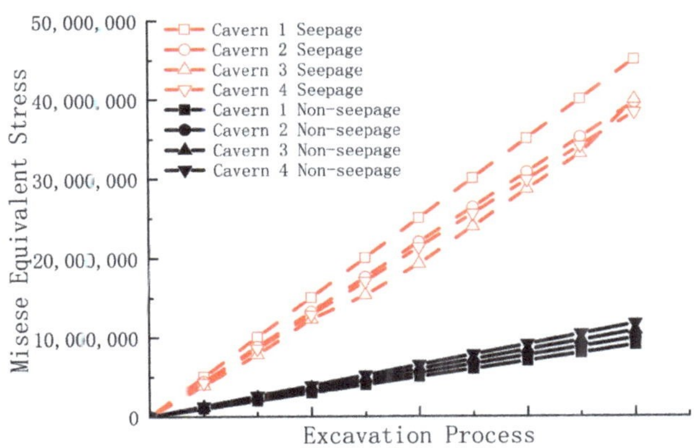

Figure 11. The average Mises stress recorded by using diverse monitoring points.

As shown in Figure 11, the average Mises equivalent stress recorded via using diverse monitoring points in seepage conditions and non-seepage conditions all rises during the excavation, indicating excavation has a high impact on the stress field for surrounding compact sandstone. Moreover, overall, the Mises equivalent stress in seepage conditions is two or three times higher than that in non-seepage conditions. It means the hydromechanical effects raise the stress of surrounding compact sandstone remarkably during excavation.

4.2.3. Plastic Deformation Analysis

According to the outcomes for numerical simulation, the distribution of plastic zones in the surrounding compact sandstone in seepage conditions and non-seepage conditions is demonstrated in Figure 12.

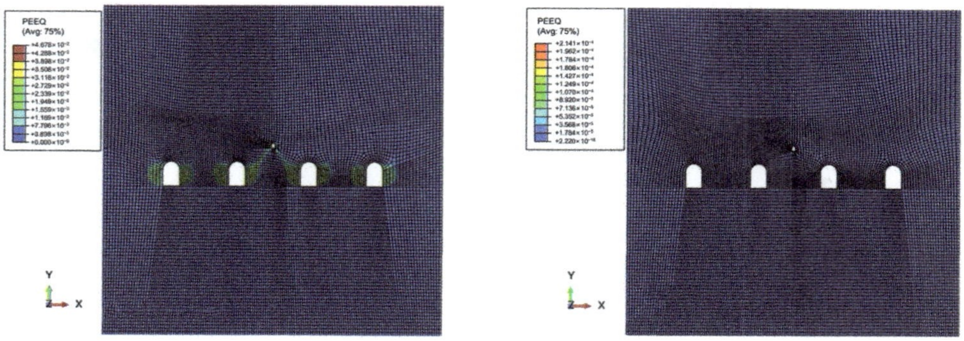

(**a**) (**b**)

Figure 12. The distribution of the plastic zones: (**a**) Seepage conditions; (**b**) Non-seepage conditions.

As shown in Figure 12, in seepage conditions, there are obvious plastic zones in the surrounding compact sandstone during excavation, mainly concentrating around the side walls of storage caverns and the water curtain channel. In comparison, in non-seepage conditions, there is no apparent plastic zone to be observed in the surrounding compact sandstone. It indicates that hydromechanical effects significantly enlarge the area of plastic zones surrounding compact sandstone during excavation.

As shown in Figure 13, in seepage conditions, the plastic deformation of the surrounding compact sandstone around the caverns gradually increases during the excavation. In comparison, in non-seepage conditions, the plastic deformation during excavation keeps stable, without notable rise. Overall, the plastic deformation in seepage conditions is remarkably larger than that in non-seepage conditions, which means the hydromechanical effects can evidentially increase the plastic deformation of surrounding compact sandstone during excavation.

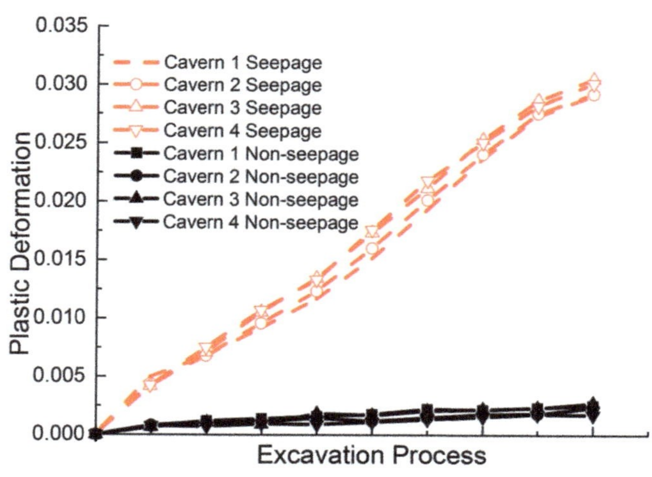

Figure 13. The average plastic deformation.

4.2.4. Stability Analysis

By using the strength reduction method, the stability safety factor of the surrounding compact sandstone (shear strength, tensile strength, etc.) during the excavation of underground oil storage caverns is calculated. The factors are 2.04 in non-seepage conditions and 1.36 in seepage conditions, which has a notable reduction due to the hydromechanical effect. Combined with the aforementioned analysis, the hydromechanical effect significantly rises the deformation, stress and the area of plastic zones in surrounding compact sandstone during the excavation of caverns. It indicates the stability analysis from the adopted hydromechanical method has evident differences compared to the non-seepage method, and for practical underground water-sealed oil storage caverns, it is suitable to apply the hydromechanical method to analyze their stability.

5. Discussion

Previous studies on compact sandstone have mainly focused on its mechanical properties, with little consideration given to the influence of fluid flow. In other words, the hydromechanical behavior of compact sandstone has rarely been investigated, and the intrinsic relationship between the permeability and porosity of compact sandstone, which is a necessary parameter for establishing numerical models that consider fluid mechanics, is extremely difficult to determine. Therefore, there have been few numerical studies on the hydromechanical response of compact sandstone. In this paper, a combined approach of experiments and numerical simulations is employed to propose a method for analyzing the hydromechanical behavior of compact sandstone.

Based on the findings presented in the paper, it can be concluded that the proposed method for simulating the hydromechanical behavior of compact sandstone using a combination of physical experiments and numerical analysis is effective. This was demonstrated through the comparison of numerical and physical results from triaxial shear experiments on sandstone under seepage conditions.

Using the validated method, the stability of underground water-sealed oil and gas storage caverns surrounding compact sandstone during excavation was analyzed. The results showed that hydromechanical effects have a significant impact on the stability of the surrounding compact sandstone during excavation. This leads to an increase in stress, deformation, and plastic deformation zones, as well as a remarkable decrease in the stability safety factor.

Several important conclusions were drawn from the study. First, the intrinsic permeability of compact sandstone has a power function relationship with porosity. Second, the combination of the porous media elastic model and modified Drucker–Prager plasticity model accurately represents the mechanical properties of compact sandstone. Third, the proposed method can accurately replicate the hydromechanical response of compact sandstone under the effect of seepage.

Overall, these findings can be used to guide the implementation of underground water-sealed oil and gas storage cavern excavations in compact sandstone and highlight the importance of considering hydromechanical effects in such projects.

6. Conclusions

This paper proposes a solution to describe the hydromechanical properties of compact sandstone by combing experimental and numerical methods. Then, the effectiveness of the presented solution is validated by comparing the numerical and physical results of triaxial shear experiments on compact sandstone under seepage conditions. After that, the validated solution is adopted to analyze the stability of underground water-sealed oil storage caverns during excavations with considering the hydromechanical effects.

1. The intrinsic permeability of compact sandstone has a power function relationship with the porosity;

2. The combination of the porous media elastic model and the modified Drucker–Prager model plasticity model is able to represent the mechanical properties of compact sandstone accurately;
3. The established method can correctly replicate the hydromechanical response of compact sandstone under the effect of seepage;
4. Hydromechanical effects can significantly increase the stress, deformation and plastic deformation of surrounding compact sandstone during the excavation of underground oil storage caverns and reduce the stability safety factor remarkably.

Author Contributions: Conceptualization, Z.C. and W.Z.; methodology, Z.C. and W.Z.; software, Z.C. and P.T.; validation, J.Z., formal analysis, Z.C.; investigation, W.Z.; writing—original draft preparation, Z.C. and P.T.; writing—review and editing H.W.; visualization, Y.D.; supervision, H.W.; project administration, Z.C.; funding acquisition, Z.C. All authors have read and agreed to the published version of the manuscript.

Funding: This article was funded by the 2022 Open Project of Failure Mechanics and Engineering Disaster Prevention, Key Laboratory of Sichuan Province (FMEDP202209). It is also sponsored by the Shanghai Sailing Program (22YF1415800; 23YF1416100), the Youth Project of Shanghai Soft Science (23692119700), the General Project of Shanghai Natural Science Foundation (23ZR1426200), and the Open Subject of Key Laboratory of Estuarine and Coastal Transportation in the Ministry of Transport (KLECE202302).

Data Availability Statement: Data available upon request.

Conflicts of Interest: The authors declare no conflict of interest.

References

1. Belmokhtar, M.; Delage, P.; Ghabezloo, S.; Conil, N. Drained triaxial tests in low-permeability shales: Application to the Callovo-Oxfordian claystone. *Rock. Mech. Rock. Eng.* **2018**, *51*, 1979–1993. [CrossRef]
2. Chao, Z.M.; Ma, G.T.; Wang, M. Experimental and numerical modelling of the mechanical behaviour of low-permeability sandstone considering hydromechanics. *Mech. Mater.* **2020**, *148*, 103454. [CrossRef]
3. Chao, Z.M.; Fowmes, G. Modified stress and temperature-controlled direct shear apparatus on soil-geosynthetics interfaces. *Geotext. Geomembr.* **2021**, *49*, 825–841. [CrossRef]
4. Zeng, L.B.; Gao, C.Y.; Qi, J.F.; Wang, L.L.; Qu, X.F. The distribution rule and seepage effect of the fractures in the ultra-low permeability sandstone reservoir in east Gansu Province, Ordos Basin. *Sci. China Ser. D Earth Sci.* **2008**, *51*, 44–52. [CrossRef]
5. Feng, M.; Wu, J.; Ma, D.; Ni, X.Y.; Yu, B.Y.; Chen, Z.Q. Experimental investigation on the seepage property of saturated broken red sandstone of continuous gradation. *Bull. Eng. Geol. Environ.* **2018**, *77*, 1167–1178. [CrossRef]
6. Chao, Z.M.; Ma, G.T.; Hu, X.W.; Luo, G. Research on anisotropic permeability and porosity of columnar jointed rock masses during cyclic loading and unloading based on physical model experiments. *Bull. Eng. Geol. Environ.* **2020**, *79*, 5433–5454. [CrossRef]
7. Chao, Z.M.; Ma, G.T.; He, K.; Wang, M. Investigating low-permeability sandstone based on physical experiments and predictive modeling. *Undergr. Space* **2021**, *6*, 364–378. [CrossRef]
8. Roberts, L.A.; Buchholz, S.A.; Mellegard, K.D.; Düsterloh, U. Cyclic loading effects on the creep and dilation of salt rock. *Rock. Mech. Rock. Eng.* **2015**, *48*, 2581–2590. [CrossRef]
9. Wang, H.L.; Xu, W.Y. Permeability evolution laws and equations during the course of deformation and failure of brittle rock. *J. Eng. Mech. Asce.* **2013**, *139*, 1621–1626. [CrossRef]
10. Kohlstedt, D.L.; Evans, B.; Mackwell, S.J. Strength of the lithosphere: Constraints imposed by laboratory experiments. *J. Geophys. Res. Solid Earth* **1995**, *100*, 17587–17602. [CrossRef]
11. Hao, Y.; Smith, M.; Sholokhova, Y.; Carroll, S. CO2-induced dissolution of low permeability carbonates. Part II: Numerical modeling of experiments. *Adv. Water. Resour.* **2013**, *62*, 388–408. [CrossRef]
12. Yang, Z.M.; Yu, R.; Su, Z.X.; Zhang, Y.; Cui, D. Numerical simulation of the nonlinear flow in ultra-low permeability reservoirs. *Petrol. Explor. Dev.* **2010**, *37*, 94–98.
13. Chen, Y.F.; Hong, J.M.; Zheng, H.K.; Li, Y.; Hu, R.; Zhou, C.B. Evaluation of groundwater leakage into a drainage tunnel in Jinping-I arch dam foundation in southwestern China: A case study. *Rock. Mech. Rock. Eng.* **2016**, *49*, 961–979. [CrossRef]
14. Ju, Y.; Xi, C.D.; Zhang, Y.; Mao, L.T.; Gao, F.; Xie, H.P. Laboratory in situ CT observation of the evolution of 3D fracture networks in coal subjected to confining pressures and axial compressive loads: A novel approach. *Rock. Mech. Rock. Eng.* **2018**, *51*, 3361–3375. [CrossRef]
15. Zhang, W.B.; Shi, D.D.; Shen, Z.Z.; Wang, X.H.; Gan, L.; Shao, W.; Tang, P.; Zhang, H.W.; Yu, S.Y. Effect of calcium leaching on the fracture properties of concrete. *Constr. Build. Mater.* **2023**, *365*, 130018. [CrossRef]

16. Chen, S.K.; Yang, T.H.; Ranjith, P.G.; Wei, C.H. Mechanism of the two-phase flow model for water and gas based on adsorption and desorption in fractured coal and rock. *Rock. Mech. Rock. Eng.* **2017**, *50*, 571–586. [CrossRef]
17. Rathnaweera, T.D.; Ranjith, P.G.; Perera, M.S.A. Effect of salinity on effective CO_2 permeability in reservoir rock determined by pressure transient methods: An experimental study on Hawkesbury sandstone. *Rock. Mech. Rock. Eng.* **2015**, *48*, 2093–2110. [CrossRef]
18. Lee, M.Y.; Bauer, S.J. Development of helium-mass-spectrometry-permeameter for the measurement of permeability of near-impermeable rock. *Rock. Mech. Rock. Eng.* **2016**, *49*, 4661–4665. [CrossRef]
19. Wang, H.L.; Xu, W.Y.; Cai, M.; Xiang, Z.P.; Kong, Q. Gas permeability and porosity evolution of a porous sandstone under repeated loading and unloading conditions. *Rock. Mech. Rock. Eng.* **2017**, *50*, 2071–2083. [CrossRef]
20. Zhang, W.B.; Shi, D.D.; Shen, Z.Z.; Shao, W.; Gan, L.; Yuan, Y.; Tang, P.; Zhao, S.; Chen, Y.S. Reduction of the calcium leaching effect on the physical and mechanical properties of concrete by adding chopped basalt fibers. *Constr. Build. Mater.* **2023**, *365*, 130080. [CrossRef]
21. Chao, Z.M.; Fowmes, G. The short-term and creep mechanical behaviour of clayey soil-geocomposite drainage layer interfaces subjected to environmental loadings. *Geotext. Geomembr.* **2022**, *50*, 238–248. [CrossRef]
22. Xiong, J.; Huang, X.L.; Ma, H.L. Gas leakage mechanism in bedded salt rock storage cavern considering damaged interface. *Petroleum.* **2015**, *1*, 366–372. [CrossRef]
23. Yang, C.H.; Wang, T.T.; Qu, D.A.; Ma, H.L.; Li, Y.P.; Shi, X.L.; Daemen, J.J.K. Feasibility analysis of using horizontal caverns for underground gas storage: A case study of Yunying salt district. *J. Nat. Gas. Sci. Eng.* **2016**, *36*, 252–266. [CrossRef]
24. Liu, W.; Zhang, Z.X.; Chen, J.; Jiang, D.Y.; Wu, F.; Fan, J.Y.; Li, Y.P. Feasibility evaluation of large-scale underground hydrogen storage in bedded salt rocks of China: A case study in Jiangsu province. *Energy* **2020**, *198*, 117348. [CrossRef]
25. Xiao, N.; Liang, W.G.; Zhang, S.L. Feasibility analysis of a single-well retreating horizontal cavern for natural gas storage in bedded salt rock. *J. Nat. Gas. Sci. Eng.* **2020**, *99*, 104446. [CrossRef]
26. Jeanne, P.; Zhang, Y.Q.; Rutqvist, J. Influence of hysteretic stress path behavior on seal integrity during gas storage operation in a depleted reservoir. *J. Rock Mech. Geotech. Eng.* **2020**, *12*, 886–899. [CrossRef]
27. Ma, C.H.; Yang, J.; Cheng, L.; Ran, L. Research on slope reliability analysis using multi-kernel relevance vector machine and advanced first-order second-moment method. *Eng. Comput.* **2021**, *38*, 3057–3068. [CrossRef]
28. Ma, C.H.; Yang, J.; Zenz, G.; Staudacher, E.J.; Cheng, L. Calibration of the microparameters of the discrete element method using a relevance vector machine and its application to rockfill materials. *Adv. Eng. Softw.* **2020**, *147*, 102833. [CrossRef]
29. Liu, W.; Li, Y.P.; Yang, C.H.; Daemen, J.J.K.; Yang, Y.; Zhang, G.M. Permeability characteristics of mudstone cap rock and interlayers in bedded salt formations and tightness assessment for underground gas storage caverns. *Eng. Geol.* **2015**, *193*, 212–223. [CrossRef]
30. Ma, C.H.; Gao, Z.Y.; Yang, J.; Cheng, L.; Zhao, T.H. Calibration of adjustment coefficient of the viscous boundary in particle discrete element method based on water cycle algorithm. *Water* **2022**, *14*, 439. [CrossRef]
31. Ma, C.H.; Zhao, T.H.; Li, G.C.; Zhang, A.A.; Cheng, L. Intelligent anomaly identification of uplift pressure monitoring data and structural diagnosis of concrete dam. *Appl. Sci.* **2022**, *12*, 612. [CrossRef]
32. Ma, C.H.; Gao, Z.Y.; Yang, J.; Cheng, L.; Chen, L. Operation performance and seepage flow of impervious body in blast-fill dams using discrete element method and measured data. *Water* **2022**, *14*, 1443. [CrossRef]
33. Maldal, T.; Tappel, I.M. CO_2 underground storage for Snøhvit gas field development. *Energy* **2004**, *29*, 1403–1411. [CrossRef]
34. Xu, X.Y.; Yang, J.; Ma, C.H.; Qu, X.D.; Chen, J.M.; Cheng, L. Segmented modeling method of dam displacement based on BEAST time series decomposition. *Measurement* **2022**, *202*, 111811. [CrossRef]
35. Cheng, L.; Ma, C.H.; Yuan, X.N.; Yang, J.; Hu, L.C.; Zheng, D.J. A literature review and result interpretation of the system identification of arch dams using seismic monitoring data. *Water* **2022**, *14*, 3207. [CrossRef]
36. Pakzad, R.; Wang, S.Y.; Sloan, S.W. Numerical simulation of hydraulic fracturing in low-/high-permeability, quasi-brittle and heterogeneous rocks. *Rock Mech. Rock Eng.* **2018**, *51*, 1153–1171. [CrossRef]
37. Kirkham, D. Explanation of paradoxes in Dupuit-Forchheimer seepage theory. *Water Resour.* **1967**, *3*, 609–622. [CrossRef]
38. Zhao, X.L.; Huang, B.X.; Wang, Z. Experimental investigation on the basic law of directional hydraulic fracturing controlled by dense linear multi-hole drilling. *Rock. Mech. Rock. Eng.* **2018**, *51*, 1739–1754. [CrossRef]
39. Yan, C.; Fan, H.; Huang, D.; Wang, G. A 2D mixed fracture–pore seepage model and hydromechanical coupling for fractured porous media. *Acta Geotech* **2021**, *16*, 3061–3086. [CrossRef]
40. Oliger, J.; Sundström, A. Theoretical and practical aspects of some initial boundary value problems in fluid dynamics. *Siam J. Appl. Math.* **1978**, *35*, 419–446. [CrossRef]
41. Cerfontaine, B.; Charlier, R.; Collin, F.; Taiebat, M. Validation of a new elastoplastic constitutive model dedicated to the cyclic behaviour of brittle rock materials. *Rock. Mech. Rock. Eng.* **2017**, *50*, 2677–2694. [CrossRef]
42. Zhou, X.P.; Xiao, N. 3D numerical reconstruction of porous sandstone using improved simulated annealing algorithms. *Rock. Mech. Rock. Eng.* **2018**, *51*, 2135–2151. [CrossRef]
43. Hu, M.M.; Veveakis, M.; Poulet, T.; Regenauer-Lieb, K. The role of temperature in shear instability and bifurcation of internally pressurized deep boreholes. *Rock. Mech. Rock. Eng.* **2017**, *50*, 3003–3017. [CrossRef]
44. Shin, H.; Kim, J.B.; Kim, S.J.; Rhee, K.Y. A simulation-based determination of cap parameters of the modified Drucker–Prager cap model by considering specimen barreling during conventional triaxial testing. *Comp. Mater. Sci.* **2015**, *100*, 31–38. [CrossRef]

45. Jiang, H.; Xie, Y.L. A note on the Mohr–Coulomb and Drucker–Prager strength criteria. *Mech. Res. Commun.* **2011**, *38*, 309–314. [CrossRef]
46. Bhavikatti, S.S. *Finite Element Analysis;* New Age International: Delhi, India, 2005.
47. Liu, J.J.; Liu, X.G.; Hu, Y.R.; Zhang, Z.S. Study of fluid-solid coupling flow in low permeablable oil reservoir. *Chin. J. Rock. Mechan. Eng.* **2002**, *21*, 48–54.
48. Chaki, S.; Takarli, M.; Agbodjan, W.P. Influence of thermal damage on physical properties of a granite rock: Porosity, permeability and ultrasonic wave evolutions. *Constr. Build. Mater.* **2008**, *22*, 1456–1461. [CrossRef]
49. Meng, Q.B.; Zhang, M.W.; Han, L.J.; Pu, H.; Chen, Y.L. Acoustic emission characteristics of red sandstone specimens under uniaxial cyclic loading and unloading compression. *Rock. Mech. Rock. Eng.* **2018**, *51*, 969–988. [CrossRef]
50. Zhu, G.Y.; Liu, X.G.; Li, S.T.; Huang, Y.Z.; Hao, M.Q. A study of slippage effect of gas percolation in low permeability gas pools. *Nat. Gas. Ind.* **2007**, *5*, 44–47+150.
51. Meng, T.; You, Y.C.; Chen, J.; Hu, Y.Q. Investigation on the permeability evolution of gypsum interlayer under high temperature and triaxial pressure. *Rock. Mech. Rock. Eng.* **2017**, *50*, 2059–2069.
52. Wang, H.L.; Xu, W.Y.; Shao, J.F.; Skoczylas, F. The gas permeability properties of low-permeability rock in the process of triaxial compression test. *Mater. Lett.* **2014**, *116*, 386–388. [CrossRef]
53. Shi, H.B.; Liu, B.G. Design and seepage discharge analysis of artificial water curtains for water sealed underground petroleum storage caverns in rock. *Chin. J. Geotech.* **2010**, *32*, 130–137.
54. Baud, P.; Zhu, W.; Wong, T. Failure mode and weakening effect of water on sandstone. *J. Geophys. Res. Solid Earth* **2000**, *105*, 16371–16389. [CrossRef]
55. Bakar, M.Z.A.; Gertsch, L.S. Evaluation of saturation effects on drag pick cutting of a brittle sandstone from full scale linear cutting tests. *Tunn. Undergr. Space Technol. Tech.* **2013**, *34*, 124–134. [CrossRef]
56. Zhang, Z.; Zhang, Y.Q.; Zheng, Y.; Zhang, J.S.; Emmanuel, F.R.; Zang, W.; Ji, R.W. Power fluctuation and wake characteristics of tidal stream turbine subjected to wave and current interaction. *Energy* **2023**, *264*, 126185. [CrossRef]
57. Zhao, G.W.; Zhu, Z.; Ren, G.Z.; Wu, T.; Ju, P.; Ding, S.J.; Shi, M.; Fan, H.H. Utilization of recycled concrete powder in modification of the dispersive soil: A potential way to improve the engineering properties. *Constr. Build. Mater.* **2023**, *389*, 131626. [CrossRef]
58. Zhang, Y.; Zang, W.; Zheng, J.H.; Lorenzo, C.; Zhang, J.S.; Zheng, Y.; Fernandez-Rodriguez, E. The influence of waves propagating with the current on the wake of a tidal stream turbine. *Appl. Energy* **2021**, *290*, 116729. [CrossRef]
59. Zang, W.; Zheng, Y.; Zhang, Y.; Lin, X.; Li, Y.; Fernandez-Rodriguez, E. Numerical Investigation on a Diffuser-Augmented Horizontal Axis Tidal Stream Turbine with the Entropy Production Theory. *Mathematics* **2023**, *11*, 116. [CrossRef]
60. Zhao, G.W.; Wu, T.; Ren, G.Z.; Zhu, Z.; Gao, Y.; Shi, M.; Ding, S.J.; Fan, H.H. Reusing waste coal gangue to improve the dispersivity and mechanical properties of dispersive soil. *J. Clean Prod.* **2023**, *404*, 136993. [CrossRef]

water

MDPI

Article

Research on Risk Evolution Mechanism of Urban River Ecological Governance Project Based on Social Network Analysis

Junke Xu [1,2,3,*], Jiwei Zhu [1,2,3] and Jiancang Xie [1,2]

1 State Key Laboratory of Eco-Hydraulics in Northwest Arid Region, Xi'an University of Technology, Xi'an 710048, China
2 Research Center of Eco-Hydraulics and Sustainable Development, The New Style Think Tank of Shaanxi Universities, Xi'an 710048, China
3 Department of Engineering Management, Xi'an University of Technology, Xi'an 710048, China
* Correspondence: 1180711014@stu.xaut.edu.cn

Abstract: The evolution and transfer of risk elements of urban river ecological management projects are primarily responsible for the difficulty of risk management in these projects. In this paper, we identify 63 risk elements of urban river ecological management projects using in-depth literature reviews and brainstorming. The association among all the risk elements is constructed using an expert survey method, and the risk elements are utilized as network nodes. The relationships between these nodes are then used as network edges (i.e., paths) to construct a complex network model. By using the network visualization and analysis tool anaconda3, we analyze the overall and local characteristic parameters of the risk network. The risk transmission characteristics of the urban river ecological management project are analyzed according to the parameter characteristics to reveal the inner relationships of risk transmission inherent in the complex network. We use the Jinghe ecological management project in Jinghe New City to verify the effectiveness of the proposed model. The study demonstrates that the starting node risk needs to be controlled, and the conduction node that indirectly triggers risk propagation needs to be cut off to achieve risk prevention and control. Accordingly, the risk prevention strategy is proposed, namely, paying close attention to the starting nodes of schedule delay risk, construction cycle risk and cost overrun risk, as well as the conduction risk nodes of project complexity risk, quality assessment risk, construction accident risk and improper drawing design risk. Effective measures should be taken to control the transmission and occurrence of risks based on these two aspects. The study reveals the network evolution of risk factors, which enriches the theory of the risk factor network evolution and evaluation of urban river ecological management projects.

Keywords: urban river ecological management project; complex network; risk identification; risk transfer

check for updates

Citation: Xu, J.; Zhu, J.; Xie, J. Research on Risk Evolution Mechanism of Urban River Ecological Governance Project Based on Social Network Analysis. *Water* **2023**, *15*, 2012. https://doi.org/10.3390/w15112012

Academic Editor: Caterina Valeo

Received: 24 March 2023
Revised: 7 May 2023
Accepted: 24 May 2023
Published: 25 May 2023

1. Introduction

With the continuous development of the Chinese economy, the water conservancy industry is currently in a stage of rapid development. An increasing number of single and comprehensive water conservancy projects are continuously being constructed and applied, providing an important guarantee for the social stability and sustained economic development of the country. Recent statistics indicate that the total cumulative investment in water conservancy construction reached two trillion Yuan in the "water conservancy reform and development" "Twelfth Five-Year Plan" period. During the "Thirteenth Five-Year Plan" period, the national water conservancy investment reached 3.58 trillion Yuan, which increased 57% over the previous "Twelfth Five-Year Plan" period. The number of water conservancy projects and investment funds is increasing quickly.

Compared with conventional engineering, water conservancy projects are characterized by high capital demand, high construction difficulty, long construction periods and a high impact on the surrounding ecological environment [1]. With the development of water conservancy project construction, the construction mode is diversified, the construction management technology is diversified, the construction management environment is more complex, and social benefits are more multi-targeted. All of these risk factors will inevitably lead to many uncertainties encountered during the construction process of water conservancy projects. These uncertainties are referred to as "risks" [2]. The above-mentioned reasons result in various risky accidents in the construction process of water conservancy projects. Increasingly more attention is paid by researchers to the evaluation of various risks involved in water conservancy projects [3,4]. Researchers around the world have applied traditional analysis methods to evaluate the risks inherent in water conservancy engineering construction projects, including the evidence-based reasoning method [5], safety checklist method [6], fuzzy comprehensive evaluation method [7], accident tree analysis method [8], cloud model evaluation method [9] and material element topological evaluation method [10]. These studies primarily focus on the level of risk evaluation research, and there is a lack of research that focuses on the influence relationship between project risk factors.

Complex network theory can be used to describe realistic systems and reveal the correlation between objective facts. Recently, researchers from various fields conducted numerous network topology and empirical statistical applications of actual complex systems based on constructions of models. For example, Xu et al. [11], Hua and Zheng [12], and Ma [13] constructed causal models of railroad accidents with complex networks; they highlighted the key factors and the correlation among them by analyzing the statistical characteristics of complex networks, such as node degree and mediators. Xiao [14] analyzed the risk evolution process of amphibious seaplane takeoff and landing safety based on the complex network construction model. Liu [15] constructed an accident causation network of highways through town sections by complex network theory and analyzed the network by using the GN algorithm of association division. Meng et al. [16] constructed an unweighted directed network for a pipeline system leakage evolution system and investigated the shortest path of accident risk transmission by analyzing node access degree and clustering coefficient. Li [17] constructed a DEMATEL-ISM explanatory structure model to study the hierarchical structure between causal risk factors and the corresponding complex network model. There are also productive results of complex networks in the natural field, social sciences, biology, engineering, etc.

Social Networks: Yao et al. [18] studied network nodes in undirected and directed graphs using proximity centrality, mesoscopic centrality, node importance and PageRank as evaluation metrics. These metrics proved effective for identifying key nodes, confirming the algorithm's reliability. Sun et al. [19] used a network evolution game model to simulate the impact of social networks and market environment factors on advertising strategies in complex social networks. Technology Network: Yu et al. [20] used a scale-free network to model the diffusion of electric energy substitution technology. They constructed an evolutionary game model that incorporated indicators such as the penalty strength of coal-fired enterprises, government subsidies and transformation costs. The study analyzed the impact of various parameters on the diffusion of the technology. Ecological Network: Williams et al. [21] conducted an empirical study on the topology of the seven largest food chain networks in the world. They found that with a short average path length, the degree distribution of the network obeyed a smaller power–law distribution. Moreover, Montoya et al. [22] and Camacho et al. [23] also obtained similar results. Biological Networks: Jeong et al. [24] found that all the outgoing and incoming degrees obeyed the curtain–law distribution in the metabolic system of 43 living organisms. Fell et al. [25] not only obtained the curtain–law distribution but also found that the corresponding undirected network has a small-world property. Transportation Network: Yuan et al. [26] identified the vulnerable points and domains of the complex network from the risk resistance of the urban metro network system (UMNS). Liu et al. [27] calculated the complex network characteristics of a communication network based on the topological characteristics and investigated the node protection strategies to reduce the occurrence of accidents using centrality analysis.

In the past years, scholars have already studied the problem of risk contagion in engineering projects and obtained rich results. However, there are difficulties in obtaining data, uncertainty in determining network structure, the potential impact on model reliability and prediction accuracy due to assumptions and simplifications, and limitations in the real-time application for engineering risk management and decision-making. From the perspective of the research paradigm, there is a lack of excavation on the mechanism of risk contagion generation and evolution, as well as a lack of providing corresponding interpretation on the mechanism of project risk contagion generation, accumulation and contagion process. Moreover, from the perspective of the research object, most of the current research focus on project combinations such as computer, technology research and development or supply chain. To the best of our knowledge, there are few applications of complex networks in the field of engineering projects, especially in the urban river ecological management project. Based on the engineering characteristics of the urban river ecological management project, we applied the complex network to discuss the evolution of risk in the present paper.

Risk in urban river ecological management projects is the result of multiple risk factors interacting with each other. The risk factors show nonlinear change characteristics, so the complex network is beneficial for analyzing the correlation and interaction between risk factors and their risk transfer process. In order to solve the risk transmission problem of complex engineering projects, it is crucial to investigate the transmission path and process between risk nodes. Once the correlation between risk nodes in the urban river ecological management project was generated, it formed a complex network. Therefore, it is advantageous to employ complex networks to investigate the correlation between risk elements in urban river ecological management projects. In this study, in the urban river ecological management project, the risk elements are regarded as network nodes and a complex network model is constructed to study the inherent laws of risk transmission. In addition, through the use of network visualization analysis tools, the overall and local characteristics of the risk network parameters are analyzed, and strategies for controlling risk transmission are proposed. This method is relatively less commonly used in the application of urban river ecological management projects. Therefore, the novelty of this type of research lies in providing a new way and method to identify risk elements, evaluate risk transmission patterns, and formulate risk prevention strategies, which has important practical significance for risk management in urban river ecological management projects

The remainder of this paper is organized as follows. The research design and methodology are in Section 2. The construction of the risk network is in Section 3. A case study is presented in Section 4. We present a discussion of the results in Section 5. Finally, we summarize the study in Section 6.

2. Research Design and Methodology

2.1. Study Design

Brainstorming, reviewing the literature and expert consultation are used to extract the set of risk factors for urban river ecological management projects and construct a novel risk relationship model of an urban river ecological management project based on complex network theory. Firstly, the transfer interaction of risk factors was obtained, the core nodes and starting nodes in the transfer process were identified, and then the process and mechanism of risk transfer in the urban river ecological governance project were analyzed (please see Figure 1 for the conceptual framework diagram of this study). Our results could provide a useful reference and help organizations to make better decisions to avoid major damage caused by risks and guarantee the smooth implementation of the project.

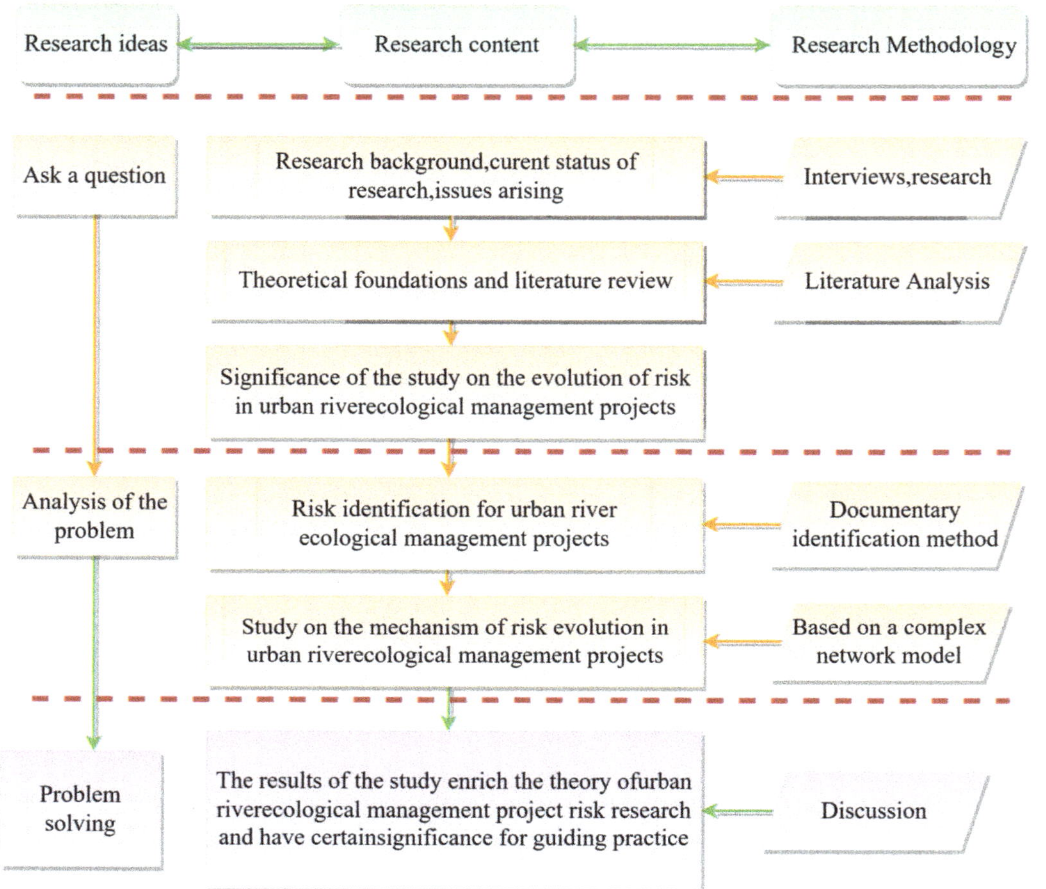

Figure 1. Project risk research framework.

2.2. Risk Node Identification

Risk nodes, as one of the components of complex networks, are the main tools for describing the research object. Network nodes refer to the risk elements that occur in urban river ecological management projects. In this work, we searched relevant academic studies from 2011 to 2021 using the CNKI and Vipul databases as the primary data source. We made a combination of queries based on keywords such as urban rivers [28–30], risk evaluation [31–33], risk identification [34–36], evolutionary mechanism, risk control [37,38] and complex network. Based on the principle of project decomposition structure and the current situation of urban river ecological management projects in China, we summarized all the possible risk elements in these types of projects, counted the frequency of each risk element, and then filtered and organized them. The risk items with similar meanings were summarized and organized. Table 1 shows the network nodes of the identified complex network model by selecting the risk elements with more frequent occurrences. We obtained the final list of risk elements of urban river ecological management projects (Table 1).

Table 1. Risk of river ecological treatment projects.

Stage	Level 1 Risk	Level 2 Risk
project plan	1 Political risks	11 Project Approval Risks 12 Government credit risk 13 Legal and regulatory risks
	2 Economic risks	21 Funded risk 22 Inflation risk 23 Interest rate change risk 24 Financing risk
	3 Natural environment risks	31 Hydrological and geological risks 32 Meteorological condition risk 33 Natural force majeure risk 34 Ecological environment risk
	4 Social risks	41 Social and cultural risks 42 Resident negotiate land acquisition risk 43 Social security risk 45 Public opinion risk 46 Land change risk
	5 Project decision risks	51 Risk of basic acceptance before implementation 52 Decision error risk 53 Approval of work procedure compliance risk 54 Incomplete collection of basic data risk
Project standard Preparation stage	6 Bidding risks	61 Bidder technical and management risk 62 Project complexity risk 63 Competitive risk 64 Bid evaluation risk 65 Calibration risk 66 Risk of information leakage 67 Risk of bidding documents 68 Contract perfection risk 69 Risk of contract signing deviating from bidding content 610 Legality risk of contract signing procedure 611 Contract dispute risk
	7 Planning and design risks	71 Qualification risk of design unit 72 Design progress lag risk 73 Design has defects, errors, omissions and frequent changes 74 Improper standard selection 75 Survey accuracy risk
	8 Preparation before construction risks	81 Construction site layout and technical preparation risk 82 Risk of insufficient supply of substances (materials) and materials 83 Risk of illegal commencement
Project construction stage	9 Construction personnel risks	91 Low technical level 92 Weak safety awareness 93 Qualification risk of employees 94 Risk of slowdown of construction personnel
	10 Construction technology risks	101 Risk of improper drawing design 102 Engineering technology risk 103 Risk of construction equipment 104 Risk of cross operation 105 Construction accident risk

Table 1. *Cont.*

Stage	Level 1 Risk	Level 2 Risk
	11 Construction management risks	111 Safety management risk 112 Coordination risks of participants 113 Rationality of construction organization design 114 Plan adjustment and engineering change 115 Contract management and execution risk 116 The organizational structure setting is chaotic 117 Management authority risk
	12 Risk factors of construction period	121 Certificate cycle 122 Construction period 123 Risk of construction delay
Project completion acceptance stage	13 Quality assessment risk	131 Risk of unqualified project acceptance and putting into use 132 Risk of file transfer not in place 133 Quality assessment risk 134 Audit risk 135 Cost overrun

2.3. *Network Characterization*

2.3.1. Overall Network Analysis

The topology of complex networks is complex and irregular. However, it was found that the overall characteristics of complex networks also exhibit some prevalent characteristics. In this study, we analyzed the following three types of indicators for the overall characteristics of the network.

(1) Network density

The "network density" metric in complex networks refers to the density of interconnected edges between nodes within the network. It is often applied in social networks to measure the density of social relationships and their evolutionary trends. The network density of a network with N nodes and L actual connected edges is:

$$d(G) = \frac{2L}{N(N-1)} \tag{1}$$

(2) Average path length

The average path length, denoted by L, is the mean value of the shortest path between any two nodes. The average path length in a small-world network is the corresponding order of magnitude of its network size.

$$L = \frac{\sum\limits_{i \neq j} d_{ij}}{N(N-1)} \tag{2}$$

where N is the total number of network nodes, and d_{ij} is the shortest distance between nodes i and j.

(3) Clustering coefficient

The clustering coefficient is an important parameter for measuring the degree of network grouping. The clustering coefficient of node i in a network represents the ratio of the number of existing edges E_i to the maximum number of possible edges in the subnet formed by all nodes directly connected to node i, denoted as C_i. Assuming that node i has

k_i nearest neighbors, then, at most, $k_i(k_i-1)/2$ edges may exist in these nearest neighbors, and the clustering coefficient of node i can be expressed as:

$$C_i = \frac{2E_i}{k_i(k_i-1)} \tag{3}$$

2.3.2. Local Network

(1) Degree distribution

The degree of a node in a directed network comprises the out-degree and in-degree. The out-degree of node i is expressed as the number of edges of node i pointing to other nodes, and the in-degree of node i is expressed as the number of edges of other nodes pointing to node i in the network. The degree k_i of node i is the sum of the in-degree k_i^{in} and out-degree k_i^{out} of node i. This is defined as:

$$k_i^{in} = \Sigma_{j=1}^{n} a_{ji}, \qquad k_i^{out} = \Sigma_{j=1}^{n} a_{ji}, \qquad k_i = k_i^{in} + k_i^{out}. \tag{4}$$

The degree distribution $p(k)$ denotes the probability that a node in the network is specified at random, and the degree of the node is equal to k. $p(k)$ is defined as the ratio of the number of nodes of degree k in the network n_k to the total number of nodes N

$$p(k) = n_k/N \tag{5}$$

(2) Medio centricity

The mediator of a node indicates the role of the node within the network. If nodes i and j, which are not adjacent to each other in a complex network, can be reached by some mediator node (i.e., if the mediator node m is more active in the network and can serve as a bridge for the relationship between node pairs), then the role of node m as a mediator is more important in the network, and its importance can be expressed by the mediator number B_m. This is defined as

$$B_m = \sum_{i,j \in N i \neq j} \frac{n_{ij(m)}}{n_{ij}} \tag{6}$$

where n_{ij} is the number of shortest paths between nodes i and j, $n_{ij(m)}$ is the number of times node m appears in all the shortest paths between i and j as a mediator role, and N is the total number of network nodes.

(3) Proximity centrality

The proximity centrality index refers to the degree of proximity between node i and other nodes j. If the node i is nearer to other nodes, the dependence of node i on other nodes to disseminate information becomes less, and the restriction by other nodes becomes weaker. The proximity centrality of a node is the inverse of the sum of the shortest paths based on node i to all other nodes j in the network, as in the following equation

$$C(i) = \frac{1}{\Sigma_y d(y,x)} \tag{7}$$

where $C(i)$ represents the proximity centrality of node i, and $\Sigma_y d(y,x)$ is the sum of distances from node i to all other nodes.

(4) PR value

PageRank (PR) is a comprehensive index that measures the number of internal and external links and the quality of the links on a web page. The higher the PR value of a web page, the higher the number of pages linking to it. If a web page is linked by another with a high PR value, its PR value increases accordingly. Thus, the nodes with a higher PR value in the project risk network are more important.

$$\text{PageRank}(p_i) = \frac{1-d}{N} + d \sum_{pj \in m(p_i)} \frac{\text{PageRank}(p_i)}{L(pj)} \tag{8}$$

(5) Eigenvector centrality

Eigenvector centrality is different from point centrality, where a node with high point centrality (i.e., with many connections) does not necessarily have high eigenvector centrality because the connections may have low eigenvector centrality. Similarly, high eigenvector centrality does not mean that it has high point centrality; it can also have high eigenvector centrality with a small number of important connectors.

Feature vector centrality considers the importance of a node based on both the number of its neighboring nodes (i.e., the degree of that node) and the importance of each neighboring node. Let x_i be the important measure of node i, then

$$EC(i) = x_i = c \sum_{j=1}^{n} a_{ij} x_j \tag{9}$$

where c is a proportional constant.

3. Construction of Risk Network

In this section, we construct the risk network. Figure 2 depicts the research method.

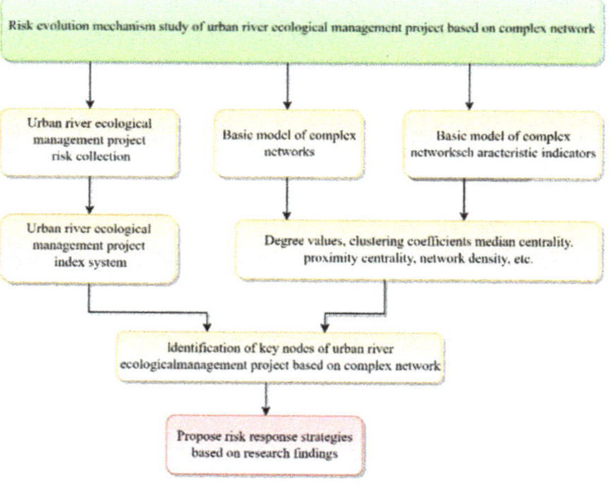

Figure 2. Research methodology process.

3.1. Risk Element Relationship Extraction

There is a close relationship between complex networks and risk networks in urban river management projects. The risk factors in these projects are often interrelated, and their relationships can be viewed as a complex network. In this complex network, there are various complex relationships between risk factors, including positive relationships, negative relationships, nonlinear relationships, etc. These relationships can be expressed by data, and different types of data are needed to represent them. By studying these relationships, the interactive relationships between risk factors in urban river management projects can be revealed, which can lead to the development of more effective risk management strategies.

The relationship of risk factors in the complex network of risk factors of urban river ecological management projects is expressed by data, and various relationships need to be represented by different types of data. Since the number of risk factors in urban river ecological management projects is large and complex, it is difficult to quantify the precise relationship with different values. Therefore, this study aimed to analyze the risk transfer mechanism of urban river ecological management projects, focusing on the causal relationship between risks. This network relationship is directional, and binary directional relationship data were utilized. The expert scoring method was used to collect their judgments on whether a correlation between risk elements exists. In this study, 11 experts were selected, and the "0, 1" method was applied. A larger number of answers were selected to maximize the risk relationship and obtain the results of the risk relationship elements.

The construction of risk network relationships for urban river ecological management projects using the expert survey method was established in the following steps.

(1) Formation of a senior expert group

Risk assessment experts from river ecological management construction units and researchers from universities were invited to form a senior expert group.

(2) Conducting expert consultation activities

A group of 11 experts, all with relevant expertise and experience, were invited to identify the relationships between engineering risk factors, and 2 rounds of questionnaires were conducted to obtain the desired results and avoid interference. The expert opinions were summarized to form the final risk factor relationships.

3.2. Construction of Complex Network Model

The topology of a complex network can be divided into the multi-core, single-core, no-core, star, lumped line, ring and mesh network. Network topologies such as coreless, star, and convergence lines cannot fully describe the complex interrelationships and interactions in a risk network. Coreless networks lack a clear central node, making it difficult to describe the important core components and key nodes in a risk network. Star and convergent wire networks lack stability and reversibility, which can easily lead to unstable and uneven information transmission and integration, and do not fully reflect the characteristics of risk networks. Therefore, single-core ring networks have important topological characteristics, such as a high degree of stability and reversibility. At the same time, single-core ring networks have efficient information transfer and integration capabilities and can effectively describe the relationships and interactions between risk factors. Single-core ring networks are better suited to describe the topology and complexity of risk networks than other types of networks.

In this study, according to the risk factor relationship data matrix defined in Table 2, columns represent the emitters of the relationship (i.e., causers), while rows represent the affected parties of the relationship (i.e., effectors). The "1" indicates the existence of the relationship, while the "0" indicates the non-existence of the relationship. Let n be risk nodes in the risk element set A. Let $A_h = (R_1, R_2, \cdots, R_h)$ be the row risk element set and $A_m = (R_1, R_2, \cdots, R_m)$ be the column risk element set, respectively. We denote b_{ij} as the binary relationship data with $i, \ j = 1, \ 2, \cdots, n$.

Table 2. Risk relationship matrix of urban river ecological treatment project.

Risk Factor	A_1	A_2	A_3	...	A_n
A_1	0	b_{12}	b_{13}	...	b_{1n}
A_2	b_{21}	0	b_{23}	...	b_{2n}
A_3	b_{31}	b_{32}	0	...	b_{3n}
⋮	⋮	⋮	⋮	0	b_{4n}
A_n	b_{n1}	b_{n2}	b_{n3}	b_{n4}	0

$b_{ij} = 1$ means that the risk element in row i affects the risk element in column j.

$b_{ij} = 0$ means that the risk element in row i does not affect the risk element in column j.

Since the risk factor cannot be causally related to itself, the risk factor's impact on itself is "0", and the value on the diagonal is "0". The above data are expressed in the form of a matrix to derive the risk factor relationship matrix A. Since the relationship between two risk factors is not necessarily causal, the adjacency matrix A is not necessarily symmetric. Therefore, the adjacency matrix A is an asymmetric matrix.

4. Case Study: Application in Jinghe River Management

The Jing River is the mother river of Jinghe New City, and it carries the history of the community. This ecological environment comprehensive management project is an important part of the establishment of the Jinghe New City waterfront landscape belt, which enhances the cultural quality of the city, strengthens the livable urban environment and promotes the Jinghe New City developed in high-speed standards.

In Jinghe New City, the upstream of the Jinghe River flood control and ecological management project (referred to as the "Jinghe comprehensive management project") is about 1.0 km from the Jinghe River Jiyuan Bridge (formerly Xiushi Du Bridge). The downstream of the project is down to the Jinghe River Xiantong railroad bridge. This project invests 3.8 billion Yuan and includes the Jinghe River embankment construction project, the Jinghe River beach management and ecological restoration project, and the Jinghe River outside the ecological protection project.

4.1. Mapping of Risk Network

The social network analysis software is selected to draw and analyze the risk network diagram. The risk network adjacency matrix is imported into the software to obtain a visualization of the risk network (Figure 3), where R_i represents the risk element number and the one-way arrow indicates the causal relationship existing between the risk elements.

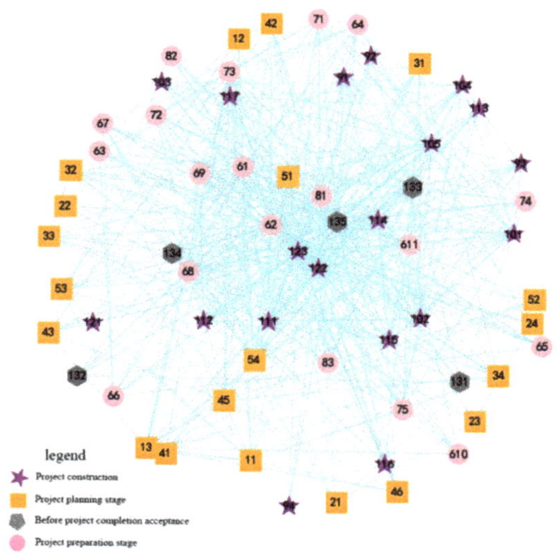

Figure 3. Project risk model.

4.2. Risk Network Parameter Analysis

In this work, we employed the anaconda3 software to analyze a large amount of network parameter data of an urban river ecological management project. Based on the analysis of the data for the overall and local networks, we could obtain the overall characteristics of the network and the key risk factors in the process of project risk transmission. Firstly, we analyzed the parameters for the overall characteristics of the complex network and obtained the basic characteristics of network density, network means path and network aggregation coefficient of the risk network of urban river ecological management project with the aid of Equations (1)–(3). The parameters of the local characteristics of the network were then analyzed. By using Equations (4)–(9), we obtained the network degree and degree distribution, proximity centrality, intermediary centrality, feature vector centrality and PR value, respectively. Finally, we obtain the most critical risk initiation nodes and conduction nodes in the complex network of the project.

4.2.1. Analysis of Overall Network Parameters

(1) Network density

Network density is a parameter used to determine the connectivity between nodes in a complex network. A higher network density indicated more connected paths among nodes and stronger network connectivity. In this study, the average network density of the complex network of urban river ecological management risk was calculated to be 0.2209 by anaconda3 software, indicating that the network is more closely connected.

(2) Clustering coefficient

The clustering coefficient size of a complex network reflects the degree of clustering and connectivity of the network. The connectivity and clustering of the network are stronger in the region where the network clustering is larger. In this study, the average clustering coefficient of the network was computed by anaconda3 software as 0.5559. The distribution range of the clustering coefficients of nodes was 0.2408–1.0000, which shows that the overall clustering and connectivity of the network is at a low level, but the uniform distribution of the clustering coefficients of nodes from the low to the middle level indicates the existence of small association distribution within the network.

The network nodes with large clustering coefficients have a strong correlation with their neighboring factors. According to the calculation in the network, construction workers' denial risk (95), weak safety awareness (92) and natural force majeure risk (33) clustering coefficient values were 1, 0.88 and 0.82, respectively. The clustering coefficient values were large if the neighboring risk factors also had problems, resulting in a chain effect caused by the overall project risk. Therefore, in the risk complex network, such nodes with high clustering coefficients should be controlled to block the occurrence of a chain reaction in the risk network.

(3) Average path length

The average path length of a network is the average of the shortest path lengths among all the pairs of nodes within the network. If the average path length of the network is shorter, the risk nodes need to pass through fewer intermediate nodes, indicating that the rate of risk propagation in the network is faster. By using the anaconda3 software, the average path length of the network of the urban river ecological management project was calculated to be 1.8815. This means that any risk factor within the network needs to pass through 1.8815 edges on average to cause changes in non-neighboring other factors. Therefore, the relationship between risks is relatively "tight".

(4) Cohesive analysis

Conducting nodal component analysis is the most common method for cohesive subgroup analysis. A component analysis is divided into strong component analysis and weak component analysis. Strong component analysis requires a bidirectional connection between any two risk elements within the network. Moreover, weak component analysis has lower requirements; a one-way connection between any two risk elements is sufficient. Therefore, this study applies weak component analysis to analyze the small group coefficients and factional indicators of urban river ecological management project risk factors. The top five small group coefficients are shown in Table 3.

Table 3. Small groups of risk factors.

Clique	Risk Name	Size	Cohesion Index
Clique74	122, 123, 135, 66, 61, 63, 64, 65	8	2.558
Clique50	122, 123, 135, 32, 31, 33, 75, 105, 114	9	2.418
Clique49	122, 123, 135, 32, 31, 33, 75, 105, 111	9	2.371
Clique4	122, 123, 135,62, 10, 101, 114, 611, 112, 113, 115	11	2.335
Clique51	122, 123, 135, 32, 31, 33, 34	7	2.333

The present study investigated the risk factors associated with urban river ecological management projects and analyzed the robustness of the risk factions using a K-cluster analysis. According to Table 3, the risk factors 122, 123 and 135 demonstrated an overlap in small groups with a frequency of 5, indicating their significance as important risks for such projects. Conversely, some risk factor nodes did not appear in the faction, implying their relative lack of connection to the overall risk network and their identification as non-core factors. For the K-cluster analysis, the results indicated that a K value of 2 and a group size of 2 accurately reflected the state of the cohesive subgroups in the risk network of urban river ecological quality engineering projects. The results of the K-cluster coefficient analysis identified three explicit cohesive subgroups in the risk network of the project, as shown in Table 4.

Table 4. Condensed subgroup analysis of risk factors.

G-Clique	Risk Name	Size	Density
G-Clique23	72, 73, 54, 114, 115, 62, 122, 123, 101, 102, 611, 135	12	0.985
G-Clique24	72, 113, 54, 114, 115, 62, 122, 123, 101, 102, 611, 135	12	0.985
G-Clique26	112, 113, 54, 114, 115, 62, 122, 123, 101, 102, 611, 135	12	0.985

4.2.2. Network Local Parameter Analysis

(1) Degree and degree distribution

Degree, one measure of the centrality of network nodes, is a simple but important concept for describing individual nodes. A higher degree value of a particular node indicated more nodes connected to it and a greater influence ability of this node in the network. For directed networks, the degree is divided into three types, namely, entry degree, exit degree and degree. The entry degree is the sum of the number of neighboring edges in a node, and a node with a higher degree of entry is more likely to be influenced by other nodes. The node out-degree is the sum of the number of neighboring edges connected outward from a node, and a node with a larger out-degree is more likely to influence other nodes. The sum of in-degree and out-degree for the node is referred to as the degree. The degree values of each node in the urban river ecological management project are obtained by the anaconda3 software. The degree value is shown in Table 5, and the degree chart is shown in Figure 4.

Table 5. Risk node degree value.

Node Number	The Degree of Input	The Degree of Output	Degree	Node Number	The Degree of Input	The Degree of Output	Degree
11	12	8	20	105	27	24	51
12	7	7	14	81	12	13	25
13	17	12	29	111	26	19	45
31	14	9	23	41	7	3	10
32	12	7	19	72	12	14	26
34	9	10	19	101	24	20	44
45	9	16	25	114	18	22	40
73	18	16	34	134	10	20	30
75	14	12	26	54	20	19	39
24	5	4	9	71	20	15	35

Table 5. *Cont.*

Node Number	The Degree of Input	The Degree of Output	Degree	Node Number	The Degree of Input	The Degree of Output	Degree
42	9	9	18	83	9	13	22
52	5	5	10	93	10	10	20
53	11	8	19	103	9	6	15
43	7	8	15	131	4	8	12
46	5	8	13	132	5	6	11
51	10	21	31	61	14	16	30
68	20	21	41	62	29	26	55
69	23	23	46	67	11	10	21
610	6	6	12	113	14	14	28
611	13	20	33	115	17	16	33
74	14	11	25	133	27	24	51
21	5	3	8	64	14	15	29
22	4	4	8	65	9	6	15
23	4	4	8	66	5	9	14
82	9	4	13	104	11	18	29
135	47	47	94	112	16	11	27
63	10	10	20	116	9	7	16
122	47	50	97	117	7	10	17
123	46	52	98	92	6	5	11
33	10	8	18	94	4	3	7
91	12	14	26	121	2	3	5
102	21	21	42				

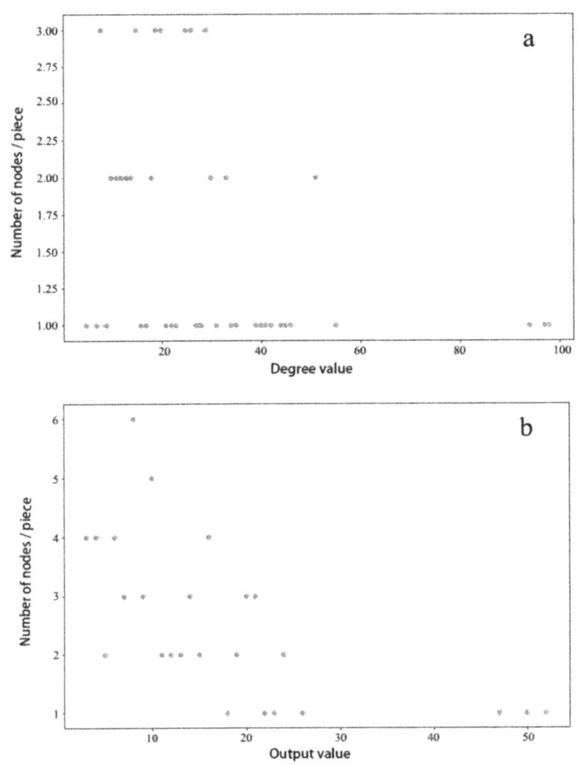

Figure 4. *Cont.*

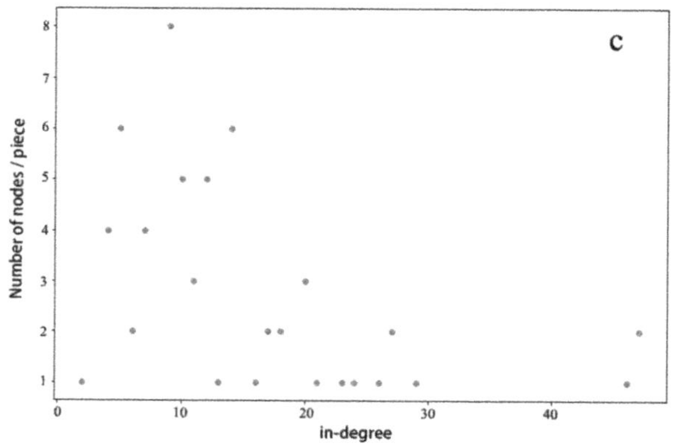

Figure 4. Degree distribution of project risk network: (**a**) degree distribution; (**b**) out-degree distribution; (**c**) in-degree distribution.

In this study, the nodes with high-risk out-degree value and low-risk in-degree value are the starting nodes in the risk transmission process. The control of such nodes should be increased in the urban river ecological management project to reduce the impact of risk transmission. The top 10 nodes in this study are the risk of delay, construction cycle, cost overrun, project complexity, quality assessment, construction accident, contract deviation from the bidding content, plan adjustment and engineering changes, engineering technology and the risk of basic acceptance before implementation.

The higher entry degree value of a node risk indicated a larger influence by other nodes. In the complex engineering project risk network constructed, nodes with both risks out-degree and in-degree values are important conduction nodes in the network. Thus, several nodes with higher out-degree in the project risk network were selected for further analysis. The top −10 risk entry degree nodes in the sequence are as follows: construction cycle, cost overrun, delay risk, project complexity risk, quality assessment risk, construction accident risk, safety management risk, improper design risk, contract signing deviation from the bidding content risk and engineering technology risk.

The nodes with higher degree values in the sequence are schedule delay risk, construction cycle, cost overrun, project complexity risk, quality assessment risk, construction accident risk, design schedule lag, safety management risk, improper drawing design risk and engineering technology risk. These particular types of risks have a more dominating impact than other risk factors.

(2) Proximity to the center

It is equal to the reciprocal sum of all the paths between the node and other nodes. The larger the value is, the smaller the sum of the paths from one node to other nodes can be (i.e., the closer the node is to the center of the network). Therefore, the proximity to the center of the project risk network is used as the basis for determining whether the node is the "center of gravity" of the overall network. According to Table 6, the top ten nodes of proximity to the center are, in order, cost overrun, construction cycle, delay risk, project complexity risk, construction accident risk, safety management risk, quality assessment risk, improper design risk, contract signing deviation from the bidding content risk and design unit qualification risk. These are closest to the center of the risk factor network and are extremely important in the risk network.

Table 6. Local parameters of project risk nodes.

Ranking	Risk Name	Near Centrality	Risk Name	Intermediary Centrality	Risk Name	Centrality of Eigenvector	Risk Name	Clustering Coefficient	Risk Name	Pagerank
1	135	0.81	135	698.31	122	0.29	95	1.00	135	0.05
2	122	0.81	132	600.86	132	0.29	92	0.88	122	0.05
3	132	0.79	122	557.77	135	0.28	33	0.82	132	0.05
4	62	0.63	62	146.86	105	0.22	32	0.77	105	0.03
5	105	0.62	133	115.06	62	0.22	117	0.76	62	0.03
6	111	0.62	69	82.86	111	0.21	93	0.75	111	0.03
7	133	0.61	13	77.95	101	0.20	64	0.74	133	0.03
8	101	0.61	73	73.66	102	0.18	82	0.73	101	0.02
9	69	0.60	105	67.54	133	0.18	112	0.72	102	0.02
10	71	0.59	45	63.70	69	0.17	91	0.71	69	0.02

(3) Intermediation centrality

Intermediation centrality measures the size of the transport capacity of nodes in a complex network. A higher intermediation centrality indicated more influence on the nodes Thus, nodes with large intermediary centrality values in the project risk network have a strong risk transmission capability. As shown in Table 6, the top ten risks are cost overrun delay risk, construction cycle, project complexity risk, quality assessment risk, contract signing deviation from the bidding content risk, laws and regulations risk, design defects errors, omissions, frequent design case changes, construction accident risk and social opinion risk.

(4) Eigenvector centrality

Eigenvector centrality is a parameter that measures the influence of a node on the network and can be used to describe the criticality of a node. A greater eigenvector centrality of a node in the project risk network indicated that the node in the network is more critical. Similar to the degree value, the eigenvector centrality also reflects the importance of a node to a certain extent. Moreover, it takes into account the importance of the node's neighboring nodes and the differences of the neighboring nodes, rather than treating the neighboring nodes "equally". This enables it to reflect the actual important nodes of the risk network more objectively. According to Table 6, the top 10 risks are construction cycle, delay risk, cost overrun risk, construction accident risk, project complexity risk, safety management risk, improper design risk, engineering risk, quality assessment risk and contract deviation from the bidding content risk.

(5) PR value

The higher PR value of a web page indicated a higher number of connected pages. Additionally, if a web page is linked by another web page with a high PR value, its PR value increases accordingly. Thus, nodes with higher PR values in the project risk network have a stronger risk transmission effect. According to Table 6, the top ten risks are as follows: cost overrun, construction cycle, delay risk, construction accident risk, project complexity risk, safety management risk, quality assessment risk, improper drawing design risk, engineering technology risk and contract signing deviation from the bidding content risk.

4.2.3. Critical Risk Determination

Four indicators are considered to determine the importance of nodes, namely, the degree value of nodes, proximity to the center, PR value and feature vector centrality. The top ten risk elements of these indicators in the project risk network are selected and listed in Table 7. The risk elements with a frequency higher or equal to 3 are identified as key nodes in the network. The key transmission nodes are judged based on the size of the outgoing and incoming degree values, the centrality of the characteristic vector and the PR value. The results are shown in Table 8.

Table 7. List of the top 10 risk nodes.

Serial Number	Degree	Near Centrality	Feature Vector	PageRank
1	132	135	122	135
2	122	122	132	122
3	135	132	135	132
4	62	62	105	105
5	133	105	62	62
6	105	111	111	111
7	72	133	101	133
8	111	101	102	101
9	101	69	133	102
10	102	71	69	69

Table 8. Key nodes of the risk network.

Node Type	Node	Risk Name	Production Stage	Judgment Basis
Initial risk node	132	Risk of construction delay	Project construction stage	Ranking first in terms of exit, third in terms of entry and first in terms of value
	122	Construction period risk	Project construction stage	The output ranked second, the entry ranked first and the value ranked second
	135	Cost overrun risk	Project completion acceptance stage	The output ranks third, the input ranks second and the value ranks third
Conduction risk node	62	Project complexity risk	Project preparation stage	The output ranks fourth, the PR value ranks top (fifth), the feature vector (fifth) and the intermediary centrality (fourth)
	133	Quality assessment risk	Project completion acceptance stage	The ranking of degree value, PR value (seventh), intermediary centrality (fifth) and proximity centrality (seventh)
	105	Construction accident risk	Project construction stage	The ranking of degree value is high (sixth), PR is high (fourth), intermediary centrality is high (ninth) and close to centrality (fifth)
	111	Safety management risk	Project construction stage	The ranking of degree value is high (eighth), the ranking of PR value is high (sixth), the ranking of intermediary centrality is high (fifth) and close to centrality (seventh))
	101	Risk of improper drawing design	Project construction stage	The degree value ranks first (ninth), the PR value ranks first (seventh), the centrality of the eigenvector ranks first (seventh) and the near centrality ranks eighth
	69	Risk of contract signing deviating from bidding content	Project preparation stage	The output value ranks high, the PR value ranks high (tenth), the intermediary centrality ranks high (sixth) and the feature vector centrality ranks high

5. Discussion

In contrast to similar research literature [39,40], in terms of research content, this study provides a more comprehensive analysis of the risk network of water conservancy projects, starting from the four stages of water conservancy project construction. In terms of research methodology, this study adds a cohesive analysis of risk in urban river ecological management projects based on complex networks, further elaborating the aggregation effect of risk factors in risk networks. In terms of findings, the results of this study are partially consistent with the focused risk nodes obtained from studies in the literature [39,40], justifying the results of this study.

This study constructs a complex network model and analyzes the overall and local characteristic parameters of the risk network. Moreover, we analyzed the risk transfer characteristics of the urban river ecological management project according to the parameter characteristics, and we also revealed the inner law of risk transfer of the complex network in the case of the Jinghe ecological management project in Jinghe New City. The results can improve the risk management awareness and ability of the main body of the urban river ecological management project to effectively achieve the goal of risk management. In particular, this paper is based on the case study of the risk of the Jinghe ecological management project. A detailed analysis of engineering risk management to guarantee the quality of the project and reduce the occurrence of engineering risks ensures the optimization of the ecological environment of Jinghe, promoting the high-quality development of Jinghe and improving the lives of the people.

Firstly, this study enriched the theory and practice of urban river ecological management engineering risk management research. With the rapid development of urban river ecological management engineering in recent years, it requires a higher level of engineering risk management. This paper investigates the urban river ecological management engineering risk evolution process, and it is meaningful for river management engineering, risk management theory and multi-link complex theory. Secondly, it can improve the risk management level of urban river ecological management project participants. Urban river ecological management project is extremely complex, involving many management factors and technical factors, including personnel, materials, management, environment and technical methods. Managers need sufficient safety theory and management level. By exploring the problem of dynamic engineering risk evolution, this study found the risk links in the construction process, we controlled the transformation conditions by judging the important links, and we also replaced the remedial measures after the occurrence of the risk with preventive management in advance, to improve the management's emergency response-ability to the risk. Thirdly, this study can help the managers of urban river ecological management understand the law of accident risk evolution by preventing the occurrence of risks and improving engineering economic efficiency. These research results help managers formulate corresponding preventive measures to reduce the probability of accidents and increase the economic benefits of enterprises and society.

This study comes with several limitations. The occurrence of the risk and its causes are complex, and 63 risk points and risk factor data cannot fully reflect the law of urban river ecological management project risk evolution. Therefore, the model slightly deviates from reality. It is crucial to expand on existing data in further research and improve the complex network model to further confirm the research results. Additionally, this study only considered the static risk network. In the future, we can consider adding the state of risk nodes and other factors to construct a dynamic risk network model. Moreover, the network constructed in this study considers the relationship between nodes as directed and unweighted, but the relationship between risk elements in engineering practice is more complex, and the weight of edges should be taken into account in future studies to make the conclusions more applicable to the actual situation, and the realistic risk network of complex engineering projects is dynamic and variable, and the risk elements in it may increase, decrease or change with time, but this study only analyzes the static network structure, and the dynamic changes of risks can be included in the study afterward.

6. Conclusions

Urban river ecological management projects are typically characterized by large capital investment, many involved parties, long construction and operation cycles, and pervasive risk factors. This makes their project risks difficult to control, and the risk of any one link may lead to the risk of the overall project. Risk prevention and control need to control direct risks and cut off the transmission path of indirectly triggered risks. This study draws the following conclusions by constructing a risk factor network, analyzing the statistical characteristics of the network and comprehensively evaluating the importance of the risk factors involved.

From the perspective of the risk generation phase, key risks in the project risk network primarily appear in the second preparation phase and the third project construction phase phases of the project. They are also the most critical phases of the project, with the longest time span and the strongest interaction among the parties involved. Therefore, the number of risk elements in these two particular phases grows extremely fast, and the interaction among the nodes increases obviously. The analysis demonstrates that schedule delay risk, construction cycle risk, and project complexity risk are the most important risks in these two phases, which interact with each other. This shows that they have a key role in the risk transfer process.

The cost overrun risk in the completion and acceptance stage is also an important transmission node that cannot be ignored. The high entry value of this risk indicates that the occurrence of risks in other stages of the project leads to the occurrence of this risk, and it will also react to other risks.

The risk factor network has complex characteristics, small-world characteristics and scale-free characteristics. It reflects that the interaction among risk factors leads to the occurrence of risk events, and the risk factors have a close and prompt influence on each other. The risk factors of the conduction node and the starting node within the network have a dominant position. The results obtained demonstrate that schedule delay risk, construction cycle risk and cost overrun risk are the starting risk factors of urban river ecological management projects. Furthermore, the project complexity risk, quality assessment risk and construction accident risk have important conduction roles. Therefore, according to the key nodes of the risk network, searching for approaches to prevent risk generation is a crucial part of the project risk control work.

Author Contributions: Conceptualization, J.X. (Junke Xu); Methodology, J.X. (Junke Xu); Software, J.X. (Junke Xu); Validation, J.X. (Jiancang Xie).; Formal analysis, J.X. (Junke Xu); Investigation, J.X. (Junke Xu); Writing—original draft, J.X. (Junke Xu); Visualization, J.X. (Junke Xu); Project administration, J.Z. All authors have read and agreed to the published version of the manuscript.

Funding: This research was supported by Jinghe New City Jinghe Flood Control and Ecological Management Engineering Consulting Research Project (No. 107-441220098).

Data Availability Statement: The data used to support the findings of this study are available from the corresponding author upon request.

Conflicts of Interest: The authors declare that they have no conflict of interest.

References

1. Wang, S.; Wu, J.; Wang, S.; Xie, X.; Fan, Y.; Lv, L.; Huang, G. Copula-based multivariate simulation approach for flood risk transfer of multi-reservoirs in the Weihe River, China. *Water* **2022**, *14*, 2676. [CrossRef]
2. Pashayan, N.; Antoniou, A.C.; Lee, A.N.; Wolfson, M.; Chiquette, J.; Eloy, L.; Eisen, A.; Stockley, T.L.; Nabi, H.; Brooks, J.D.; et al. Should age-dependent absolute risk thresholds be used for risk stratification in risk-stratified breast cancer screening? *J. Pers. Med.* **2021**, *11*, 916. [CrossRef]
3. Ma, C.; Yang, J.; Zenz, G.; Staudacher, E.J.; Cheng, L. Calibration of the microparameters of the discrete element method using a relevance vector machine and its application to rockfill materials. *Adv. Eng. Softw.* **2020**, *147*, 102833. [CrossRef]
4. Ma, C.H.; Yang, J.; Cheng, L.; Ran, L. Research on slope reliability analysis using multi-kernel relevance vector machine and advanced first-order second-moment method. *Eng. Comput.* **2021**, *38*, 3057–3068. [CrossRef]
5. Wang, A.; Gao, X.D. A variable scale case-based reasoning method for evidence location in digital forensics. *Future Gener. Comput. Syst.* **2021**, *122*, 209–219. [CrossRef]
6. Kaleh, H.; Mohammadian, F.; Pouyakian, M. Developing a safety audit checklist for in-operation administrative buildings. *Work* **2021**, *70*, 785–794. [CrossRef]
7. Zhang, W.P.; Li, B.; Liu, Z.J.; Zhang, B.B. Application of improved fuzzy comprehensive evaluation method in karst groundwater quality evaluation: A case study of Cengong county. *Earth Sci. Inform.* **2021**, *14*, 1101–1109. [CrossRef]
8. Ahn, Y.J.; Yu, Y.U.; Kim, J.K. Accident cause factor of fires and explosions in tankers using fault tree analysis. *J. Mar. Sci. Eng.* **2021**, *9*, 844. [CrossRef]
9. Cheng, Y.; Wang, Z.Z.; Xu, C.D.; Cheng, H.; Zhu, X.L. Safety grade evaluation of aqueduct structure based on fuzzy cloud theory analysis. *Teh. Vjesn.—Tech. Gaz.* **2020**, *27*, 874–882. [CrossRef]
10. Kang, J.F.; Dong, E.C.; Li, X.D.; Guo, Z.; Shi, L.; Li, D.C.; Wang, L. Topological design and biomechanical evaluation for 3D printed multi-segment artificial vertebral implants. *Mater. Sci. Eng. C—Mater. Biol. Appl.* **2021**, *127*, 112250. [CrossRef]
11. Wei, X.; Shiwei, H.; Zhaohui, L.; Yidong, W.; Mengyao, W.; Weiwen, M. Construction and analysis of railway accident causal network based on association rules. *Railw. Transp. Econ.* **2020**, *42*, 72–79. [CrossRef]

12. Lingling, H.; Wei, Z. Analysis of railway accident causes based on complex network theory. *Chin. J. Saf. Sci.* **2019**, *29*, 114–119. [CrossRef]
13. Xin, M. Risk assessment of railway dangerous goods transportation based on complex network and fuzzy analytic hierarchy process. *Transp. Technol. Econ.* **2018**, *20*, 29–33+39. [CrossRef]
14. Qin, X.; Fan, L. Evolution of safety risks for amphibious seaplane takeoff and landing based on complex networks. *Complex Syst. Complex. Sci.* **2019**, *16*, 19–30. [CrossRef]
15. Wenheng, L. Analysis of Traffic Accident Characteristics and Accident Cause Depth in Road Sections Passing through Villages and Towns. Master's Thesis, Beijing Jiaotong University, Beijing, China, 2018.
16. Xiangkun, M.; Guoming, C.; Hongwei, Z. Complex network analysis of submarine pipeline leakage risk evolution. *China's Saf Prod. Sci. Technol.* **2017**, *13*, 26–31. [CrossRef]
17. Xiaopeng, L.; Cunbin, L.; Ding, L.; Xiaokun, S. Analysis of power CPS accident modeling based on DEMATEL-ISM. *J. North China Electr. Power Univ. Nat. Sci. Ed.* **2018**, *45*, 67–77.
18. Jing, Y.; Tongzhou, Z. Research on the importance of nodes in complex social networks. *Comput. Digit. Eng.* **2016**, *44* 80–82+87. [CrossRef]
19. Jingchun, S.; Chenyu, W. Research on simulation of corporate advertising strategies based on complex social networks. *Calculate* **2021**, *40*, 62–68.
20. Haoran, Y.; Sipeng, H. Research on the diffusion of power substitution technology in terminal enterprise clusters based on complex network evolutionary game theory. *Power Supply Consum.* **2022**, *39*, 48–53+60. [CrossRef]
21. Williams, R.J.; Berlow, E.L.; Dunne, J.A.; Barabasi, A.-L.; Martinez, N.D. Two degrees of separation in complex food webs. *Proc. Natl. Acad. Sci. USA* **2002**, *99*, 12913–12916. [CrossRef]
22. Montoya, J.M.; Sol, R.V. Small world patterns in food webs. *J. Theor. Biol.* **2002**, *214*, 405–412. [CrossRef] [PubMed]
23. Camacho, J.; Guimera, R.; Nunes Amaral, L.A. Robust patterns in food web structure. *Phys. Rev. Lett.* **2002**, *88*, 228102. [CrossRef]
24. Jeong, H.; Tombor, B.; Albert, R.; Oltvai, Z.N.; Barabasi, A.L. The large-scale organization of metabolic networks. *Nature* **2000**, *407*, 651–654. [CrossRef] [PubMed]
25. Fell, D.A.; Wagner, A. The small world of metabolism. *Nat. Biotechnol.* **2000**, *18*, 1121–1122. [CrossRef] [PubMed]
26. Jingfeng, Y.; Qiming, L.; Ruoyu, J.; Zhiru, W. Vulnerability analysis of urban metro network system operation. *Chin. J. Saf. Sci.* **2012**, *22*, 92–98. [CrossRef]
27. Dachen, L.; Xingpei, J.; Bo, W.; Fei, T. Topological vulnerability analysis and countermeasures of electric power communication network based on complex network theory. *Power Grid Technol.* **2015**, *39*, 3615–3621. [CrossRef]
28. Qinglin, Z. Study on risk decision model of water conservancy engineering bid evaluation based on life cycle theory. *Water Conserv. Plan. Des.* **2019**, *01*, 80–83. [CrossRef]
29. Ziye, Y.; Yakun, Z.; Zekun, W.; Wenzhe, T. A partnership based water conservancy project construction management model: A case study of ningxia water conservancy project. *J. Hydropower* **2022**, *41*, 35–41.
30. Chunsheng, C.; Ting, Z.; Yunge, L.; Xiaoning, X.; Wenjie, Z. Study on risk assessment of large water conservancy project based on entropy weighted matter element. *Pract. Underst. Math.* **2021**, *51*, 47–55.
31. Wei, G.; Dingbin, L.; Xichen, Z.; Ji, L.; Jianyou, W.; Fengyao, Z. Atastrophe evaluation method for evaluating the risk of land acquisition and resettlement in large linear projects. *Yellow River* **2019**, *41*, 119–121+130. [CrossRef]
32. Xin, J.; Xuelian, L.; Jinghan, W.; Wei, L.; Dongli, L.; Fan, F. SD model for social stability risk evolution of sensitive water conservancy projects. *Chin. J. Saf. Sci.* **2021**, *31*, 18–26. [CrossRef]
33. Qixiang, F.; Peng, L.; Pengcheng, W.; Zhilin, W.; Xianrong, L.; Jianye, Y. Occurrence Mechanism and Management Countermeasures of Safety Accidents in Hydropower Projects. *Chin. J. Saf. Sci.* **2019**, *29*, 144–149. [CrossRef]
34. Hongyan, Z.; Peng, G. Risk analysis and countermeasures for safety management of water conservancy and hydropower projects. *China's Saf. Prod. Sci. Technol.* **2017**, *13*, 89–92.
35. Li, Z. Risk assessment of long span bridge project based on fuzzy comprehensive theory. *Highw. Transp. Technol. Appl. Technol. Ed.* **2011**, *7*, 270–272.
36. Xuan, Q.; Lei, J. Assessment and analysis of life cycle risk factors for green buildings: A questionnaire-based exploration. *J. Civ. Eng.* **2013**, *46*, 123–135. [CrossRef]
37. Guixiang, C.; Hongwei, H.; Jianxin, Y. Research on life cycle risk management of metro projects. *J. Undergr. Space Eng.* **2005**, *01*, 47–51.
38. Yunli, G.; Hongnan, L.; Guojun, Z. Dynamic fuzzy evaluation of engineering project risk based on cooperation. *J. Dalian Univ. Technol.* **2010**, *50*, 404–408.
39. Jiwei, F.; Kaichang, S.; Xin, Y. Human risk analysis of hydraulic engineering based on complex network theory. *People's Yangtze River* **2022**, *53*, 155–159. [CrossRef]
40. Changsheng, L.; Wenzhe, T.; Tengfei, W. Risk management of water conservancy project: Taking Ningxia Water Conservancy Project as an example. *J. Tsinghua Univ. Nat. Sci. Ed.* **2023**, *63* 233–241. [CrossRef]

water

MDPI

Article

Intelligent Inversion Analysis of Hydraulic Engineering Geological Permeability Coefficient Based on an RF–HHO Model

Wei Zhao [1,2], Qiaogang Yin [1] and Lifeng Wen [1,*]

1 State Key Laboratory of Eco-Hydraulics in Northwest Arid Region, Xi'an University of Technology, Xi'an 710048, China; zw349250197@163.com (W.Z.); yinqg96@163.com (Q.Y.)
2 Shaanxi Province Institute of Water Resources and Electric Power Investigation and Design, Xi'an 710001, China
* Correspondence: wenlifeng@xaut.edu.cn

Abstract: The permeability of the natural geology plays a crucial role in accurately analyzing seepage behavior in the project area. This study presents a novel approach for the inverse analysis of the permeability coefficient. The finite element model (FEM) combined with orthogonal experimental design is used to construct a sample set of permeability coefficient inversion. The established random forest (RF) algorithm surrogate model is applied to determine the optimal values of permeability parameters in the project area using the Harris hawk optimization (HHO) algorithm. This method was used to explore and verify the distribution of natural seepage fields for the P hydropower station. The results showed that the RF model outperformed the classical CART and BP models at each borehole regarding performance evaluation indices. Furthermore, the water head prediction results were more accurate, and the RF model performed admirably in terms of prediction, anti-interference, and generalization. The HHO algorithm effectively searched for the optimal permeability coefficient of the geology. The maximum value of the relative error of the borehole water head inverted was 1.11%, and the accuracy met engineering standards. The initial seepage field distribution pattern calculated followed the basic distribution pattern of the mountain seepage field.

Keywords: permeability coefficient; inversion analysis; orthogonal experimental design; RF; HHO

check for updates

Citation: Zhao, W.; Yin, Q.; Wen, L. Intelligent Inversion Analysis of Hydraulic Engineering Geological Permeability Coefficient Based on an RF–HHO Model. *Water* **2023**, *15*, 1993. https://doi.org/10.3390/w15111993

Academic Editor: Bommanna Krishnappan

Received: 17 April 2023
Revised: 18 May 2023
Accepted: 22 May 2023
Published: 24 May 2023

1. Introduction

In recent years, China has built a considerable number of new high dams and large reservoirs, pumped storage power plants, water transfer projects, and other large-scale water conservancy projects, encouraging the sustainable and healthy growth of the economy and society. To ensure the safe construction and normal operation of the project, it is vitally necessary to master the seepage distribution in the project area. As a result, numerical simulation methods are commonly employed to judge the seepage properties. However, this procedure is based on the determination of the permeability coefficient of the natural stratum [1]. The inversion analysis based on onsite observation data can more accurately obtain permeability parameters than indoor and in situ experiments [2], making it one of the most effective ways to determine the permeability coefficients of materials [3]. Many inversion methods [4–7] have been proposed. However, these classic methods necessitate many calls to the seepage analysis forward model, which is computationally time-consuming and inefficient. Therefore, practical approaches must be adopted to improve the calculation efficiency of inversion methods.

As a method to improve the efficiency of the inversion, surrogate models based on machine learning algorithms have been widely used in recent years. The response surface technique [8], radial basis function (RBF) [9], extreme learning machine (ELM) [10], backpropagation (BP) neural network [11], support vector regression (SVR) [12], multiple adaptive regression splines (MARS) [13], and others are commonly used as surrogate models. These models generate learning samples using FEM. Then, the mathematical model is used to build the relationship between the permeability coefficient (i.e., input variable) and the water head (i.e., output variable). This can then be used to replace the seepage forward model to solve the problem quickly.

Applying the surrogate model can enhance the computational efficiency of the inversion, but its essence remains the standard inversion method of "forward problem inverse calculation". The parameters inverted are not optimal and have low accuracy; hence, further research into novel seepage inversion theories and methods is required. Chi et al. [9] studied the inversion of permeability coefficients of the high-core rockfill dam using the RBF model and the particle swarm optimization (PSO) algorithm. Ni et al. [14] established a surrogate model based on SVR. They searched for the optimal permeability coefficients of the partition inverted for the Nuozhadu high-core rockfill dam using the PSO algorithm. Li et al. [15] established an inversion analysis model based on the relevance vector machine model and cuckoo search algorithm. They determined the permeability coefficient of each stratum in the study area. Shu et al. [16] used ELM and an optimization algorithm to perform a back study on the permeability coefficient of the dam anti-seepage curtain and determined its optimal value. These studies showed that it is feasible to invert the optimal values of seepage parameters using an optimization algorithm combined with the surrogate model, whereby the inversion accuracy is improved.

In the process of nonlinear modeling of the permeability coefficient, the above model exhibits several problems, including low prediction accuracy, poor robustness, slow convergence speed, limited generalization ability, and quickly falling into a local minimum. Additionally, the model assumes the medium to be isotropic during inversion, disregarding the impact of its anisotropy on seepage behavior. Given this, this study introduces the random forest (RF) algorithm, considers the anisotropic characteristics of the medium, and constructs a surrogate model for seepage analysis. Then, Harris hawk optimization (HHO) is introduced to establish a new inversion model that can intelligently optimize the stratum permeability coefficient and lay the foundation for safety analysis of seepage properties in the project area. Lastly, the rationality and validity of the model proposed are verified using an engineering example.

2. Methodology

2.1. Random Forest (RF)

RF [17] is an ensemble learning algorithm widely used for nonlinear regression problems [18,19]. The algorithm's core is to use the ensemble learning method (i.e., bootstrap aggregating) to model the decision tree for each sample set extracted using the bootstrap method. A single decision tree with overfitting and local convergence problems becomes multiple "forests", improving the model's performance. The implementation steps are as follows:

Step 1: T training sets $S_1, S_2, \ldots, S_T$ are randomly generated using the bootstrap method. Let the set of samples be $S = \{x_1, x_2, \cdots, x_n\}$; there is putback from the set of samples S to draw n times, forming a new set of samples S_i (i = 1, 2, ... , T). The total samples of S_i are the same, but it contains only about 62.3% of samples in the original set.

Step 2: According to each training set S_i (i = 1, 2, ... , T), the corresponding decision tree models $C_1, C_2, \ldots, C_T$ are generated.

Step 3: The corresponding result C_1 (X), C_2 (X), ... , C_T (X) is obtained for the test set of samples X using each of the generated decision tree models.

Step 4: The output results of T decision trees are averaged as the result of the X calculation.

The RF model contains three critical parameters: the number of decision trees N_{tree}, the number of random characteristics of node splitting M_{try}, and the minimum number of samples of leaf nodes $Node_{size}$. Generally, compared with N_{tree} and M_{try}, $Node_{size}$ has less impact on the performance and efficiency of the RF model, and the default value is generally chosen.

2.2. Harris Hawk Optimization (HHO)

The HHO algorithm is a novel global search algorithm constructed by Heidari et al. [20] on the basis of the predation patterns of the Harris hawk group and Levy flight characteristics. This algorithm has the characteristics of few adjustment parameters, an optimal random search path, easy implementation, and high calculation accuracy [21]. It has been widely applied to mechanical multi-objective optimization [22], dam deformation prediction [23], environmental drought index prediction [24], landslide prediction [25,26], and other fields. This paper introduces the HHO algorithm to search for the optimal permeability coefficient value. The optimization process consists of the exploration phase, the transition from exploration to exploitation, and the exploitation phase. The specific process is described below.

(1) Exploration phase

In this phase, HHO simulates the Harris hawks' behavior of tracking prey. Their positions are updated according to Equation (1).

$$X(t+1) = \begin{cases} X_{rand}(t) - r_1|X_{rand}(t) - 2r_2X(t)| & q \geq 0.5 \\ (X_{rabbit}(t) - X_m(t)) - r_3(\mathrm{LB} + r_4(\mathrm{UB} - \mathrm{LB})) & q < 0.5 \end{cases}, \tag{1}$$

where $X(t)$ and $X(t+1)$ are the positions of hawks in the t-th iteration and $t+1$-th iteration, $X_{rand}(t)$ is a randomly selected Harris hawk individual, $X_{rabbit}(t)$ is the current location of the prey, and r_1, r_2, r_3, r_4, and q are random numbers in the range [0, 1]. UB and LB are the upper and lower bounds of the search space. $X_m(t)$ is the average position of hawks and is computed as shown in Equation (2).

$$X_m(t) = \frac{1}{N}(\sum_{i=1}^{N} X_i(t)), \tag{2}$$

where N is the number of Harris hawks, and $X_i(t)$ is the location of each hawk in the t-th iteration.

(2) Transition from exploration to exploitation

HHO conducts the transition from exploration to exploitation according to the escape energy of the prey. It is defined as shown in Equation (3).

$$E = 2E_0(1 - \frac{t}{T}), \tag{3}$$

where E is the escape energy of the prey, E_0 is a random initial energy between $(-1, 1)$, t is the number of iterations, and T is the maximum number of iterations.

When $|E| \geq 1$, the hawks track the prey in different regions, and HHO executes the exploration phase; when $|E| < 1$, HHO executes the exploitation phase in the neighborhood of the solution, and hawks encircle, approach, and attack the prey.

(3) Exploitation phase

In this phase, when $|E| \geq 0.5$, the hawks conduct a soft besiege on the prey; when $|E| < 0.5$, the hawks conduct a hard besiege on the prey. In addition, the chance of the prey escaping is measured using r. When $r \geq 0.5$, the prey has enough energy to escape; when $r < 0.5$, the prey does not have enough energy to escape. Therefore, the hawks have the following four strategies to catch the prey.

When $r \geq 0.5$ and $|E| \geq 0.5$, the hawks execute the soft besiege strategy and update their positions according to Equation (4).

$$X(t+1) = X_{rabbit}(t) - X(t) - E|JX_{rabbit}(t) - X(t)|, \tag{4}$$

where J is the random jumping strength of the prey in the process of escape, when $J = 2(1 - r_5)$, and r_5 is a random number in the range (0,1).

When $r \geq 0.5$ and $|E| < 0.5$, the hawks execute the hard besiege strategy, and their positions are updated as represented by Equation (5).

$$X(t+1) = X_{rabbit}(t) - E|X_{rabbit}(t) - X(t)|. \quad (5)$$

When $|E| \geq 0.5$ and $r < 0.5$, the hawks execute the soft besiege strategy of progressive rapid dives and update their positions using Equation (6).

$$X(t+1) = \begin{cases} Y, f(Y) < f(X(t)) \\ Z, f(Z) < f(X(t)) \end{cases}, \quad (6)$$

where f is the fitness function. Y and Z are obtained using Equations (7) and (8).

$$Y(t) = X_{rabbit}(t) - E|J \cdot X_{rabbit}(t) - X(t)|, \quad (7)$$

$$Z = Y + S \times LF(D), \quad (8)$$

where D is the problem dimension, S is a D-dimension random variable, with its elements being random numbers in the range [0, 1], and LF is the Levy flight function [19].

When $|E| < 0.5$ and $r < 0.5$, the hawks execute the hard besiege strategy of progressive rapid dives, and their positions are updated as described in Equations (9) and (10).

$$X(t+1) = \begin{cases} Y, f(Y) < f(X(t)) \\ Z, f(Z) < f(X(t)) \end{cases}, \quad (9)$$

where Y and Z are calculated using Equations (10) and (8).

$$Y(t) = X_{rabbit}(t) - E\left|J \cdot X_{rabbit}(t) - \frac{1}{N}(\sum_{i=1}^{N} X_i(t))\right|. \quad (10)$$

2.3. Calculation Principle of the Three-Dimensional Stable Seepage Field

According to the continuity equation of water flow, the primary differential Equation [3] of steady seepage is expressed as shown in Equation (11).

$$\frac{\partial}{\partial x}\left(k_x \frac{\partial H}{\partial x}\right) + \frac{\partial}{\partial y}\left(k_y \frac{\partial H}{\partial y}\right) + \frac{\partial}{\partial z}\left(k_z \frac{\partial H}{\partial z}\right) = 0, \quad (11)$$

where H is the water head function, and k_x, k_y, and k_z are the permeability coefficients in x-, y-, and z-directions, respectively.

According to the common boundary conditions [3] and the variational principle, Equation (11) can be solved using Equation (12).

$$I(H) = \iiint_{\Omega} \frac{1}{2}\left[k_x\left(\frac{\partial H}{\partial x}\right)^2 + k_y\left(\frac{\partial H}{\partial y}\right)^2 + k_z\left(\frac{\partial H}{\partial z}\right)^2\right] dxdydz - \iint_{S_2} qHds \Rightarrow \min, \quad (12)$$

where Ω is the calculation area.

By discretizing the seepage calculation area, the basic finite element equation for solving a three-dimensional seepage field can be obtained [27] from Equation (12) as shown in Equation (13).

$$[\boldsymbol{K}] \cdot \{\boldsymbol{H}\} = \{\boldsymbol{F}\}, \quad (13)$$

where $[\boldsymbol{K}]$ is the overall penetration matrix, $\{\boldsymbol{H}\}$ is node head array, and $\{\boldsymbol{F}\}$ is the corresponding flow matrix.

2.4. The Fitness Function of the RF–HHO Model

The essential purpose of seepage field inversion is to find the optimal combination of permeability coefficients within a given range. Therefore, the essence of the permeability coefficient inversion is an optimization problem. This study used HHO to find the optimal permeability coefficient for the geology, and an inversion model based on RF–HHO was established. The model's fitness function was the minimum mean square error (MSE) between the measured and calculated water head of boreholes, as shown in Equation (14). The pseudo-code of the RF–HHO model is shown in Algorithm 1.

$$f = \text{minMSE} = \min\left[\frac{1}{q}\sum_{k=1}^{q}\left(H_k - H_k'\right)^2\right], \tag{14}$$

where H_k is the measured water head value of the k-th borehole, H_k' is the calculated water head of the k-th borehole, and q is the number of boreholes.

The constraints of the objective function are shown in Equation (15).

$$\text{s.t. } x_m^d \le x_m \le x_m^u \ (m = 1, 2, \ldots, M), \tag{15}$$

where x_m is the permeability coefficient inverted, x_m^u and x_m^d are its upper and lower limits, respectively, and M is the total number of permeability coefficients inverted.

Algorithm 1: Pseudo-code for RF–HHO implementation.

```
Input: training examples and range of permeability coefficient values
Output: Optimal combination of permeability coefficients for the geology of the project area
Initialize N and T, generate the initial population X_{N×D}, and calculate the fitness value f;
While t < T
   Set X_rabbit as the prey (best location), and update E0, E, and r for each hawk (X_i)
   If |E| ≥ 1
 Use Equation (1) to update the population;
   if |E| < 1
if r ≥ 0.5 and |E| ≥ 0.5
      Use Equation (4) to update the population;
if r ≥ 0.5 and |E| < 0.5
      Use Equation (5) to update the population;
if r < 0.5 and |E| ≥ 0.5
      Use Equation (6) to update the population;
if r < 0.5 and |E| < 0.5
      Use Equation (9) to update the population;
end
   end
  Calculate the fitness value of the new individual, and update their positions and optimal
fitness value;
  t = t + 1;
end
```

2.5. Establishment of a Permeability Coefficient Inversion Model for the Dam Site Area

Using the RF and HHO algorithms, this paper established an inverse model of geology permeability coefficients. It can intelligently search for the optimal permeability coefficients of the dam foundation. Figure 1 shows the modeling procedure. The main steps are as follows:

Step 1: The orthogonal test design method is used to construct the combination of permeability coefficients. Then, the borehole water head under the corresponding combination is obtained by FEM. Finally, the inversion sample set is established.

Step 2: The permeability coefficient is selected as the input variable of the RF model, and the corresponding borehole head calculation value is selected as the output variable of the model.

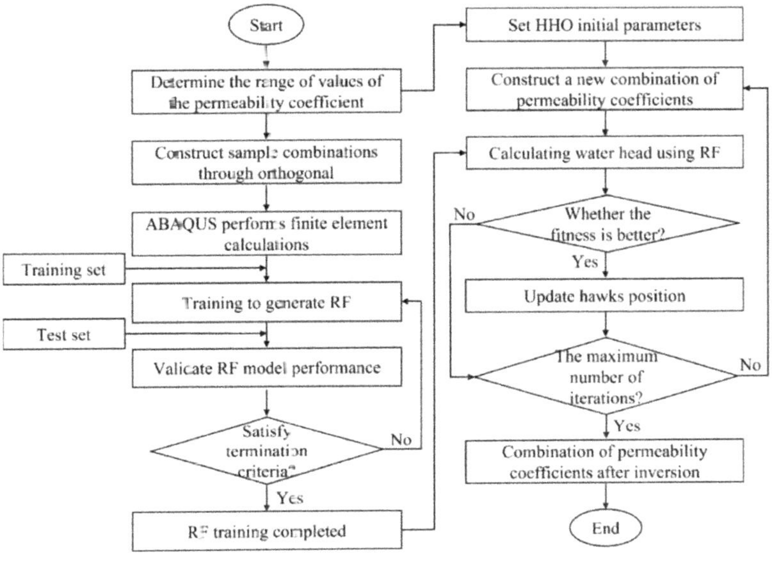

Figure 1. Execution process of the inversion model of the geological permeability coefficient.

Step 3: The sample set is split into 80% for training and 20% for testing. Simultaneously, to eliminate the impact of variable magnitude on the RF model, the sample data are preprocessed using the normalization Equation (16) to limit the data to [0, 1].

$$x'_m = \frac{x_m - x_m^d}{x_m^u - x_m^d}, \quad (16)$$

where x'_m is the normalized permeability coefficient value.

Step 4: Tenfold cross-validation is used to determine the optimal parameters M_{try} and N_{tree} of the RF model. Then, the inversion surrogate model is trained.

Step 5: It is verified whether the accuracy of the surrogate model obtained in Step 4 meets the error threshold. If it meets the standards, the procedure proceeds to the next step. Otherwise, steps 4 and 5 are repeated.

Step 6: After denormalizing the output results, the surrogate model established is evaluated using the following statistical indices [18,19,28–30]: mean absolute error (MAE), mean absolute percentage error (MAPE), root-mean-square error (RMSE), and goodness of fit (R^2). The calculated expressions are shown in Equations (17)–(20), respectively.

$$\text{MAE} = \frac{1}{n}\sum_{i=1}^{n}|y_i - \hat{y}_i|, \quad (17)$$

$$\text{MAPE} = \frac{100}{n}\sum_{i=1}^{n}\left|\frac{y_i - \hat{y}_i}{y_i}\right|, \quad (18)$$

$$\text{RMSE} = \sqrt{\frac{1}{n}\sum_{i=1}^{n}(y_i - \hat{y}_i)^2}, \quad (19)$$

$$R^2 = 1 - \frac{\sum_{i=1}^{n} (y_i - \hat{y}_i)^2}{\sum_{i=1}^{n} (y_i - \overline{y})^2}, \tag{20}$$

where y_i and $\hat{y}_i$ are the measured and predicted values, respectively, $\overline{y}$ and $\overline{\hat{y}}_i$ are the averages of the measured and predicted values, respectively, and n is the number of measured values.

Step 7: When employing HHO to find the optimal permeability coefficients, a new combination of permeability coefficients is constructed. Their corresponding borehole head is calculated using the established RF model. If the fitness under this combination is less than the existing optimal fitness, the existing combination and fitness are update. Otherwise, the procedure proceeds to step 8.

Step 8: If HHO does not reach the maximum number of iterations, step 7 is repeated. Otherwise, the current combination and its fitness are updated, along with the final inversion result.

3. Case Study

3.1. Basic Information on the Project Area

The reservoir of P hydropower station is a class III medium-sized reservoir. The pivotal project is composed of left and right bank water-retaining dam sections, riverbed overflow dam sections, bottom outlets, and water-intake dam sections. The dam is a roller-compacted concrete gravity dam with a maximum height of 59.50 m, a crest length of 217.00 m, and a crest width of 8 m. The normal storage and the design flood level of the reservoir are 946.5 m, the check flood level is 946.95 m, the total storage capacity is 15.81×10^6 m^3, and the backwater length is 6.082 km. The dam is a grade III structure, the design flood standard is a 50 year return period, and the corresponding peak flow is 808 m^3/s; the check flood standard is a 500 year return period, and the corresponding peak discharge is 2110 m^3/s. The seismic design intensity is VII.

The project area is located in a low hill and gully landscape with a high north and low south terrain; the river valley gully is relatively developed, and its cross-section is "U"-shaped. The main stratum of the project area is the Triassic Ermaying Formation. The lithology is mainly gray-green coarse-grained feldspathic sandstone, sandstone mudstone interbedding, siltstone, and silty mudstone. The rock stratum is stable and about 200 m thick. The stratum in the project area is divided into a strongly weathered, weakly weathered, and slightly new strata. The stratum is weakly permeable with good bottom sealing. The groundwater is mainly bedrock fissure phreatic water and quaternary pore phreatic water. Both reservoir banks are relatively wide and thick, without a thin watershed distribution, large structure, and fault distribution. Two groups of conjugate high-angle shear joints, NWW and NNE, are common, with a fracture spacing of 0.5~2.0 m.

3.2. Establishment of the Finite Element Model

According to the geological survey report and relevant hydrogeological data, a three-dimensional finite element seepage model in the engineering area was established. The model's boundary was determined by the seepage field and the site's terrain. The surrounding was taken from the mountain watershed, and the thickness of the rock around the reservoir was not less than 3–5 times the dam height or the excavation span [31,32]. When the distance from the surface was less than five times the excavation span, it was calculated to the surface to ensure the rationality of the model calculation boundary. The upstream and downstream boundaries of the established model were 7.90 and 10.01 times the dam height from the dam site, respectively; the left and right boundaries were 3.79 and 9.29 times the dam height from the midpoint of the dam axis, respectively. Moreover, the left boundary was along the channel on the left bank of the downstream. Figure 2 shows the model's scope, measuring 1284.67 m in length and 1183.87 m in width.

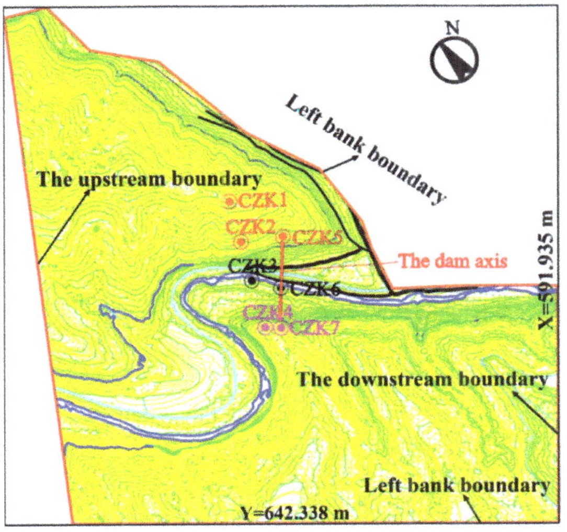

Figure 2. Calculation scope of seepage inversion analysis model in the project area.

In order to make the calculation model consistent with the actual hydrogeological conditions in the project area, a detailed simulation of the strata located on both banks and the river valley was conducted on the basis of the geological profile of the dam axis, as illustrated in Figure 3. The thickness of the strongly and weakly weathered strata was 13.5 m and 24.2 m on the left bank, 3.9 m and 26.1 m on the river valley, and 3.8 m and 20.6 m on the right bank, respectively. Furthermore, the survey of adits on both banks revealed the presence of a wide fissure, measuring approximately 8 cm in width, and a long fissure, roughly 1.0 cm in width, on the right bank. On the left bank, a weak rock belt composed of sandy mudstone was observed, with a maximum thickness of 5 cm. These findings were simulated in the study.

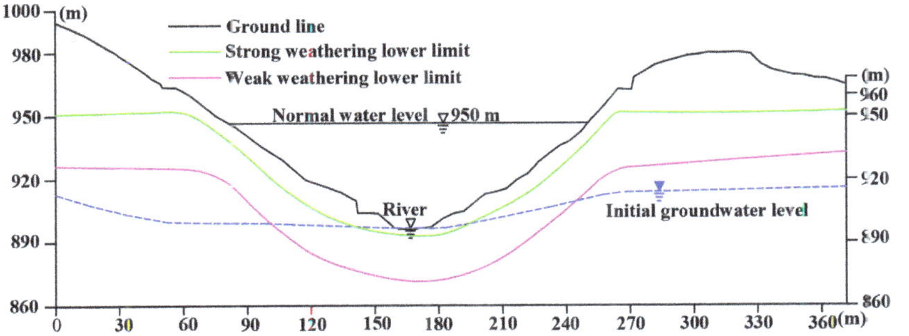

Figure 3. Engineering geological profile of dam axis at dam site.

In conclusion, the calculation model built using ABAQUS is shown in Figure 4. The bottom elevation of the model was 507.1 m, and the top elevation was the corresponding surface elevation. The four-node isoparametric element (C3D4P) was used to mesh the model. The numbers of nodes and units were 54,552 and 300,618, respectively. In the calculations, the right boundary, the left boundary, and the middle channel of the model used the measured groundwater level, the corresponding ditch bottom water level, and the natural river valley level as their constant water head boundaries, respectively; the upstream, downstream, and bottom were impermeable boundaries.

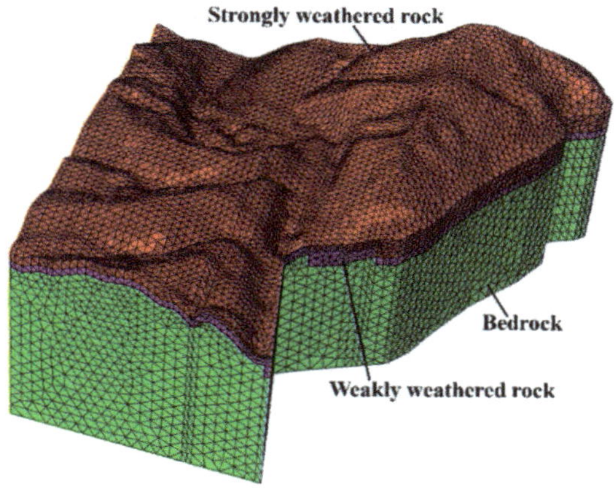

Figure 4. The inversion calculation model of the seepage field.

The study used the proposed inversion model to invert the dam foundation permeability coefficients of the P hydropower station using the measured water levels from seven boreholes (i.e., CZK1, CZK2, CZK3, CZK4, CZK5, CZK6, and CZK7) in the natural seepage field.

3.3. Sample Construction Based on Orthogonal Design

The permeability coefficient of various strata and fractures was an independent variable in the inversion. Since each stratum was mostly sandstone and mudstone interbedded with uneven lithology, their permeability coefficients were considered anisotropic. According to the geological survey data and combined with general engineering experience, the range of permeability coefficients was determined, as shown in Table 1, where k_1 and k_2 are the tangential permeability coefficient values, and k_3 is the normal permeability coefficient value.

Table 1. Value range of permeability coefficients for each stratum and fracture.

Rock Stratum	Measured Range of k_1 and k_2 (m/s)	Measured Range of k_3 (m/s)
Strongly weathered stratum	$[4.32, 9.86] \times 10^{-5}$	$[1.12, 6.57] \times 10^{-5}$
Weakly weathered stratum	$[3.57, 9.25] \times 10^{-6}$	$[1.63, 7.49] \times 10^{-6}$
Bedrock	$[5.61, 8.94] \times 10^{-7}$	$[1.38, 4.67] \times 10^{-7}$
Fracture	$[5.35, 8.58] \times 10^{-6}$	$[1.24, 4.27] \times 10^{-6}$

Constructing a reasonable permeability coefficient sample is crucial for developing a surrogate model. To ensure the RF model precision without increasing the workload of finite element forward analysis, this study adopted the orthogonal design method [33] to arrange a representative combination scheme of permeability coefficients. Orthogonal tables are essential to orthogonal test sample preparation. The scheme is generally expressed as L_n (t^c), where L is the orthogonal table, n is the total number of tests, t is the level number of factors, and c is the maximum number of factors arranged. The four factors of the test were the permeability coefficients of strongly weathered stratum, weakly weathered stratum, bedrock, and fracture. Table 2 shows each factor's nine level numbers. The 81 permeability coefficient combination schemes were generated by SPSS using the orthogonal table L_{81} (9^4). Subsequently, the corresponding water head was determined through finite element forward analysis, and 81 sets of inversion samples were constructed. A total of 65 groups of samples (the first 80% of the data) were chosen for training the RF model, while 16 groups of samples (the remaining 20%) were used for verifying the model.

Table 2. Values of parameters at different levels based on orthogonal design.

Level	Strongly Weathered Layer ($\times 10^{-5}$ m/s)		Weakly Weathered Layer ($\times 10^{-6}$ m/s)		Bedrock ($\times 10^{-7}$ m/s)		Fissure ($\times 10^{-5}$ m/s)	
	$k_1 = k_2$	k_3	$k_1 = k_2$	k_3	$k_1 = k_2$	k_3	$k_1 = k_2$	k_3
1	4.32	1.12	3.57	1.63	5.61	1.38	5.35	1.24
2	5.01	1.80	4.28	2.36	6.03	1.79	5.75	1.62
3	5.71	2.48	4.99	3.10	6.44	2.20	6.16	2.00
4	6.40	3.16	5.70	3.83	6.86	2.61	6.56	2.38
5	7.09	3.85	6.41	4.56	7.28	3.03	6.97	2.76
6	7.78	4.53	7.12	5.29	7.69	3.44	7.37	3.13
7	8.48	5.21	7.83	6.03	8.11	3.85	7.77	3.51
8	9.17	5.89	8.54	6.76	8.52	4.26	8.18	3.89
9	9.86	6.57	9.25	7.49	8.94	4.67	8.58	4.27

3.4. Determination of the RF Model Parameters

The establishment of the RF model relies heavily on setting the optimal Mtry and Ntree values, as demonstrated in Section 2.1. The variable control method was applied to analyze and determine RF parameter settings using the sample data from Section 3.3. The ideal RF parameters were selected using a tenfold cross-validation based on the model's least MSE, effectively reducing model unpredictability.

To examine the model's generalization ability under different N_{tree}, M_{try} was set to the default value (i.e., one-third of the number of variables), and the N_{tree} range was set to [50, 1000] with a step size of 50. Figure 5a depicts the model's MSE change curve for different N_{tree}. It is evident that the model's MSE stabilized when N_{tree} reached 400, indicating that increasing the number of model trees did not significantly affect the model error when M_{try} was fixed. Therefore, to enhance the computational efficiency and ensure the accuracy of the RF model, this study selected N_{tree} = 400. Subsequently, the model's generalization ability was evaluated under different M_{try} levels, with N_{tree} set to 400 and M_{try} set to [1, 20] with a step size of 1. Figure 5a depicts the model's MSE change curve for different M_{try}. It shows that the model's MSE increased gradually with an increase in M_{try}, with the smallest MSE observed at M_{try} = 4. Accordingly, N_{tree} = 150 and M_{try} = 4 were identified as the optimal parameters for the RF model, leading to improved performance and computational efficiency.

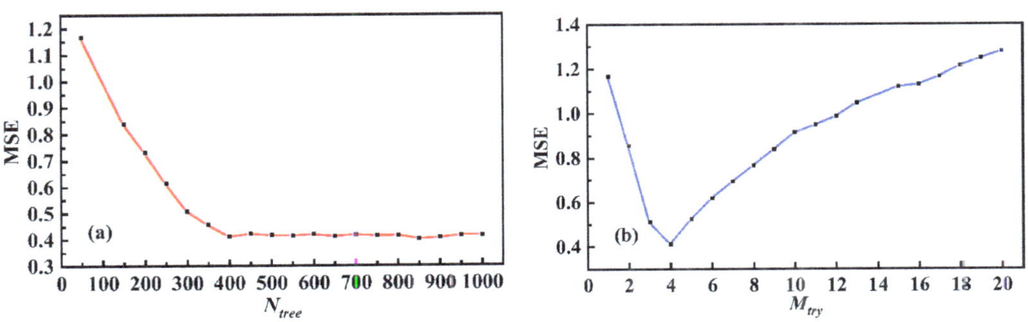

Figure 5. The effect of the parameters on RF model performance: (**a**) N_{tree}, and (**b**) M_{try}.

4. Results and Analysis

4.1. Performance Validation of the RF Model

The study evaluated the ability of the RF model to invert permeability parameters at borehole CZK1, comparing it to the CART model [18] and BP neural network model [19]. The RF model was configured with parameters N_{tree} and M_{try} set to 400 and 4, respectively; the CART model was configured with parameters minLeafSize and smoothingK set to

6 and 20, respectively; the BP model's hidden layer consisted of seven neurons, and the maximum number of iterations was set to 200. Figure 6 displays the prediction results of the RF, CART, and BP models for the borehole water level. Table 3 presents the generalization ability and forecast accuracy of the three models according to different evaluation indices.

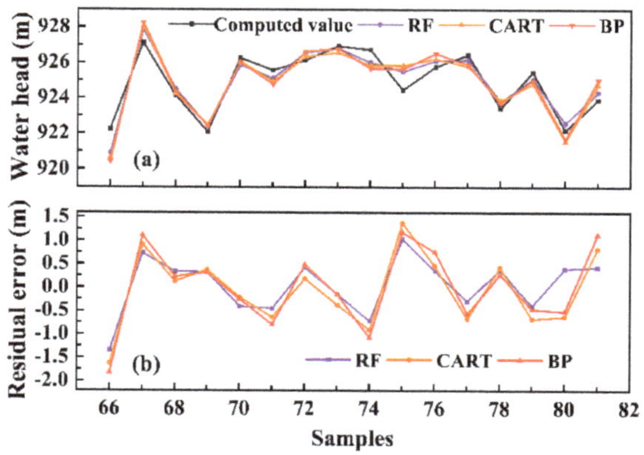

Figure 6. Model accuracy verification results at borehole CZK1: (**a**) water head, and (**b**) residuals.

Table 3. Evaluation indices of prediction results of RF, CART, and BP models at borehole CZK1.

Models	Model Training				Model Verification			
	MAE (m)	MAPE (%)	RMSE (m)	R^2	MAE (m)	MAPE (%)	RMSE (m)	R^2
RF	0.290	0.031	0.353	0.962	0.255	0.055	0.589	0.879
CART	0.357	0.039	0.416	0.948	0.325	0.070	0.764	0.797
BP	0.385	0.042	0.457	0.937	0.695	0.075	0.827	0.762

Figure 6a shows that the prediction curves of the three models were close to the finite element calculation curves, and the prediction errors were negligible. However, Figure 6b illustrates that the overall prediction residuals of the RF model were smaller than those of the CART and BP models. Table 3 proves that the RF model outperformed the CART and BP models in all evaluation indices. Notably, the R^2 values for the training and test sets were 0.962 and 0.879, respectively, and other indices also achieved low values, indicating that the RF model fit was highly significant. Consequently, the prediction accuracy of the RF model was higher, and the proxy model based on the RF model accurately predicted the water head of borehole CZK1.

The water heads of boreholes CZK2, CZK3, CZK4, CZK5, CZK6, and CZK7 were utilized to assess the generalization ability of the RF model. Figure 7 illustrates the water head prediction results of the RF, CART, and BP models at these boreholes. Table 4 presents a quantitative evaluation of the prediction results.

Figure 7 indicates that the RF model outperformed the other two models regarding anti-interference and generalization. Specifically, the RF model's R^2 values at boreholes CZK2, CZK3, CZK4, CZK5, CZK6, and CZK7 were 0.864, 0.898, 0.896, 0.922, 0.909, and 0.922, respectively, which were superior to those of the CART (R^2 = 0.797, 0.818, 0.860, 0.871, 0.869, and 0.903) and BP (R^2 = 0.675, 0.802, 0.807, 0.836, 0.844, and 0.868) models, as reported in Table 4. Moreover, the other statistical indices of the RF model outperformed those of the CART and BP models, indicating its excellent generalization ability. Overall, the RF model accurately predicted the water head at all seven boreholes, making it a suitable surrogate model for inverting engineering seepage parameters with high accuracy and robustness.

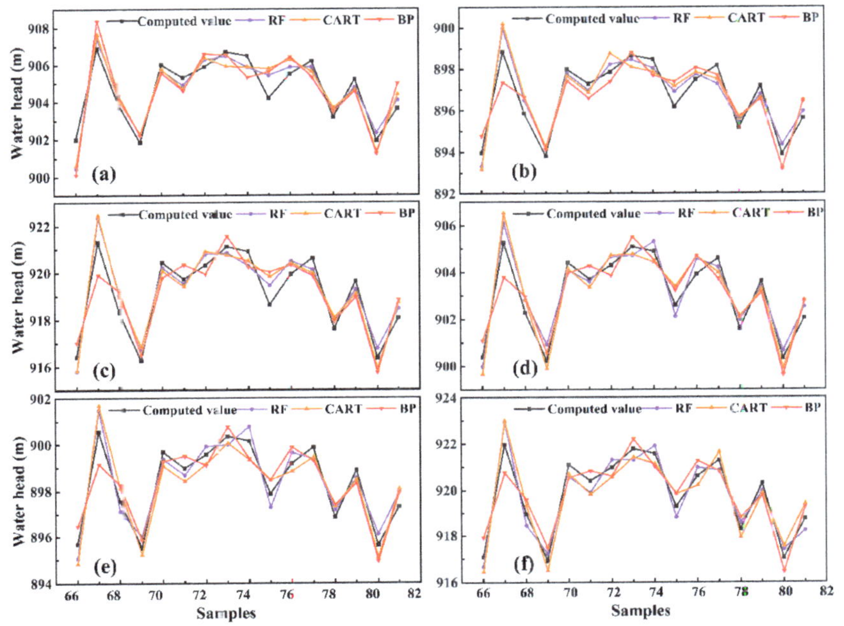

Figure 7. Model accuracy verification results at different boreholes: (**a**) CZK2, (**b**) CZK3, (**c**) CZK4, (**d**) CZK5, (**e**) CZK6, and (**f**) CZK7.

Table 4. Evaluation indices of prediction results of RF, CART, and BP models at different boreholes.

Measuring Points	Models	Model Training				Model Verification			
		MAE (m)	MAPE (%)	RMSE (m)	R^2	MAE (m)	MAPE (%)	RMSE (m)	R^2
CZK2	RF	0.304	0.034	0.366	0.960	0.262	0.058	0.626	0.864
	CART	0.366	0.041	0.430	0.944	0.341	0.075	0.764	0.797
	BP	0.385	0.043	0.457	0.937	0.844	0.093	0.967	0.675
CZK3	RF	0.339	0.038	0.413	0.948	0.241	0.054	0.543	0.898
	CART	0.373	0.042	0.437	0.942	0.338	0.075	0.724	0.818
	BP	0.388	0.043	0.449	0.939	0.687	0.077	0.755	0.802
CZK4	RF	0.326	0.035	0.377	0.957	0.252	0.055	0.548	0.896
	CART	0.362	0.039	0.405	0.951	0.294	0.064	0.635	0.860
	BP	0.419	0.046	0.477	0.931	0.672	0.073	0.745	0.807
CZK5	RF	0.339	0.038	0.379	0.957	0.221	0.049	0.473	0.922
	CART	0.362	0.040	0.407	0.950	0.280	0.062	0.609	0.871
	BP	0.422	0.047	0.453	0.938	0.633	0.070	0.688	0.836
CZK6	RF	0.350	0.039	0.384	0.955	0.242	0.054	0.510	0.909
	CART	0.426	0.047	0.449	0.939	0.285	0.063	0.610	0.869
	BP	0.482	0.054	0.516	0.920	0.626	0.070	0.669	0.844
CZK7	RF	0.403	0.044	0.452	0.938	0.224	0.049	0.475	0.922
	CART	0.406	0.044	0.424	0.946	0.250	0.054	0.527	0.903
	BP	0.471	0.051	0.490	0.927	0.587	0.064	0.617	0.868

4.2. Inversion of Permeability Coefficients Based on the HHO Algorithm

Section 4.1 verified that the constructed surrogate model predicted borehole water heads well. Utilizing this model, the HHO algorithm was employed in global optimization within the range of permeability coefficients (i.e., Table 1). The initial number of hawks was set to N = 50, the maximum number of iterations was T = 200, and the fitness value was calculated using the trained RF in Section 4.1. To validate the HHO algorithm's optimization capacity, classic optimization algorithms such as the genetic algorithm (GA) [34] and PSO [35] were introduced for comparison. During calculation, the PSO algorithm's parameters were set to 100 for the number of particle swarms, 1.5 for the individual learning factor, 2 for the social learning factor, 1 for the maximum inertia factor, 0.8 for the minimum inertia factor, and 200 for the maximum number of iterations. The GA algorithm's parameters were set to 100 for the population size, 0.85 for the generation gap, 0.6 for the crossover probability, 0.01 for the mutation probability, and 200 for the maximum genetic algebra. Figure 8 illustrates the convergence process of the HHO, PSO, and GA algorithms in searching for the optimal solution.

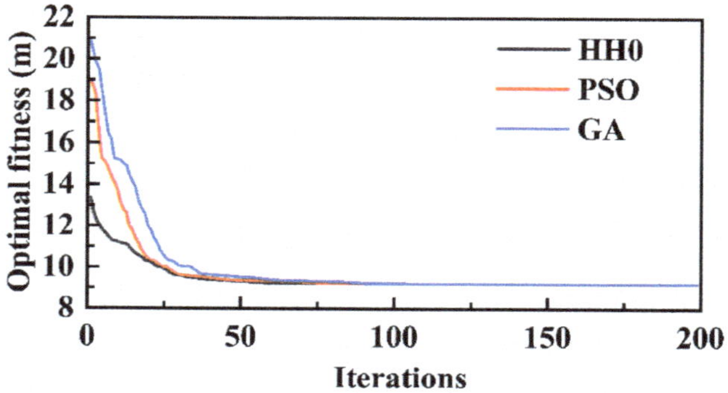

Figure 8. Convergence process of HHO, PSO, and GA algorithms.

Figure 8 illustrates that the HHO algorithm exhibited a smoother convergence curve, a superior initial optimization result, and a faster convergence rate than the PSO and GA algorithms. The HHO, PSO, and GA algorithms achieved optimal results at the 66th, 96th, and 105th generations, respectively. The convergence rates of all three algorithms were relatively flat, indicating their strong global search capabilities. However, the HHO algorithm required fewer initial parameters, with only N and T needing to be determined, thus simplifying programming. Table 5 presents the optimal permeability coefficient for each stratum and fracture.

Table 5. Inversion results of permeability coefficient of each stratum and fracture.

Rock Stratum	Calculated Value of k_1 and k_2 (m/s)	Calculated Value of k_3 (m/s)
Strongly weathered stratum	8.43×10^{-5}	5.47×10^{-5}
Weakly weathered stratum	7.81×10^{-6}	6.14×10^{-6}
Bedrock	7.46×10^{-7}	3.69×10^{-7}
Fracture	7.29×10^{-6}	2.87×10^{-6}

To confirm the validity of the permeability coefficient obtained, we applied the inverted permeability coefficients of each stratum and fracture to the FEM. This allowed us to calculate the water head of each borehole and compare it with the measured water level. The results are presented in Table 6. The calculation formulas for absolute error e and relative error e_r are presented in Equations (21) and (22), respectively:

$$e = H' - H, \tag{21}$$

$$e_r = \left[(H' - H)/H\right] \times 100\%, \tag{22}$$

where H' and H are the calculated and measured water levels of the borehole, respectively.

Table 6. Comparison of measured and inverted water levels at each borehole.

Borehole	Calculated Water Level (m)	Measured Water Level (m)	Absolute Error (m)	Relative Error (%)
CZK1	920.83	928.74	−7.91	0.85
CZK2	905.67	902.24	3.43	0.38
CZK3	899.83	895.97	3.86	0.43
CZK4	922.75	912.62	10.13	1.11
CZK5	904.56	899.64	4.92	0.55
CZK6	898.32	895.84	2.48	0.28
CZK7	921.41	913.51	7.90	0.87

Several studies [3,31,32] suggested that the inverted permeability coefficient is suitable when $e < 10$ m and $e_r < 5\%$ at each borehole. Table 6 shows that CZK4 had the highest absolute error of 10.13 m and the highest relative error of 1.11%, while the remaining boreholes had minor absolute and relative errors that met the accuracy control requirements. The disparity between the calculated and measured water level at CZK4 was due to a decline in the groundwater level caused by the construction of the exploration adit near the borehole, which resulted in a significant difference between the measured and true water level for the borehole. Therefore, in future research, the borehole water level must be corrected near the adit, or one should avoid using similar borehole information before inversion. However, the disparity between the calculated and measured water level in the remaining boreholes was slight, indicating that the geological permeability coefficient inverted was close to its true value, and the seepage properties of the project area calculated could be used as its natural seepage field distribution. In conclusion, the permeability coefficients obtained using the RF–HHO model met the project's requirements for accuracy. They can be applied to analyze and calculate seepage properties under different situations during the construction and operation of the project area.

For seepage calculation analysis, X = 591.935 m and Y = 642.335 m in the calculation model were used as typical profiles, as shown in Figure 2. Figures 9 and 10 show the total water head and pressure water head distribution calculated for the two profiles. The left and right banks of the X profile are the upstream and downstream, respectively; the left and right of the Y profile are the right and left banks, respectively.

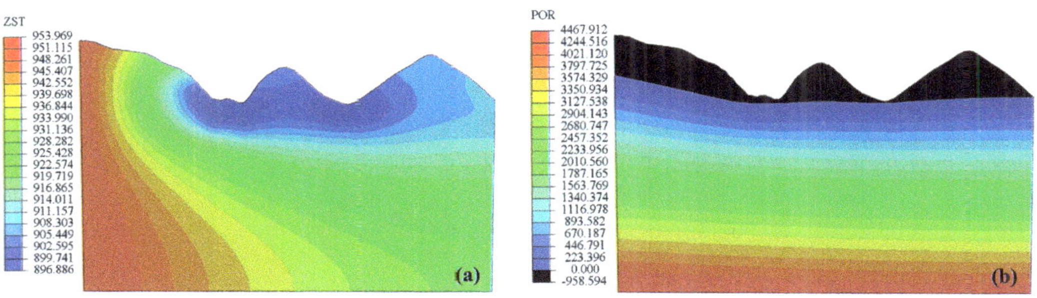

Figure 9. Calculation results of Y = 642.335 m profile (unit/m): (a) total water head, and (b) pore water pressure.

The results indicated that the upstream reservoir area of the initial seepage field exhibited a higher water head than the downstream area, validating the principle of seepage supplement from upstream to downstream. The upstream water head line was distributed more densely, and the hydraulic gradient was relatively large. Similarly, the water head on the right side of the initial seepage field model was higher than the left side, indicating the law of seepage supplement from the mountain to the river valley. The water head line was distributed more densely on the right side, and the hydraulic gradient was relatively large. The distribution pattern of the initial seepage field computed aligned with the pattern revealed by the borehole, which was consistent with the distribution law of the general mountain seepage field. Moreover, the distribution of the total head and pore water pressure was essentially reasonable.

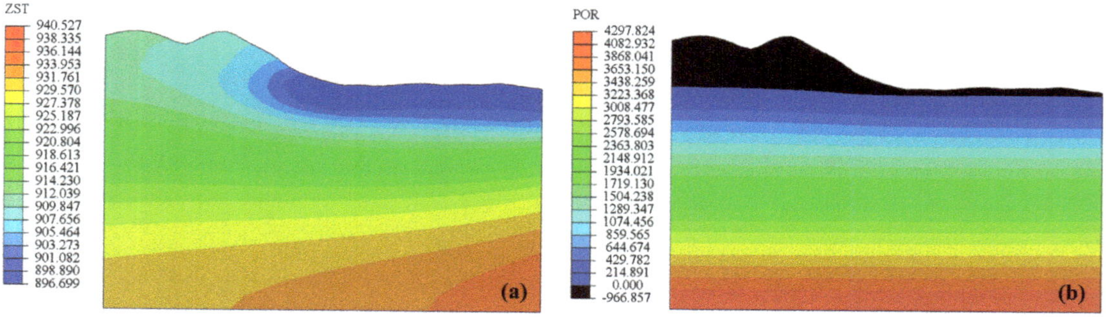

Figure 10. Calculation results of X = 591.935 m profile (unit/m): (**a**) total water head, and (**b**) pore water pressure.

5. Conclusions

The geological permeability coefficient is a critical parameter in hydraulic engineering for analyzing the safety of three-dimensional seepage fields. However, estimating its actual value is a challenge due to the interaction of various elements. Therefore, it is crucial to establish a high-precision model to quickly obtain the permeability coefficient's "true value". In this study, we applied an RF–HHO-based model for permeability coefficient inversion to the dam foundation of the P hydropower station, which revealed the overall distribution characteristics of its natural seepage field. On the basis of our research, we can draw the following conclusions:

(1) The RF model showed promising potential for engineering seepage parameter inversion. Compared to other models, the RF model's water level prediction at all boreholes was closer to the calculated value of the FEM, with its evaluation index being the smallest, indicating its greater prediction accuracy and generalization ability. The RF-based surrogate model can replace the FEM for seepage calculation, avoiding the time-consuming process of FEM seepage calculation and improving the efficiency of the inversion process.

(2) The HHO algorithm demonstrated remarkable proficiency in conducting global searches. As evidenced by the convergence curve of parameter optimization, the HHO method surpassed the PSO and GA algorithms regarding optimization efficiency and initial setting parameters. It could rapidly identify the optimal solution by determining the population and maximum iterations.

(3) The inversion model constructed provided a solid foundation for the numerical study of the natural seepage field in the project area. The RF–HHO model successfully determined the optimal permeability coefficient of the geology for the P hydropower station, and then calculated the water head for each borehole using FEM. The absolute and relative errors between the calculated and measured water levels in the borehole were small. Additionally, the calculated distribution pattern of the initial seepage field was consistent with the distribution law of the mountain seepage field. These results indicate that the inversion model was reasonable and met the engineering requirements for accuracy.

However, it is a deficiency that the study only used single-borehole data modeling to invert the optimum permeability coefficient for the geology without considering the linkage between different boreholes. Boreholes are frequently situated near one another and possess interrelated characteristics that can be utilized for parametric inversion modeling.

As such, our forthcoming research endeavors will maximize the linkages between the different boreholes to enhance the current model.

Author Contributions: Conceptualization, W.Z. and L.W.; methodology, Q.Y.; software, Q.Y.; validation, Q.Y.; formal analysis, Q.Y.; investigation, Q.Y.; resources, W.Z.; data curation, L.W.; writing—original draft preparation, W.Z.; writing—review and editing, L.W.; visualization, Q.Y.; supervision, L.W. All authors have read and agreed to the published version of the manuscript.

Funding: This research received no external funding.

Data Availability Statement: The data presented in this study are available on request from the corresponding author. The data are not publicly available due to confidentiality of data.

Conflicts of Interest: The authors declare no conflict of interest.

References

1. Mao, X.; Pan, S.; Bai, Z. Back analysis of initial seepage field of complex dam foundation at Shuangjiangkou Hydropower Station. *Rock Soil Mech.* **2008**, *29* (Suppl. S1), 135–139.
2. Dong, W.; Yuan, H.; Xu, W.; Zhang, B.; Yu, Y. Dynamic back-analysis of material parameters of Nuozhadu high earth-rock-fill dam. *J. Hydroelectr. Eng.* **2012**, *31*, 203–208.
3. Xu, Z.; Cao, C.; Li, K.; Junrui, C. Analysis of part seepage control scheme of an upper reservoir of a pumped-storage power station. *Chin. J. Appl. Mech.* **2018**, *35*, 417–422+459.
4. Mao, C. *Seepage Calculation, Analysis and Control*; Water Resources and Electric Power Press: Beijing, China, 1990.
5. Sheng, J.; Su, B.; Zhan, M. Back analysis of 3d seepage problem and its engineering application. *Chin. J. Rock Mech. Eng.* **2003**, *22*, 203–207.
6. Zhu, B. A new method for the back analysis of seepage problem. *J. Hydraul. Eng.* **1994**, *39*, 42–46.
7. Duan, B.; Zhang, L.; He, J.; Fu, W.; Chen, G. Back analysis of natural seepage field in complicated fractured rock mass. *J. Hydroelectr. Eng.* **2012**, *31*, 188–193.
8. Lu, P.; Wang, X.; Wu, B.; Cheng, Z. Seepage parameter inversion based on Bayesian theory and entropy-blind numbers. *J. Hydroelectr. Eng.* **2019**, *38*, 108–118.
9. Chi, S.; Ni, S.; Liu, Z. Back analysis of the permeability coefficient of a high core rockfill dam based on a RBF neural network optimized using the PSO algorithm. *Math. Probl. Eng.* **2015**, *2015*, 124042. [CrossRef]
10. Xu, L.; Shen, Z. Inversion model of permeability coefficient for complex earth rock dam based on ELM-GA. *Water Resour. Power* **2021**, *39*, 86–90.
11. Tang, S.; Xiong, W.; Wan, X.; Luo, Z.; Wan, S.; Wang, Q. Multi-objective inversion analysis method for dam permeability coefficient based on GA-BP. *China Rural. Water Hydropower* **2020**, *62*, 213–216.
12. Chen, H.; Teng, Y.; Wang, J. Methods of estimation of hydraulic conductivity with genetic algorithm-support vector regression machine. *Hydrogeol. Eng. Geol.* **2011**, *38*, 14–18.
13. Yu, H.; Wang, X.; Ren, B.; Zeng, T.; Lv, M.; Wang, C. An efficient Bayesian inversion method for seepage parameters using a data-driven error model and an ensemble of surrogates considering the interactions between prediction performance indicators. *J. Hydrol.* **2022**, *604*, 127235. [CrossRef]
14. Ni, S.; Chi, S. Back analysis of permeability coefficient of high core rockfill dam based on particle swarm optimization and support vector machine. *Chin. J. Geotech. Eng.* **2017**, *39*, 727–734.
15. Li, Y.; Yang, J.; Cheng, L.; Ma, C. Inversion Analysis on Permeability Coefficient of Stratum in Engineering Area Based on RVM-CS. *J. Yangtze River Sci. Res. Inst.* **2020**, *37*, 121–127.
16. Shu, Y.; Shen, Z.; Xu, L.; Zhang, K.; Yang, C. Inversion analysis of impervious curtain permeability coefficient using calcium leaching model, extreme learning machine, and optimization algorithms. *Appl. Sci.* **2022**, *12*, 3272. [CrossRef]
17. Breiman, L. Random forests. *Mach. Learn.* **2001**, *45*, 5–32. [CrossRef]
18. Chen, X.; Zhao, X.; Tahmasebi, P.; Luo, C.; Cai, J. NMR-data-driven prediction of matrix permeability in sandstone aquifers. *J. Hydrol.* **2023**, *618*, 129147. [CrossRef]
19. Zhao, X.; Chen, X.; Huang, Q.; Lan, Z.; Wang, X.; Yao, G. Logging-data-driven permeability prediction in low-permeable sandstones based on machine learning with pattern visualization: A case study in Wenchang A Sag, Pearl River Mouth Basin. *J. Pet. Sci. Eng.* **2022**, *214*, 110517. [CrossRef]
20. Heidari, A.; Mirjalili, S.; Faris, H.; Aljarah, I.; Mafarja, M.; Chen, H. Harris hawks optimization: Algorithm and applications. *Future Gener. Comput. Syst.* **2019**, *97*, 849–872. [CrossRef]
21. Yu, J.; Kim, C.H.; Rhee, S.B. The comparison of lately proposed Harris hawks optimization and jaya optimization in solving directional overcurrent relays coordination problem. *Complexity* **2020**, *2020*, 3807653. [CrossRef]
22. Abbasi, A.; Firouzi, B.; Sendur, P. On the application of Harris hawks optimization (HHO) algorithm to the design of microchannel heat sinks. *Eng. Comput.* **2021**, *37*, 1409–1428. [CrossRef]

23. Li, Y.; Yin, Q.; Zhang, Y.; Qiu, W. Prediction of long-term maximum settlement deformation of concrete face rockfill dams using hybrid support vector regression optimized with HHO algorithm. *J. Civ. Struct. Health Monit.* **2023**, *13*, 371–386. [CrossRef]
24. Malik, A.; Tikhamarine, Y.; Sammen, S.S.; Abba, S.I.; Shahid, S. Prediction of meteorological drought by using hybrid support vector regression optimized with HHO versus PSO algorithms. *Environ. Sci. Pollut. Res.* **2021**, *28*, 39139–39158. [CrossRef] [PubMed]
25. Moayedi, H.; Osouli, A.; Nguyen, H.; Rashid, A.S.A. A novel Harris hawks' optimization and k-fold cross-validation predicting slope stability. *Eng. Comput.* **2021**, *37*, 369–379. [CrossRef]
26. Bui, D.T.; Moayedi, H.; Kalantar, B.; Osouli, A.; Pradhan, B.; Nguyen, H.; Rashid, A.S.A. Harris hawks optimization: A novel swarm intelligence technique for spatial assessment of landslide susceptibility. *Sensors* **2019**, *19*, 3590. [CrossRef] [PubMed]
27. Xu, K.; Lei, X.; Meng, Q. Application of seepage back analysis to a hydropower engineering. *Eng. J. Wuhan Univ.* **2011**, *44*, 37–39.
28. Barati, R. Application of excel solver for parameter estimation of the nonlinear Muskingum models. *KSCE J. Civ. Eng.* **2013**, *17*, 1139–1148. [CrossRef]
29. Hosseini, K.; Nodoushan, E.J.; Barati, R.; Shahheydari, H. Optimal design of labyrinth spillways using meta-heuristic algorithms. *KSCE J. Civ. Eng.* **2016**, *20*, 468–477. [CrossRef]
30. Alizadeh, M.J.; Shahheydari, H.; Kavianpour, M.R.; Shamloo, H.; Barati, R. Prediction of longitudinal dispersion coefficient in natural rivers using a cluster-based Bayesian network. *Environ. Earth Sci.* **2017**, *76*, 86. [CrossRef]
31. Sun, C.; Chai, J.; Xu, Z.; Qin, Y. Numerical simulation and assessment of seepage control effects on surrounding fractured rocks of underground powerhouse in Jinchuan Hydropower Station. *Chin. J. Geotech. Eng.* **2016**, *38*, 786–797.
32. Xu, Z.; Liu, Y.; Huang, J.; Wen, L. Performance assessment of the complex seepage control system at the LuDila hydropower station in China. *Int. J. Geomech.* **2019**, *19*, 05019001. [CrossRef]
33. Li, Y.; Li, S.; Ding, Z.; Tu, X. The sensitivity analysis of Duncan-Chang E-B model parameters based on the orthogonal test method. *J. Hydraul. Eng.* **2013**, *44*, 873–879.
34. Haupt, R.L.; Haupt, S.E. *Practical Genetic Algorithms*; John Wiley & Sons: Hoboken, NJ, USA, 2004.
35. Mirjalili, S.; Dong, J.; Lewis, A.; Sadiq, A.S. Particle swarm optimization: Theory, literature review, and application in airfoil design. In *Nature-Inspired Optimizers*; Springer: Berlin/Heidelberg, Germany, 2020; pp. 167–184.

water

MDPI

Article

Parameter Optimization of Centrifugal Pump Splitter Blades with Artificial Fish Swarm Algorithm

Qidi Ke [1,*], Lingfeng Tang [1], Wenbin Luo [1] and Jingzhe Cao [2]

[1] School of Mechanical Engineering, Anhui Polytechnic University, Wuhu 241000, China; tlfwym@ahpu.edu.cn (L.T.); lwbyt08@163.com (W.L.)
[2] Tonglu County Forestry Water Conservancy Bureau, Hangzhou 310000, China; qq1025572164@163.com
* Correspondence: 2210110117@stu.ahpu.edu.cn

Abstract: Low specific speed centrifugal pumps typically suffer from low efficiency and severe backflow; adding optimally structured splitter blades can play a role. In this paper, the distribution of pressure and velocity in the flow channel is analyzed using CFD simulation for a low specific speed centrifugal pump. The geometric parameters of the splitter blade are optimized using an orthogonal test and an artificial fish swarm algorithm; then the optimal splitter blade structure is obtained. Results showed that the splitter blade not only effectively solves the backflow of the flow channel and compresses the range of the trailing vortex, but it also alleviates the cavitation at the inlet of the main blade. When considering the best head, the order of influence of each factor is: Splitter blade thickness > Splitter blade inlet diameter > Splitter blade inlet width. At this time, the thickness of the splitter blade is 4.5 mm, splitter blade inlet diameter is 155 mm (0.775) and Splitter blade inlet width is 23 mm. Through the closed pump experimental system, it is confirmed that hydraulic performance has been improved.

Keywords: centrifugal pump; splitter blades; CFD simulation; artificial fish swarm algorithm (AFSA)

Citation: Ke, Q.; Tang, L.; Luo, W.; Cao, J. Parameter Optimization of Centrifugal Pump Splitter Blades with Artificial Fish Swarm Algorithm. *Water* **2023**, *15*, 1806. https://doi.org/10.3390/w15101806

Academic Editors: Jie Yang, Lin Cheng and Chunhui Ma

Received: 25 March 2023
Revised: 3 May 2023
Accepted: 5 May 2023
Published: 9 May 2023

1. Introduction

Pumping systems consume about 9% of the world's electricity; therefore, high efficiency has always been a constant requirement for pump geometry design [1–3]. The low specific speed centrifugal pump is widely used in small and micro feeding, as well as conveying occasions, because of its small flow and high head characteristics [4]. However, due to its narrow channel structure, high disc friction is unsuitable. Therefore, the low specific speed centrifugal pump should have a special design method suitable for its characteristics [5].

In order to improve performance, scholars have tried many methods and achieved good results. Installing splitter blades is the most popular research direction for dealing with channel diffusion; the principle is to increase the number of blades while reducing the inlet crowding, which can effectively increase the head and broaden the high efficiency area [6,7]. The slotting treatment of the blade can also improve the flow channel condition; the purpose is to make the boundary layer flow more stable by blowing the passive jet. The improvement brought by the slotted blade is not as good as that of the splitter blade, but they can be used together. Orthogonal experiments are often used to design slit shapes, due to their practicality. However, at present, the related research is not sufficient [8–13]. As a slotted blade, the clearance blade cannot increase the head, but it greatly improves efficiency [14]. For the main blade, the optimized logarithmic spiral line is superior to other lines, such as the double arc and the B-spline curve [15]. The variation blade with a maximum thickness of 4% chord length is better than the equal thickness blade [16]. In order to make the numerical simulation more accurate, scholars have used supercomputers to calculate the real-time flow field of centrifugal pumps using large eddy simulation (LES), and they confirmed that it has more advantages than other Reynolds average methods (RANS) [17–20]. In addition, in order to ensure the life of the centrifugal pump, leakage, cavitation and vibration need to be controlled. Cavitation number (pressure difference), Reynolds number (turbulence level) and thermodynamic parameters (temperature) have a comprehensive inhibitory effect on hydrodynamic cavitation intensity and spatiotemporal characteristics [21–24]. There is a strong correlation between cavitation and vibration. After cavitation occurs, high amplitude vibration begins to appear and concentrates in the range of 1200∼1400 Hz. Under the same cavitation number, the sound power on the suction side of the blade is greater than that on the pressure side. With a decrease in cavitation number, the fluctuating radiation noise of the cavitation volume is the main noise source for the increased sound power [25–30].

In recent years, with the development of big data and computing power, the innovation and application of algorithms is unprecedented. Scholars have used genetic algorithms, artificial neural networks and other optimization algorithms to optimize the structural parameters of centrifugal pumps, and they have achieved good results [31–35]. The artificial fish swarm algorithm (AFSA) has the advantages of flexibility, fast convergence and insensitivity to initial parameter settings; it occupies the second application rate of swarm intelligence algorithms (SI) all year round [36].

In this study of splitter blades, this paper considers the case that the inlet width of splitter blades is larger than that of main blades for the first time. In this study, Section 2 uses CFD simulation to analyze the characteristics of a low specific speed centrifugal pump. In Section 3, the influence of the geometric parameters of the splitter blade on the flow field is studied using an orthogonal test. In Section 4, the artificial fish swarm algorithm is used to optimize the structure. Finally, the accuracy is verified through experiments. This study provides a reference for the shape selection of splitter blades.

2. Numerical Method

2.1. Research Model

Figure 1 shows the fluid domain of the low specific speed centrifugal pump used in this study. The design parameters are shown in Table 1. The impeller specific speed is 42 and the working fluid is liquid water at 25°. Note that the blade outlet width should have been calculated by Formulas (1) and (2); however, in the actual production process, in order to obtain higher head and stronger flow capacity, the blade width is often stretched as needed. In the numerical model, in order to make the water flow at the impeller inlet more stable and reduce the influence on the backflow of the blade channel, the length of the inlet pipe is appropriately extended.

$$b_2 = k_b \sqrt[3]{\frac{Q}{n}} \tag{1}$$

$$k_b = 0.64 k_{b_2} \left(\frac{n_s}{100}\right)^{\frac{5}{6}} \tag{2}$$

Table 1. Major Design Parameters.

Parameters	**Flow** $Q\left(m^3/h^{-1}\right)$	**Head** H **(m)**	**Speed** n **(r/min)**
value	10	12.5	1440

In Equation (2), k_{b_2} is the correction coefficient of k_b, which is related to the form and specific speed of the pump. So k_{b_2}= 1.536. Calculated b_2= 6.1 mm and broadened to 20 mm in this article.

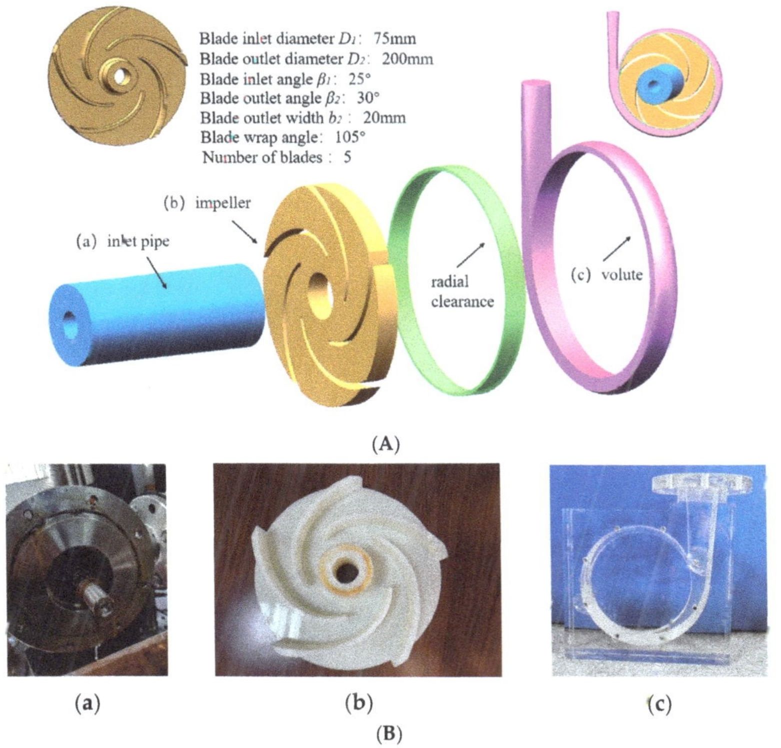

Figure 1. Fluid domain view of centrifugal pump. (**A**) Numerical model; (**B**) Physical model: (**a**) Inlet pipe; (**b**) Impeller; (**c**) Volute.

2.2. Turbulence Model and Boundary Conditions

The fluid in the centrifugal pump continuously exchanges energy through turbulen In the CFD flow field calculation, selecting the appropriate calculation model plays a de sive role in simulation results. In this simulation, the focus is on the external characteristi The real-time turbulence development of the internal flow field is not concerned; the fore, the steady-state Reynolds average method is selected. Nowadays, the *Realizablek* − model and the $SSTk-\omega$ model are widely used. They can not only adapt to most turb lence conditions, but they also have the advantages of easy convergence. In this pap considering the characteristics of the impeller with high speed rotation, $RNGk-\varepsilon$ w selected, as it is more suitable for the separation flow in the near wall region in the hi speed rotating domain [37–39]. The specific expression is as follows:

$$\begin{cases} \frac{\partial(\rho k)}{\partial t}+\frac{\partial(\rho k u_i)}{\partial x_i}=\frac{\partial}{\partial x_j}\left[\alpha_k \mu_{eff}\frac{\partial k}{\partial x_j}\right]+P_k-\rho\varepsilon \\ \frac{\partial(\rho\varepsilon)}{\partial t}+\frac{\partial(\rho\varepsilon u_i)}{\partial x_i}=\frac{\partial}{\partial x_j}\left[\alpha_\varepsilon \mu_{eff}\frac{\partial \varepsilon}{\partial x_j}\right]+C_{1\varepsilon}\frac{\varepsilon}{k}P_k-C_{2\varepsilon}\frac{\varepsilon^2}{k}\rho \end{cases}$$

In the formula:

$$\mu_{eff}=\mu+\mu_t$$

$$\mu_t=\rho C_\mu\frac{k^2}{\varepsilon}$$

where: $C_\mu=0.0845$; $\alpha_k=\alpha_\varepsilon=1.39$; $C_{1\varepsilon}=1.42$; $C_{2\varepsilon}=1.68$.

The inlet boundary condition is the velocity inlet, the set value is 2.76 m/s and the turbulence intensity is set to 5%. The boundary condition of the outlet is Mass Flow Rate, which is calculated as 2.78 kg/s according to the pump flow. The impeller center is set as the rotation origin and the impeller speed is 1440 r/min. The blade and the front and rear cover plates are set to Rotating Wall; the value is 0 rev/min. The wall motion type of volute and inlet and outlet extension section is set as static wall, and the shear condition is set as non-slip wall [40].

2.3. Mesh

The turbulence model affects the separation form of the flow field in the centrifugal pump. The quality of mesh determines whether the model has the ability to accurately capture the existence of small vortices and vortex boundaries. In order to balance the mesh quality and computational efficiency, a mesh independence analysis is carried out. The results are shown in Figure 2. When the number of meshes increased to the fourth group, the two indicators gradually stabilized, and the final number of meshes was 7,991,379. The calculation formula of efficiency is shown in Equation (6).

$$\eta = \frac{\rho g Q H}{M\omega} \tag{6}$$

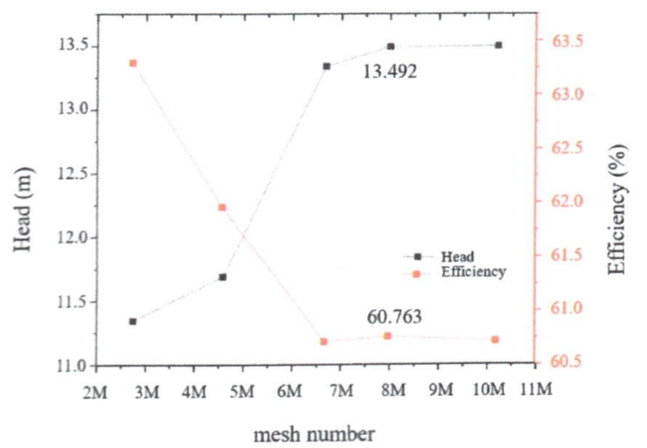

Figure 2. Mesh independence.

While ensuring the number of meshes, y^+ is used to judge whether the mesh of the wall can accurately reflect the flow of the boundary layer. The definition is:

$$y^+ = \frac{\Delta y u_\tau}{v} \tag{7}$$

In the formula, y^+ is the distance from the center of mass of the first grid to the wall, u_τ is the friction velocity and v is the kinematic viscosity.

For different turbulence models, the required y^+ range is different; the $RNGk-\varepsilon$ model requires 0~300. The y^+ calculated by this model is as shown in Figure 3; basically below 50, which meets the quality requirements of the near-wall mesh [41,42].

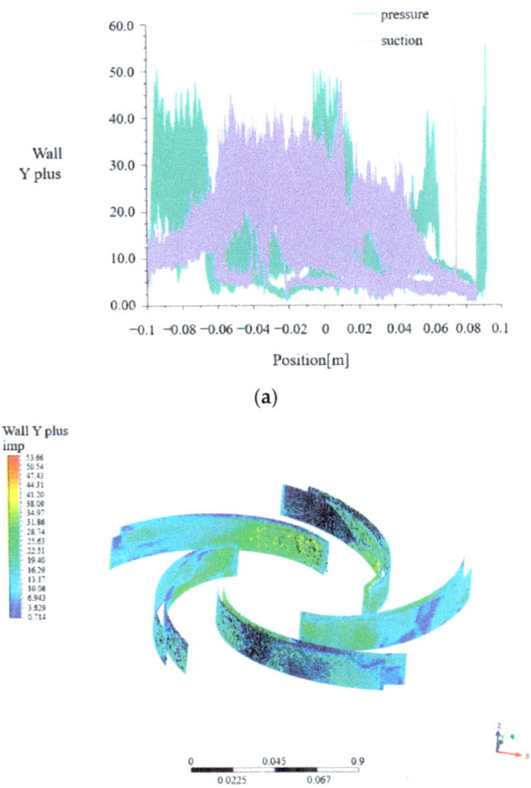

(a)

(b)

Figure 3. Wall y^+ distribution. (**a**) Scatter distribution graph; (**b**) Equivalent distribution map.

2.4. Flow Field Analysis

The CFD simulation is carried out on the rated working condition of the centrifugal pump. Figures 4 and 5 show the pressure and velocity contours inside the flow channel.

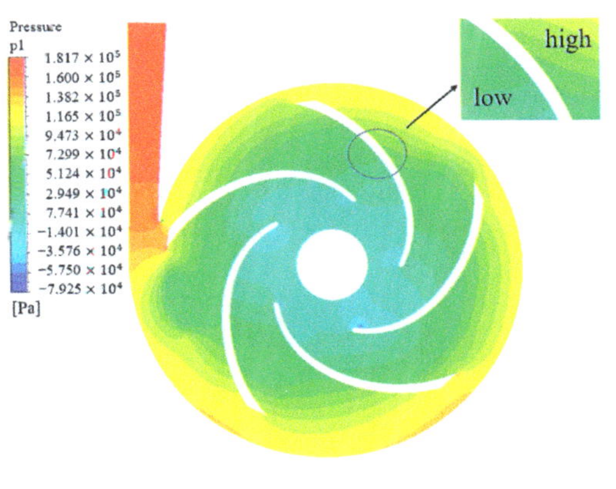

Figure 4. Velocity contour.

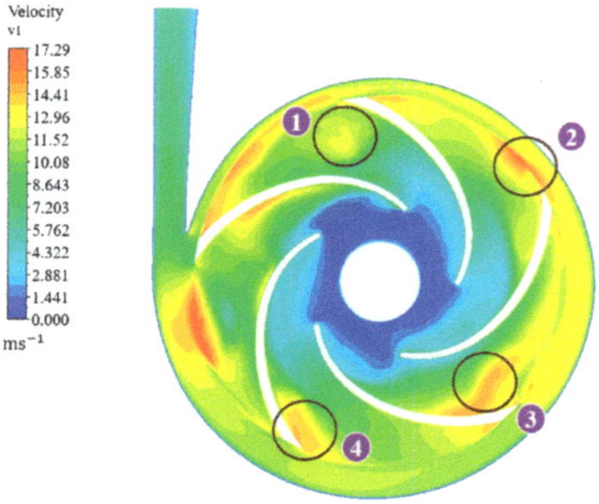

Figure 5. Pressure contour.

In Figure 4, the pressure of the impeller channel is a positive pressure gradient along the flow direction. There is an obvious pressure difference between the pressure and the suction surface. At the same radius of the same blade, the pressure on the pressure surface is higher than that on the suction surface. Near the tongue, a low-pressure zone appears at the intersection of the impeller flow channel and the volute base circle, which is the result of the combined action of diffusion at the outlet of the low specific speed pump flow channel and backflow at the tongue.

Figure 5 is the velocity contour map of the flow field. In the figure, the fluid at the outlet of the pressure surface acts violently with the suction surface through the radial gap, resulting in a backflow phenomenon at the tail of each blade, resulting in different degrees of vortices, only slightly easing near the tongue. The wake vortex at ① falls off completely, hides in the flow channel and occupies the center of the outlet of the flow channel. The trailing vortex at ②–④ is located at the tail of the blade and is continuously pulled by the outlet fluid of the suction surface. The backflow causes the outlet of the flow channel to block and consumes a lot of kinetic energy, which reduces the performance of the centrifugal pump.

The overall flow channel distribution of the model is good, which is consistent with the actual situation.

The Reynolds number is usually used to determine the flow state of viscous fluids; it means the ratio of the inertial force to the viscous force. The ideal flow state inside the impeller should be a stable laminar flow along the flow channel. However, as the flow develops to the trailing edge, the Reynolds number increases, the thickness of the laminar boundary layer gradually increases, the flow disturbance begins to develop and the boundary layer flow becomes unstable. After the transition point, the boundary layer flow becomes completely turbulent. A coordinate system is established at the entrance of the main blade; along the impeller from the inlet to the outlet direction is set to the x-axis positive direction, and perpendicular to the blade direction is set to the y-axis positive direction, assuming that the steady incompressible fluid, the $N - S$ equation is:

$$\begin{cases} \frac{\partial u}{\partial x} + \frac{\partial v}{\partial y} = 0 \\ u\frac{\partial u}{\partial x} + v\frac{\partial u}{\partial y} = -\frac{1}{\rho}\frac{\partial p}{\partial x} + v\left(\frac{\partial^2 u}{\partial x^2} + \frac{\partial^2 u}{\partial y^2}\right) \\ u\frac{\partial v}{\partial x} + v\frac{\partial v}{\partial y} = -\frac{1}{\rho}\frac{\partial p}{\partial y} + v\left(\frac{\partial^2 v}{\partial x^2} + \frac{\partial^2 v}{\partial y^2}\right) \end{cases} \tag{8}$$

Ignoring the smaller order of magnitude, Equation (8) is simplified as:

$$\frac{\partial u}{\partial x} + \frac{\partial v}{\partial y} = 0 \tag{9}$$

$$u\frac{\partial u}{\partial x}+v\frac{\partial u}{\partial y}=-\frac{1}{\rho}\frac{\partial p}{\partial x}+v\frac{\partial^2 u}{\partial y^2} \tag{10}$$

$$0=\frac{\partial p}{\partial y} \tag{11}$$

According to Equation (11), the pressure p of the fluid boundary layer does not change along the y direction. This indicates that the fluid disturbance is caused by the pressure change in the x-axis direction. In the diffusion process of the flow channel, the x-axis direction is a negative pressure gradient; that is, in Formula (10), $\frac{\partial p}{\partial x} < 0$. When the pressure is less than the flow resistance, the flow direction changes and turbulence begins to develop. This creates conditions for the birth of the wake vortex.

In this paper, short blades are added to the tail of the flow channel, which is intended to directly block the outlet reflux here and limit the development of the wake vortex.

3. Orthogonal Test

3.1. Experimental Design

The geometric shape of the splitter blade will strongly affect the streamline direction; therefore, it is necessary to explore its optimal value. In recent years, with the rise of various new materials, plastic centrifugal pumps have become the best choice in special scenarios due to their high corrosion resistance and low price. In order to ensure strength, the components of the plastic centrifugal pump will be thicker than the traditional metal centrifugal pump, which leads to the additional consideration of the influence of the blade thickness on the shape of the splitter blade. At the same time, different from the traditional design, this paper additionally considers the unequal width of the splitter blade and the main blade. Taking into account the number of experiments and the mastery of experimental rules, this paper designs sixteen sets of orthogonal experiments with three factors and four levels. According to the analysis of the flow channel and the structure of the centrifugal pump blade, the research factors are selected as follows: Splitter blade thickness S, splitter blade inlet diameter Di, splitter blade inlet width Bi. They are represented by A, B and C, respectively. The parameters are defined as Figure 6, and the specific values are shown in Table 2.

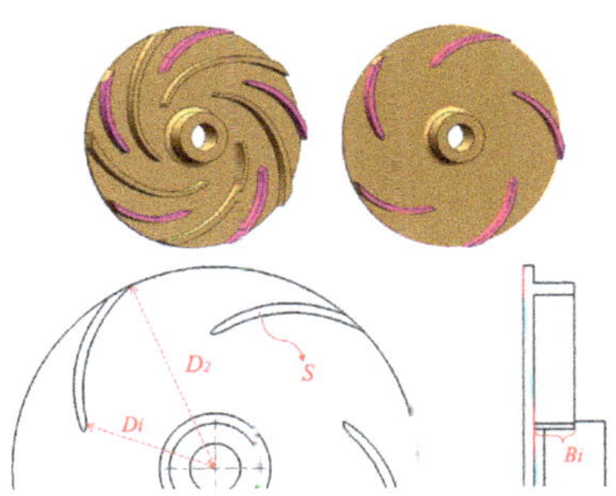

Figure 6. Geometric parameters of splitter blade.

Table 2. Factor Level Table.

Factor \ Level	A Thickness S (mm)	B Diameter Di ($Di/D2$)	C Width Bi (mm)
1	4.5	0.625	17
2	6	0.65	19
3	7.5	0.675	21
4	9	0.7	23

Table 3 is the calculation results of 16 sets of orthogonal tests under rated flow.

Table 3. Orthogonal Experiment Result.

SN	A (S)	B (*Di*)	C (*Bi*)	Head (m)	EFF (%)
1	6	0.65	23	14.25	64.95
2	6	0.625	19	15.21	62.50
3	9	0.625	23	13.52	60.03
4	7.5	0.65	17	15.35	63.20
5	7.5	0.625	21	13.56	63.87
6	4.5	0.675	23	15.17	61.96
7	4.5	0.625	17	14.98	57.36
8	4.5	0.7	19	14.35	61.80
9	7.5	0.7	23	13.88	62.23
10	4.5	0.65	21	15.02	63.60
11	9	0.65	19	14.36	61.96
12	9	0.7	21	13.31	62.79
13	6	0.7	17	14.04	60.40
14	7.5	0.675	19	14.69	66.30
15	9	0.675	17	14.96	62.15
16	6	0.675	21	14.35	64.97

3.2. Data Analysis

Figure 7 shows the results of the orthogonal test. Although the factors restrict each other, in general, adding splitter blades has a very positive effect on the performance of centrifugal pumps. This is because the splitter blade suppresses fluid separation and rotating stall, making the flow distribution at the outlet more uniform. Its essence is to double the number of impellers at the outlet of the impeller, so that the blade has a stronger control over the flow fluid when the crowding performance is not serious.

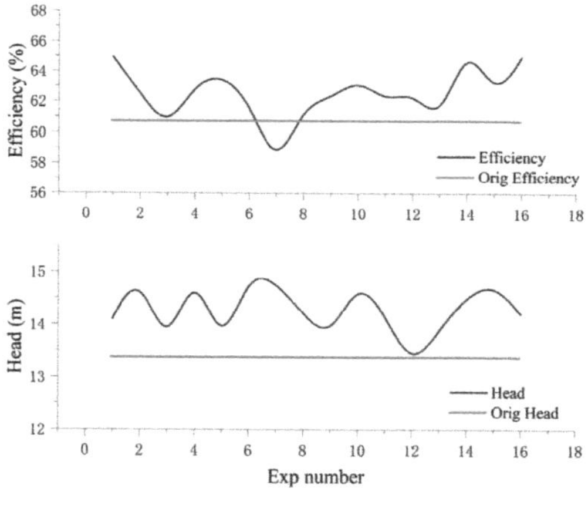

Figure 7. Results of orthogonal test.

In order to explore the influence of various factors on the external characteristics of low specific speed centrifugal pumps, extreme difference analysis was performed for different indicators. The results are shown in Tables 4 and 5.

Table 4. Head Analysis.

Index	Factor		
	A	B	C
$\overline{K_1}$	14.8825	14.165	14.85
$\overline{K_2}$	14.45	14.775	14.625
$\overline{K_3}$	14.35	14.7825	14.05
$\overline{K_4}$	13.89	13.85	14.0475
R	0.9925	0.9325	0.8025
rank	1	2	3

Table 5. Efficiency Analysis.

Index	Factor		
	A	B	C
$\overline{K_1}$	61.18	60.94	60.7775
$\overline{K_2}$	63.205	63.4275	63.14
$\overline{K_3}$	63.9	63.845	63.8075
$\overline{K_4}$	61.7325	61.805	62.2925
R	2.72	2.905	3.03
rank	3	2	1

In the Table, R is the range; the larger the value, the greater the impact of this factor on the index. This has important guiding significance for the weight setting in Section 4.

In the exergy table, factor A (thickness S) has a greater impact on the head, and factor C (inlet width Bi) has a greater impact on the efficiency. It can be seen that as long as the splitter blade exists, it can play a role in reducing the diffusion of the channel outlet, and it also limits the continuous elongation of the trailing vortex. However, too thick blades not only occupy a large number of flow channels to cause blockage, but they also inevitably lose a lot of momentum when the incoming flow impacts the blades, which limits the improvement of the centrifugal pump performance. For the inlet width Bi: the appropriate increase of the inlet width (21mm) can stir more fluid with axial clearance to improve efficiency. However, when the inlet width is stretched to 23 mm, the whole volute space is occupied, and the shaft power cannot be fully utilized.

In order to find the best combination of these three factors, intelligent algorithms are used for further analysis.

4. Artificial Fish Swarm Algorithm Optimization

4.1. Parameter Preparation

Artificial fish swarm algorithm (AFSA) is a two-dimensional swarm intelligence optimization algorithm that simulates the foraging, clustering and rear-end behavior of fish swarm. In the actual production scenario, the importance of head is far greater than efficiency, so this paper takes head as the index, that is, the Y value. X is the value obtained by weighting three factors. In this paper, the gap between the front wall of the volute and the blade is only 3 mm, which has little effect on the overall structure; therefore, the smaller weight is retained. The inlet diameter of the splitter blade is much larger than the thickness value. In order to balance the influence between the two, the weight of the thickness is set to be larger than the inlet diameter. At the same time, according to the order of the factors in Table 4, the weight is set to:

$$\omega = (\omega_1, \omega_2, \omega_3) = (0.5, 0.3, 0.2) \tag{12}$$

In order to make the two-dimensional scene wide enough, the inlet diameter Di uses a numerical form rather than a ratio. All parameters are fitted to a 6th-order Fourier polynomial (13) using Matlab:

$$\begin{aligned} Y = -72.83 + 78.18 \times cos(x \times 0.6658) - 107.6 \times sin(x \times 0.6658) \\ + 26.11 \times cos(2 \times x \times 0.6658) \\ - 6.181 \times sin(2 \times x \times 0.6658) \\ - 6.843 \times cos(3 \times x \times 0.6658) \\ + 72.18 \times sin(3 \times x \times 0.6658) \\ + 67.71 \times sin(4 \times x \times 0.6658) \\ - 33.27 \times cos(5 \times x \times 0.6658) \\ - 149.5 \times sin(5 \times x \times 0.6658) \\ + 20.37 \times cos(6 \times x \times 0.6658) \\ + 69.39 \times sin(6 \times x \times 0.6658) \end{aligned} \quad (13)$$

4.2. Parameter Optimization

The AFSA algorithm is not sensitive to the initial value and does not easily fall into local extremum. In order to describe the behavior of the fish school, relevant parameters are defined, such as in Table 6.

Table 6. AFSA Parameter Definition.

Parameter	Population Size	Maximum Iterations	Maximum Trials	Cognitive Distance	Degrees of Crowding
value	50	200	150	0.5	0.518

(1) Foraging behavior: Let the current state of the artificial fish be X_i, and randomly select a state X_j within its field of view, which is:

$$X_j = X_i + Rand(Visual) \quad (14)$$

$Rand()$ is a random number in 0~1. If the food concentration is $Y_j > Y_i$, then go further in this direction:

$$X_i^{t+1} = X_i^t + \frac{X_j - X_i^t}{\|X_j - X_i^t\|} \times Rand(Step) \quad (15)$$

Conversely, reselect X_j to ensure $Y_j > Y_i$. If the forward condition is not satisfied after t attempts, the update state is:

$$X_i^{t+1} = X_i^t + Rand(Visual) \quad (16)$$

(2) Clustering behavior: Let the current state of the artificial fish be X_i, the number of partners in the neighborhood of the field of view be n_f and the center position be X_c. If $X_c \times n_f < \delta \times X_i$, it indicates that there is more food in the center of the partner and it is not too crowded, then further to the center position X_c:

$$X_i^{t+1} = X_i^t + \frac{X_c - X_i^t}{\|X_c - X_i^t\|} \times Rand(Step) \quad (17)$$

(3) Rear-end behavior: Let the current state of artificial fish be X_i. Detect the best neighbor X_{max} in its neighborhood and the number of partners in the neighborhood of

X_{max} satisfies $Y_c/n_f < \delta \times Y_i$, indicating that there is more food near X_{max} and it is not too crowded, then go further in the direction of X_{max}:

$$X_i^{t+1} = X_i^t + \frac{X_{max} - X_i^t}{\|X_{max} - X_i^t\|} \times Rand(Step) \quad (18)$$

At the beginning of each round, the algorithm will simulate the clustering and rear-end behavior at the same time, and it will perform the behavior with higher function value. If neither behavior can find better results, the foraging behavior is performed. After all artificial fish complete the action in turn, the position state of the fish group is uniformly updated. Repeat the cycle until the end of the iteration.

In the process of taking the value, it is restricted by the actual structure of the centrifugal pump, and only part of the value range can be accepted. The constraint conditions are as follows (19).

$$\begin{cases} 4 < S < 10 \\ 5.61k_{D_2}\left(\frac{n_s}{100}\right)^{-\frac{1}{2}}\sqrt[3]{\frac{Q}{n}} < Di < 7.48k_{D_2}\left(\frac{n_s}{100}\right)^{-\frac{1}{2}}\sqrt[3]{\frac{Q}{n}} \\ 0.47k_{b_2}\left(\frac{n_s}{100}\right)^{\frac{5}{6}}\sqrt[3]{\frac{Q}{n}} < Bi < 0.81k_{b_2}\left(\frac{n_s}{100}\right)^{\frac{5}{6}}\sqrt[3]{\frac{Q}{n}} \end{cases} \quad (19)$$

In the formula, k_{D_2} and k_{b_2} are related to the form and specific speed of the pump, which are 1.125 and 1.54, respectively. The final weighted range is 41.4 < x < 57.6. The final optimization result is shown in Figure 8.

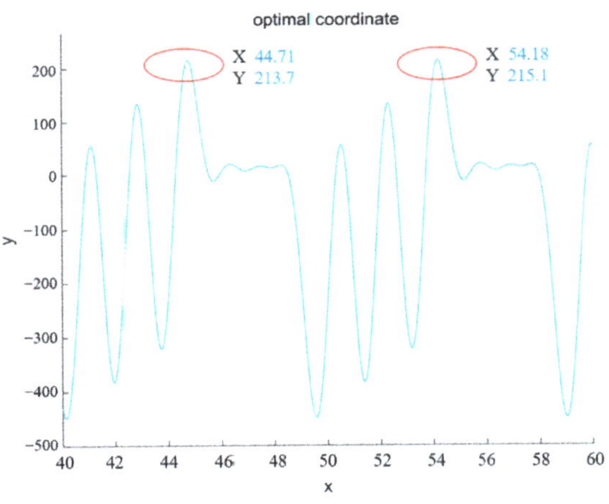

Figure 8. Optimal solution position.

It can be seen from Figure 8 that the optimal solution X obtained by artificial fish swarm optimization is 54.18. Although the effect of the other optimal solution, 44.71, is the same, the larger the X value means the larger the inlet diameter of the model. That is, the centrifugal pump with the same performance is cast with less material; therefore, the larger value 54.18 is selected. The three parameters obtained by decoupling are: thickness S is 4.5 mm, inlet diameter Di is 155 mm, inlet width Bi is 23 mm. At this time, the impeller fluid domain is shown in Figure 9, and the flow channel distribution is shown in Figure 10.

At this time, the external characteristics of the centrifugal pump are further improved: the head is 16.12 m and the efficiency is 63.45%. The head of 2.781 m and the efficiency of 2.736% are improved compared with those without splitter blades, and the optimization effect is obvious. Compared with the optimal results of the orthogonal test (group 4: 15.21 m), there are still great advantages. It can be seen that the artificial fish swarm algorithm with head as the objective function has had a good optimization effect.

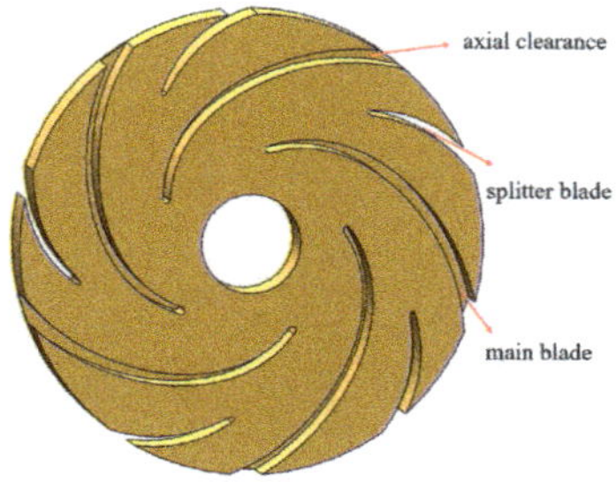

Figure 9. Impeller fluid domain.

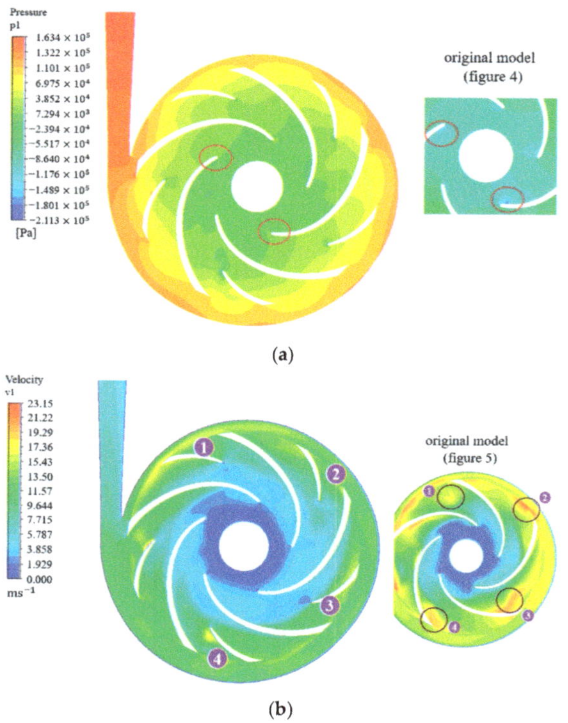

Figure 10. Flow channel distribution map. (**a**) Comparison of pressure distribution; (**b**) Comparison of velocity distribution.

Observe Figure 10. The pressure distribution of the optimized centrifugal pump is more uniform and the two low-pressure areas at the inlet of the main blade disappear, which means that the improvement of the flow at the end of the flow channel is beneficial in alleviating the cavitation at the inlet of the centrifugal pump, which has a positive effect on improving the service life of the centrifugal pump. At the same time, because the splitter blade divides the original wide outlet into two, the serious diffusion at the tail of the suction surface is limited and the outlet reflux has the tendency to be suppressed into laminar flow. In Figure 10b, the trailing vortex at ②–④ almost disappears; the area of the vortex at ① is also reduced and the overall basin composed of the impeller and the volute is well improved.

5. Experimental Verification

In order to verify the rationality and reliability of the splitter blade, physical experiments were carried out on the model.

Part of the impeller is 3D printed to obtain the object, as shown in Figure 11 (the serial number marked on the impeller is the experimental serial number, not the corresponding orthogonal test serial number).

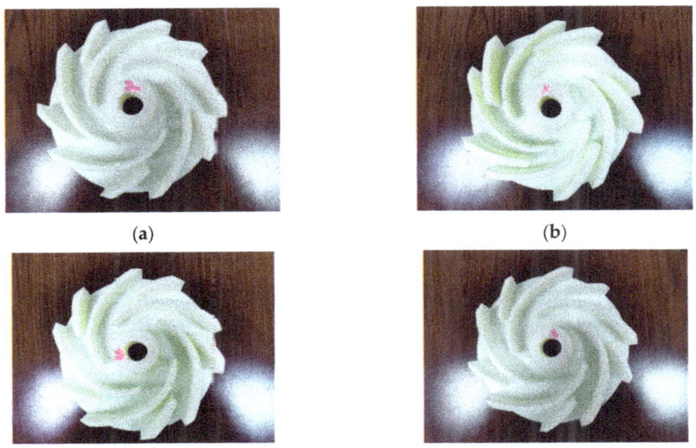

(a) (b) (c) (d)

Figure 11. Splitter blade model. (**a**) Group 1; (**b**) Group 4; (**c**) Group 11; (**d**) AFSA model.

The closed pump performance test bench is assembled, as shown in Figure 12. The experimental platform is installed and debugged by the laboratory personnel of the National Water Pump Engineering Center. The whole system meets the international standards: ≪ISO9906: 2000 Rotodynamic pumps. Hydraulic performance acceptance tests. Grades 1 and 2≫. The system accuracy can reach the highest accuracy of the corresponding standard; therefore, the numerical results are acceptable.

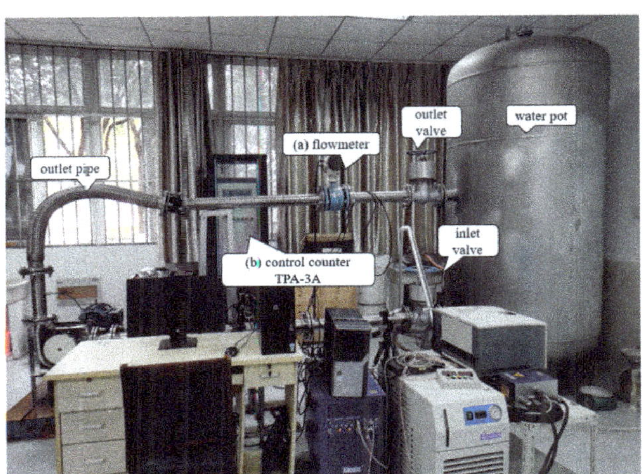

(**1**)

Figure 12. *Cont.*

(2)

(a) (b) (c)

(3)

Figure 12. Closed pump performance test bench. (**1**) Front view; (**2**) Side view; (**3**) Part of instrument: (**a**) Flowmeter; (**b**) Control counter; (**c**) Vacuum pump.

Under rated conditions (1440 r/min; 10 m^3/h), the hydraulic performance of the first three groups of blades and the optimized blades by AFSA was tested many times, and the results were averaged: The maximum head of the first three groups was 14.23 m and the head optimized by AFSA was 15.12 m. After reading the voltage and current of the pump shaft through the control panel, the total efficiency of the test system of the AFSA model is calculated to be 58.61% using Formula (20). Considering the various losses neglected in the numerical simulation, the overall performance improvement effect is better.

$$\eta = \frac{\rho g Q H}{1000UI} \quad (20)$$

6. Conclusions

(1) The flow channel of the low specific speed centrifugal pump is simulated and the reasons for the low efficiency are analyzed: The outlet of the low specific speed pump is seriously diffused. At the same time, the fluid at the tail of the blade working face does not enter the volute smoothly, but it interacts with the outlet fluid on the back, forming a tail vortex, causing the flow channel to block and consume a large amount of fluid kinetic energy.

(2) The $L_{16}(4^3)$ orthogonal table was designed and the orthogonal experiment was carried out using the geometric parameters of the splitter blade. The results obtained shower that the thickness of the splitter blade S is too thick, which will cause impact loss and reduce head. When the inlet diameter Di of the splitter blade is too small, this will cause inlet crowding and block the flow channel; when the splitter blade inlet width Bi is larger, although the head is increased, the efficiency is decreased.

Taking the level of the orthogonal table as the range and the results of the orthogonal test as the original data, the artificial fish swarm algorithm was optimized to obtain the optimal structural parameters: thickness S is 4.5 mm, inlet diameter Di is 155 mm (0.775), inlet width Bi is 23 mm. The numerical simulation confirms that the optimized model of the AFSA algorithm has more advantages. The head is 16.21 m and the efficiency is 63.45%, which is 2.781 m higher than that without splitter blades and 2.736% higher than that without splitter blades. The optimization effect is obvious, and the cavitation condition of the main blade inlet is improved.

(3) The closed water pump performance test bench was assembled, and the experimental results show that the external characteristics of the model optimized by the AFSA algorithm are effectively improved. It is proved that better performance will be achieved when the inlet width of the splitter blade is larger than that of the main blade, which breaks the traditional design idea that the two must be consistent.

Author Contributions: Conceptualization, Q.K. and L.T.; methodology, W.L.; software, Q.K.; validation, Q.K. and W.L.; formal analysis, J.C.; investigation, Q.K.; resources, L.T.; data curation, J.C.; writing—original draft preparation, Q.K.; writing—review and editing, Q.K.; visualization, Q.K.; supervision, L.T.; project administration, L.T.; funding acquisition, L.T. All authors have read and agreed to the published version of the manuscript.

Funding: This research was funded by the major projects of the "The University Synergy Innovation Program of Anhui Province (GXXT-2019-004)", by the project of the "Teaching Research Project of the Anhui Education Department(2019jyxm0229)", and by the project "Science and Technology Planning Project of Wuhu City, 2021YF58".

Data Availability Statement: The data used to support the findings of this study are included within the article.

Conflicts of Interest: The authors declare no conflict of interest.

References

1. Sorguven, E.; Incir, S.; Highgate, J. Understanding loss generation mechanisms in a centrifugal pump using large eddy simulation. *Int. J. Heat Fluid Flow* **2022**, *96*, 108994. [CrossRef]
2. Liu, M.; Tan, L.; Cao, S. Performance Prediction and Geometry Optimization for Application of Pump as Turbine: A Review. *Front. Energy Res.* **2022**, *9*, 818118. [CrossRef]
3. Gonçalves, L.d.S.; Medeiros, K.A.R.; Barbosa, C.R.H. Hydrometer Design Based on Thin-Film Resistive Sensor for Water Measurement in Residential Buildings. *Water* **2023**, *15*, 1045. [CrossRef]
4. Hu, J.; Li, K.; Su, W.; Zhao, X. Numerical Simulation of Drilling Fluid Flow in Centrifugal Pumps. *Water* **2023**, *15*, 992. [CrossRef]
5. Pokharel, N.; Ghimire, A.; Thapa, B.; Thapa, B.S. Wear in centrifugal pumps with causes, effects and remedies: A Review. *IOP Conf. Ser. Earth Environ. Sci.* **2022**, *1037*, 012042. [CrossRef]
6. Kang, C.; Teng, S.; Hu, J.; Ding, K.; Zhang, W. Mitigation of pressure fluctuation through an impeller with splitter blades for a double-volute centrifugal pump. *J. Chin. Inst. Eng.* **2023**, *46*, 293–303. [CrossRef]
7. Namazizadeh, M.; Gevari, M.T.; Mojaddam, M.; Vajdi, M. Optimization of the Splitter Blade Configuration and Geometry of a Centrifugal Pump Impeller using Design of Experiment. *J. Appl. Fluid Mech.* **2020**, *13*, 89–101. [CrossRef]
8. Ke, Q.; Tang, L. Performance Optimization of Slotted Blades for Low-Specific Speed Centrifugal Pumps. *Adv. Civ. Eng.* **2023**, *2023*, 9612947. [CrossRef]
9. Zhang, L.; Li, H.; Xu, H.; Shi, W.; Yang, Y.; Wang, W.; Zhou, L. Experimental and Numerical Investigation of Pressure Fluctuation in a Low-Specific-Speed Centrifugal Pump with a Gap Drainage Impeller. *Shock. Vib.* **2021**, *2021*, 5571178. [CrossRef]
10. Yang, Y.; Zhou, L.; Zhou, H.; Lv, W.; Wang, J.; Shi, W.; He, Z. Optimal Design of Slit Impeller for Low Specific Speed Centrifugal Pump Based on Orthogonal Test. *J. Mar. Sci. Eng.* **2021**, *9*, 121. [CrossRef]
11. Wang, H.; Long, B.; Wang, C.; Han, C.; Li, L. Effects of the Impeller Blade with a Slot Structure on the Centrifugal Pump Performance. *Energies* **2020**, *13*, 1628. [CrossRef]

12. Zhang, Z.-C.; Chen, H.-X.; Ma, Z.; He, J.-W.; Liu, H.; Liu, C. Research on Improving the Dynamic Performance of Centrifugal Pumps With Twisted Gap Drainage Blades. *J. Fluids Eng.* **2019**, *141*, 091101. [CrossRef]
13. Zhang, R.; Yun, L.-C.; Li, J.J.; Jo, H. Investigation on the effect of impeller slot jet on centrifugal pump performance. *J. Hydrodyn.* **2018**, *31*, 733–739. [CrossRef]
14. Wei, Y.; Yang, Y.; Zhou, L.; Jiang, L.; Shi, W.; Huang, G. Influence of Impeller Gap Drainage Width on the Performance of Low Specific Speed Centrifugal Pump. *J. Mar. Sci. Eng.* **2021**, *9*, 106. [CrossRef]
15. Zhang, H.; Tang, L.; Zhao, Y. Influence of Blade Profiles on Plastic Centrifugal Pump Performance. *Adv. Mater. Sci. Eng.* **2020**, *2020*, 6665520. [CrossRef]
16. Xu, Z.; Kong, F.; Tang, L.; Liu, M.; Wang, J.; Qiu, N. Effect of Blade Thickness on Internal Flow and Performance of a Plastic Centrifugal Pump. *Machines* **2022**, *10*, 61. [CrossRef]
17. Zhang, N.; Liu, X.; Gao, B.; Xia, B. DDES analysis of the unsteady wake flow and its evolution of a centrifugal pump. *Renew. Energy* **2019**, *141*, 570–582. [CrossRef]
18. Kye, B.; Park, K.; Choi, H.; Lee, M.; Kim, J.-H. Flow characteristics in a volute-type centrifugal pump using large eddy simulation. *Int. J. Heat Fluid Flow* **2018**, *72*, 52–60. [CrossRef]
19. Posa, A.; Lippolis, A.; Balaras, E. Investigation of separation phenomena in a radial pump at reduced flow rate by large-eddy simulation. *J. Fluids Eng.* **2016**, *138*, 121101. [CrossRef]
20. Stel, H.; Ofuchi, E.M.; Sabino, R.H.G.; Ancajima, F.C.; Bertoldi, D.; Neto, M.A.M.; Morales, R.E.M. Investigation of the Motion of Bubbles in a Centrifugal Pump Impeller. *J. Fluids Eng.* **2018**, *141*, 031203. [CrossRef]
21. Ge, M.; Zhang, G.; Petkovšek, M.; Long, K.; Coutier-Delgosha, O. Intensity and regimes changing of hydrodynamic cavitation considering temperature effects. *J. Clean. Prod.* **2022**, *338*, 130470. [CrossRef]
22. Ge, M.; Sun, C.; Zhang, G.; Coutier-Delgosha, O.; Fan, D. Combined suppression effects on hydrodynamic cavitation performance in Venturi-type reactor for process intensification. *Ultrason. Sonochem.* **2022**, *86*, 106035. [CrossRef] [PubMed]
23. Ge, M.; Manikkam, P.; Ghossein, J.; Kumar Subramanian, R.; Coutier-Delgosha, O.; Zhang, G. Dynamic mode decomposition to classify cavitating flow regimes induced by thermodynamic effects. *Energy* **2022**, *254*, 124426. [CrossRef]
24. Ge, M.; Petkovšek, M.; Zhang, G.; Jacobs, D.; Coutier-Delgosha, O. Cavitation dynamics and thermodynamic effects at elevated temperatures in a small Venturi channel. *Int. J. Heat Mass Transf.* **2021**, *170*, 120970. [CrossRef]
25. Babayigit, O.; Ozgoren, M.; Aksoy, M.H.; Kocaaslan, O.J.D.W.T. Experimental and CFD investigation of a multistage centrifugal pump including leakages and balance holes. *Water Treat.* **2017**, *67*, 28–40. [CrossRef]
26. Bai, L.; Zhou, L.; Jiang, X.; Pang, Q.; Ye, D. Vibration. Vibration in a Multistage Centrifugal Pump under Varied Conditions. *Shock. Vib.* **2019**, *2019*, 2057031. [CrossRef]
27. Al-Obaidi, A.R. Detection of Cavitation Phenomenon within a Centrifugal Pump Based on Vibration Analysis Technique in both Time and Frequency Domains. *Exp. Tech.* **2020**, *44*, 329–347. [CrossRef]
28. Xiao, W.; Tan, L. Design method of controllable velocity moment and optimization of pressure fluctuation suppression for a multiphase pump. *Ocean. Eng.* **2021**, *220*, 108402. [CrossRef]
29. Si, Q.; Ali, A.; Liao, M.; Yuan, J.; Gu, Y.; Yuan, S.; Bois, G. Assessment of cavitation noise in a centrifugal pump using acoustic finite element method and spherical cavity radiation theory. *Eng. Appl. Comput. Fluid Mech.* **2023**, *17*, 2173302. [CrossRef]
30. Lu, Y.; Tan, L.; Han, Y.; Liu, M. Cavitation-vibration correlation of a mixed flow pump under steady state and fast start-up conditions by experiment. *Ocean Eng.* **2022**, *251*, 111158. [CrossRef]
31. De Donno, R.; Ghidoni, A.; Noventa, G.; Rebay, S. Engineering. Shape optimization of the ERCOFTAC centrifugal pump impeller using open-source software. *Optim. Eng.* **2019**, *20*, 929–953. [CrossRef]
32. Wang, C.-N.; Yang, F.-C.; Nguyen, V.T.T.; Vo, N.T.M. CFD analysis and optimum design for a centrifugal pump using an effectively artificial intelligent algorithm. *Micromachines* **2022**, *13*, 1208. [CrossRef]
33. Thakkar, S.; Vala, H.; Patel, V.K.; Patel, R. Engineering. Performance improvement of the sanitary centrifugal pump through an integrated approach based on response surface methodology, multi-objective optimization and CFD. *J. Braz. Soc. Mech. Sci. Eng.* **2021**, *43*, 24. [CrossRef]
34. Kim, B.; Siddique, M.H.; Samad, A.; Hu, G.; Lee, D.-E. Optimization of Centrifugal Pump Impeller for Pumping Viscous Fluids Using Direct Design Optimization Technique. *Machines* **2022**, *10*, 774. [CrossRef]
35. Chen, X.B.; Zhang, R.H.; Yang, W.F. Inverse Design and Optimization of Low Specific Speed Centrifugal Pump Blade Based on Adaptive POD Hybrid Model. *J. Appl. Fluid Mech.* **2022**, *15*, 453–464. [CrossRef]
36. Pourpanah, F.; Wang, R.; Lim, C.P.; Wang, X.-Z.; Yazdani, D. A review of artificial fish swarm algorithms: Recent advances and applications. *Artif. Intell. Rev.* **2023**, *56*, 1867–1903. [CrossRef]
37. Liu, H.; Wang, Y.; Liu, D.; Yuan, S.; Wang, J. Technology. Assessment of a turbulence model for numerical predictions of sheet-cavitating flows in centrifugal pumps? *J. Mech. Sci. Technol.* **2013**, *27*, 2743–2750. [CrossRef]
38. Al-Obaidi, A.R. Effects of Different Turbulence Models on Three-Dimensional Unsteady Cavitating Flows in the Centrifugal Pump and Performance Prediction. *Int. J. Nonlinear Sci. Numer. Simul.* **2019**, *20*, 487–509. [CrossRef]
39. Shadab, M.; Karimipour, M.; Najafi, A.F.; Paydar, R.; Nourbakhsh, S.A. Effect of impeller shroud trimming on the hydraulic performance of centrifugal pumps with low and medium specific speeds. *Eng. Appl. Comput. Fluid Mech.* **2022**, *16*, 514–535. [CrossRef]

40. Gu, Y.; Li, J.; Wang, P.; Cheng, L.; Qiu, Y.; Wang, C.; Si, Q. An Improved One-Dimensional Flow Model for Side Chambers of Centrifugal Pumps Considering the Blade Slip Factor. *J. Fluids Eng.* **2022**, *144*, 091207. [CrossRef]
41. Wei, Z.; Yang, W.; Xiao, R. Pressure fluctuation and flow characteristics in a two-stage double-suction centrifugal pump. *Symmetry* **2019**, *11*, 65. [CrossRef]
42. Xu, C.; Fan, C.; Zhang, Z.; Mao, Y. Numerical study of wake and potential interactions in a two-stage centrifugal refrigeration compressor. *Eng. Appl. Comput. Fluid Mech.* **2021**, *15*, 313–327. [CrossRef]

water MDPI

Article

Application of Artificial Neural Networks for Predicting Small Urban-Reservoir Volumes: The Case of Torregrotta Town (Italy)

Biagio Saya and Carla Faraci *

Department of Engineering, University of Messina, 98166 Messina, Italy; biagio.saya@studenti.unime.it
* Correspondence: cfaraci@unime.it

Abstract: In the hydraulic construction field, approximated formulations have been widely used for calculating tank volumes. Identifying the proper water reservoir volumes is of crucial importance in order to not only satisfy water demand but also to avoid unnecessary waste in the construction phase. In this perspective, the planning and management of small reservoirs may have a positive impact on their spatial distribution and storage capacities. The purpose of this study is, therefore, to suggest an alternative approach to estimate the optimal volume of small urban reservoirs. In particular, an artificial neural network (ANN) is proposed to predict future water consumption as a function of certain environmental parameters, such as rainy days, temperature and the number of inhabitants. As the water demand is strongly influenced by such quantities, their future trend is recovered by means of the Copernicus Climate Change Service (C3S) over the next 10 years. Finally, based on ANN prediction of the future consumption requirements, the continuity equation applied to tanks was resolved through integral-discretization obtaining the time-series volume variation and the total number of crisis events.

Keywords: application of neural network; hydraulic supply system; urban reservoirs

Citation: Saya, B.; Faraci, C. Application of Artificial Neural Networks for Predicting Small Urban-Reservoir Volumes: The Case of Torregrotta Town (Italy). *Water* **2023**, *15*, 1747. https://doi.org/10.3390/w15091747

Academic Editors: Jie Yang, Chunhui Ma, Lin Cheng and Giuseppe Pezzinga

Received: 29 March 2023
Revised: 26 April 2023
Accepted: 28 April 2023
Published: 1 May 2023

1. Introduction

Among the topics of sustainable development, water resources and the services they provide are indicated as the driving forces to create social, economic and financial equity and, in turn, contribute to alleviate land exploitation and poverty and to improve health, food and energy production and environmental preservation [1]. In this perspective, reservoirs assume a key role in water management due to their functions in water supply, irrigation and soil conservation; industrial and hydropower use; as well as flood protection and control [2].

Focusing on municipal water use, in the literature, there are several examples of tank volume calculations that estimate the capacity as the sum of multiple rates [3]. In particular, water supply systems have been traditionally designed according to deterministic approaches, specifying tank components for balancing incoming and outcoming flow, fire protection and emergency storage due to potential interruptions of the supply lines [4]. Most of the available formulations for water storage determination depend on the inhabitant number and are based on the hypothesis that each user consumes a fixed water quantity [5].

Water demand, on the other hand, is highly variable, and it is influenced by a large number of factors due to its dependence on several environmental parameters [6], including weather conditions and anthropogenic aspects [7–9]. For example, [10] observed that rainfall, roof area, number of residents as well as rainwater demand strongly influence tank sizing. It must be considered that, in order to estimate water demand, time series affected by both weekly and yearly periodicity can be employed [7]. The annual periodicity strongly relates to weather conditions, while the weekly or seasonal trend can be related to the fluctuation of non-resident inhabitants.

Empirical approaches have been frequently adopted to determine the reliability of water resources for a municipal water supply system in the light of future demand needs [11]. As well, stochastic methods have been successfully applied to describe the real-time dynamics in reservoirs [12] or for analyzing the reliability of municipal storage tanks [4]. Among such approaches, Monte Carlo methods have been frequently used to tackle uncertainty analysis in complex environmental systems.

As indicated by [13], these approaches fully respect the physical constraints, evaluating the operation constraints in probabilistic terms and allowing the distribution functions and statistics to be derived through Monte Carlo simulation. Recently, ref. [14] used a Monte Carlo method to estimate the uncertainty of average monthly streamflow discharge from a hydrological model in the reservoir system, but other applications can also be found in sewage systems with optimal pumping requirements [15].

On the other hand, conceptual or physically-based models have been adopted to determine tank volume and its management rules [16]. For example, ref. [17] developed a mathematical model for assessing operative policies for a reservoir system based on a computationally efficient optimization procedure. Modern advancements in computer technology have led to the development of complex simulation models widely used in professional practice (i.e., HEC-ResSim [18,19], the WEAP model [20,21] and the CalSim model [22,23]).

Such models present mathematical bases for estimating tank outflow and storage combined with empirical operation rule curves based on historical hydrological data, which are particularly effective in water management tasks [24]. However, such approaches require professional expertise and high-resolution data about watershed characteristics and environmental parameters, and these can fail when multiple complex factors or nonlinear problems have to be faced.

Recently, the advancement in artificial intelligence (AI) has led data-driven models to become popular tools for defining reservoir operations [25]. AI models, compared to physically based ones, show significantly faster response, great learning abilities and better performance in modeling complex formulations. Among AI and machine-learning algorithms, artificial neural networks (ANNs) are black-box models, widely used in several research areas, including hydrology and hydraulics, and they are applied when it is difficult to identify the functional relationships between variables [26].

In the literature, several works deal with the application of ANN to water consumption [27–29]. For instance, Walker et al. performed an interesting study using ANN training for investigating hot water consumption [30]. Ref. [31] applied recurrent neural networks to simulate the operation of three multi-purpose reservoirs under different flow regimes—in particular, under extreme floods and droughts—demonstrating their potential applicability in practical water management, particularly for real-time reservoir operations.

Ref. [32] predicted reservoir volumes using several AI techniques, i.e., (ANN), support vector regression (SVR) and long short-term memory (LSTM), applying the aforementioned algorithms to two reservoirs in Turkey and obtaining better performances with ANN and LSTM models. In addition, ref. [33] compared different soft computing models with conventional approaches to forecast water consumption using real datasets from South Africa to understand the effects of the input parameters on the water consumption forecast.

From what is mentioned above, it is clear that the use of ANNs requires a considerable amount of data both for training and testing. The present study reports the application of an ANN to a case study in order to estimate the future consumption of the small town of Torregrotta in Sicily (Italy), using measurements of the consumption flow rate conducted with the aid of a non-remote flow meter. The main aim of this work was to recover the optimal volume to be assigned to small urban reservoirs in order to satisfy future water demand in such a way to also positively impact both the tank spatial distribution and storage capacities.

The paper is organized as follows. In Section 2, we demonstrate that the consumption trends depend on the rain events and temperature tendencies. These occurrences are related to the high percentage of houses with gardens. Due to the lack of a clear trend between the environmental parameters and the water consumption, we deemed it necessary to use an ANN. We obtain the measured water volume as output and the temperature, the number of rainy days and population as input. After the development of the ANN, as shown in Section 3.1, the neural network training leads to a good correlation for the testing and training data.

The future trends of the environmental parameters are successively recovered through the Copernicus platform, re-sampling data from a yearly scale to a monthly scale by means of an adaptive function. The obtained values are used as input for training the ANN in order to obtain future consumption volumes. To predict the tank volume, the monthly volumes are re-sampled at an hourly scale, by means of the already mentioned adaptive model functions. Finally, through the resolution of the continuity equation applied to the tank, imposing a value for the maximum volume, it is possible to obtain the future volume trend and the total number of hours of crisis. The paper ends with some conclusive remarks.

2. Materials and Methods

2.1. Study Area

The considered case study refers to the town of Torregrotta (Messina, Italy) (Figure 1). Here, in the past, several crisis events led to shortages in the water supply. The hydraulic network is served by to two tanks, as identified in Figure 1, where *Tank*-1 and *Tank*-2 are connected in series. After several interventions performed on the pipe network, it was observed that the problem persisted, meaning that the volume of the tanks was no longer adequate.

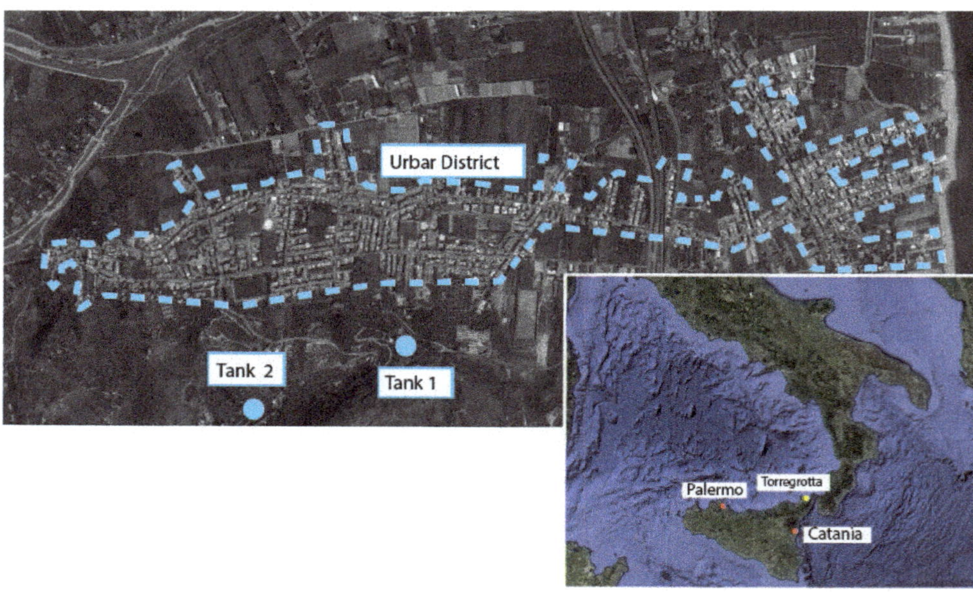

Figure 1. Study area and localization of Torregrotta Town (adapted from https://earth.google.com/, accessed on 30 September 2022).

2.2. Description of the Existing Reservoirs and Water Network

The two above-mentioned reservoirs serve the urban district shown in Figure 1. Both have the same piezometric head of 68.97 m above sea level. *Tank*-1 was built between 1972 and 1980 [34], while *Tank*-2 was built later, in the 2000s, to cope with the increasing compensation crises. The total volume of both tanks equals 1750 m^3, with *Tank*-1 contributing 1000 m^3 [34]. The water level inside the tanks can reach a maximum depth of 4.50 m. The tanks assume a compensation and reserve function and serve the urban area throughout the day. Recently, a campaign was made to investigate the flow rates in and out of the reservoirs as shown in Figure 2 for the period 1–2 January 2001.

Some peaks in the outflow are visible at fixed times of the day, i.e., during morning and evening, whereas the flow entering the reservoir assumes a random variability. It is worth pointing out that the tanks are equipped with flow meters that are not remotely controlled; thus, the hourly data series is rather limited. In order to overcome this limitation, some adaptive functions are applied in Section 4.2.

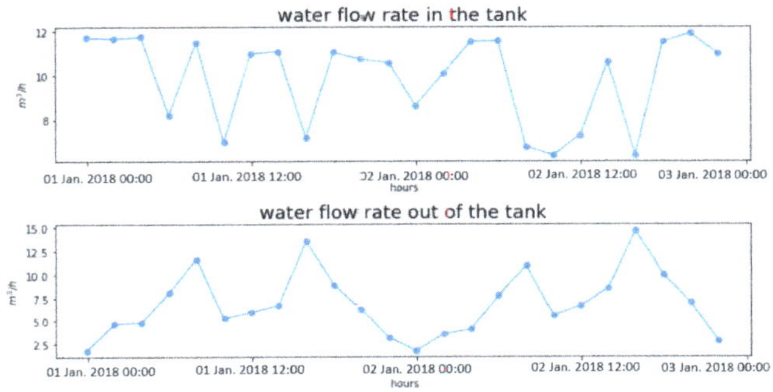

Figure 2. Trends of the water flow rates in and out of the tank measured each 120 min [35].

The water network was built right after the reservoirs were completed. It involves 10.187 m of mainly steel pipelines. Following a preliminary analysis performed in the EPANET environment, assuming the patterns shown in Figure 2 as input, the minimum piezometric head bearing on the network was estimated to be equal to 44.35 m, while the maximum reaches 55.37 m during the hour of minimum consumption.

2.3. Data Acquisition

From 1998 to today, the consortium ACAVN [36], which manages the water supply to the town, surveyed the volumes entering the network, aiming at verifying the taxed volume. In the present study, these data are used for the calculation of the reservoir volume. Figure 3 shows the trend of the measured consumption [36].

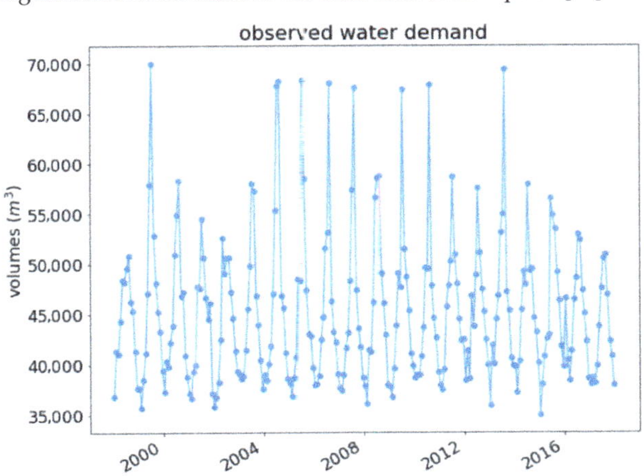

Figure 3. Monthly water consumption obtained by the surveys in the period 1998–2018 [36].

In addition, some environmental parameters, such as temperature data, number of users and number of rainy days per month were also considered, as these factors strongly influence the water demand. Temperature data, recorded by the weather station located in Messina (Geophysical Institute) were collected from [37], as they are the longest and most representative source for this site. Figure 4 shows the trend of the monthly temperatures of the last 20 years.

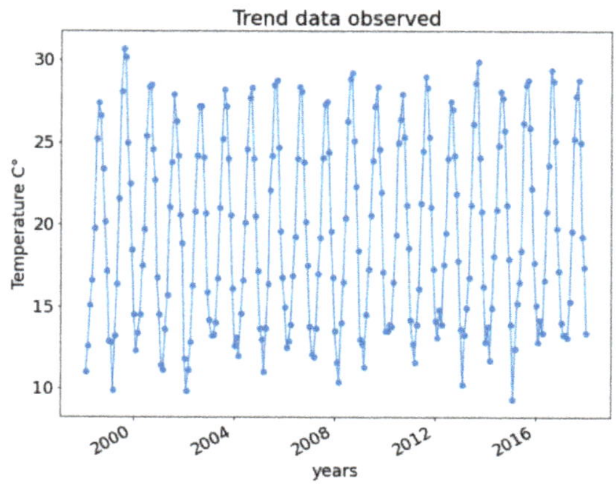

Figure 4. Temperature data in the period 1998–2018 (Hydrological Annals, Regione Sicilia, [37], http://www.sias.regione.sicilia.it, accessed on 30 September 2022).

The population varies over time due to both weekly and monthly variability. Being a touristic area, the peak of the total users occurs during the month of August, while the opposite happens during the winter, leading to greater water demand in the summer. The data relating to users were obtained from the local registry office in combination with the data collected by Istat [38]. In particular, the number of total users was obtained as the sum of the resident population and the commuter or touristic population, which was obtained by an analysis conducted by the local city administration [39]. Figure 5 shows the trend of the total network users in the investigated period, where the number of commuters ranges between 1500 and 2000 persons.

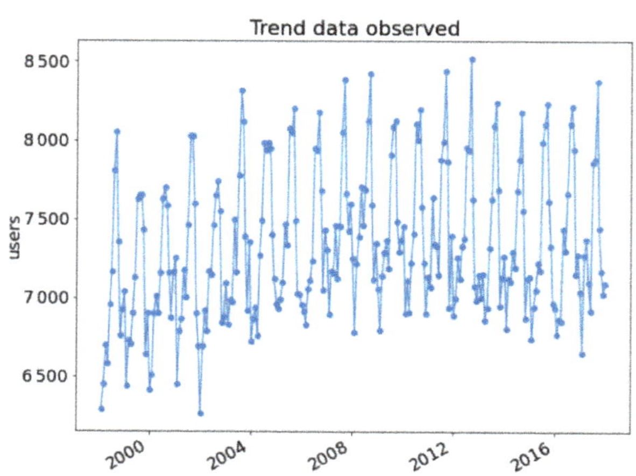

Figure 5. Users obtained by the surveys in the period 1998–2018.

Finally, Figure 6 shows the number of rainy days per month in the investigated period. It is worth pointing out that, in the present study, a day is considered rainy when the cumulative precipitation reaches at least one millimeter during the 24 h. This choice is due according to the Copernicus service, which defines a "dry day" as a day when the threshold of one millimeter of cumulative precipitation is not exceeded in 24 h [40].

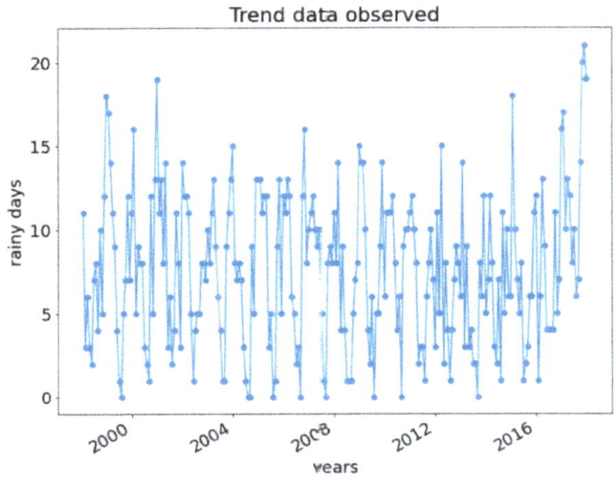

Figure 6. Rainy day monthly surveys in the period 1998–2018.

In order to verify the hypothesis of dependence between water consumption with the previously mentioned environmental parameters, the correlation matrix in Figure 7, reports the correlation factors between the water demand with the rainy days per month, the population and the temperature. It is possible to note that, though being correlated with water consumption, the previous environmental parameters cannot be expressed with an analytical function of the first variable.

This leads to considering the possibility to use a neural network approach in order to forecast the water supply as a function of all the parameters that can influence the water demand. In the following sections, the possibility to adopt an artificial neural network to define the water supply for the considered case study will be explored. Data coming from the trained ANN will serve as a benchmark to obtain the needed tank volume for satisfying the consumption.

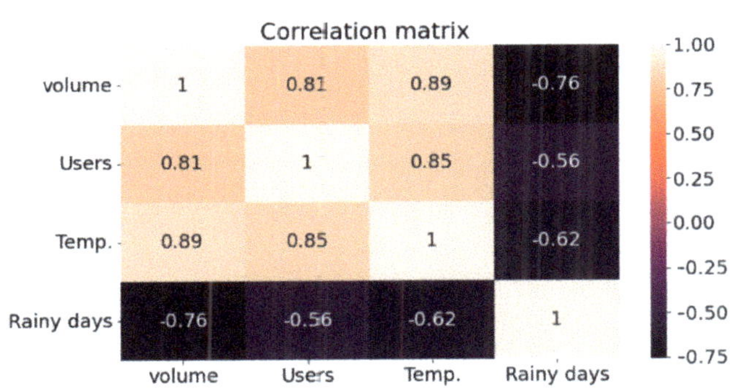

Figure 7. Correlation matrix of parameters and volume consumption.

3. Development of the ANN Model

3.1. Data Arrangement

The ANN model employed here is a multi-layer perception, i.e., a model comprising an input layer, hidden layers and an output layer [41]. The input layer allows the environmental parameters influencing the process to be considered. Domestic water usage over time is obtained as output. The hidden neurons are arranged into hidden layers, whose configuration depends on the problem to be faced and can be optimized as a function of the available dataset. In this specific case, the input layers are the three datasets of temperature, rainy days and population. First, the data were normalized according to the range between the minimum and maximum values:

$$\tilde{x} = \frac{x - x_{max}}{x_{max} - x_{min}} \tag{1}$$

where the x_{max} and x_{min} are, respectively, the maximum and minimum input values, while x is the generic input value. $\tilde{x}$ is the normalized value. Similarly, as far as the output is concerned:

$$\tilde{y} = \frac{y - y_{max}}{y_{max} - y_{min}} \tag{2}$$

where the y_{max} and y_{min} are, respectively, the maximum and minimum output values. The $\tilde{y}$ is the normalized value.

As mentioned before, in the present study, three input datasets were considered, specifically, the temperature, the monthly rainy days and the users served, while the total volume consumed over the month is the output.

3.2. Performance Evaluation

In order to develop the ANN Model, 70% of the N data were used for training and 30% for testing. Then, the root mean squared error $RMSE$, the Pearson correlation coefficient r and the determination coefficient R^2 were employed as performance indices of the models, which are shown and represented in the following.

$$RMSE = \sqrt{\frac{\sum_{k=1}^{N}(P_k - T_k)^2}{N}} \tag{3}$$

$$r = \frac{N\left(\sum_{k=1}^{N} T_k\, P_k\right) - \left(\sum_{k=1}^{N} T_k\right)\cdot\left(\sum_{k=1}^{N} P_k\right)}{\sqrt{\left(N\sum_{k=1}^{N} T_k^2 - \left(\sum_{k=1}^{N} T_k\right)^2\right)\cdot\left(N\sum_{k=1}^{N} T_k^2 - \left(\sum_{k=1}^{N} P_k\right)^2\right)}} \tag{4}$$

$$R^2 = \frac{\left|\sum_{k=1}^{N}\left(T_k - \bar{T}_k\right)\cdot\left(P_k - \bar{P}_k\right)\right|^2}{\sum_{k=1}^{N}\left(T_k - \bar{T}_k\right)\cdot\left(P_k - \bar{P}_k\right)} \tag{5}$$

where P and T are the predicted and target values, the bar indicates the average value, and S is the total number of training or testing samples. For building the neural network reported in this section, Tensorflow [42] and Keras [43] libraries were used.

3.3. ANN Efficiency

The efficiency of ANN models depends on the architecture of the neural network i.e., the number of hidden layers and the number of neurons. A single hidden layer architecture with a different number of neurons in the hidden layer was adopted for the ANN model as this structure generally leads to better results. It should be noted that, for a typical ANN algorithm, it is necessary to optimize the overall ANN structure before deciding the final number of hidden nodes.

Repeated training and testing should be performed with different ANN structures. On the one hand, if the number of hidden layers and/or hidden neurons is too high, there is a risk of over-fitting [44]; on the other hand, if their number is too low, this leads to under-fitting [45]. For the previous reasons, each model was run three times; then, the average values of $RMSE$, R^2 and r were determined for each ANN model. Table 1 shows the performance indices of the ANN models.

Table 1. Performance of the ANN model for different architectures.

	Data Training			Data Testing		
Numbers of Neurons	r	$RMSE$ m³	R^2	r	$RMSE$ m³	R^2
5	0.824	1423.20	0.732	0.798	1501.80	0.694
8	0.931	1364.22	0.821	0.907	1411.82	0.817
13	0.961	1234.50	0.920	0.947	1253.78	0.947
15	0.908	1345.10	0.897	0.929	1378.90	0.831

As can be seen in the table, selecting the number of 13 neurons can lead to better results by having the highest values of r and R^2 and the lowest $RMSE$ value. Moreover, the network shows a good correlation between the training and testing data as reported in the following Table 2.

Table 2. Performance of the ANN model.

	Data Training			Data Testing		
Max Iteration	r	$RMSE$ m³	R^2	r	$RMSE$ m³	R^2
2000	0.961	1234.50	0.920	0.947	1253.78	0.947

As shown in Figure 8, for the highest y values, the correlation becomes slightly worse, likely due to the difficulty in estimating the fluctuating population. From what is shown above, it is possible to confirm that the volumes of consumed water strongly depend on the temperature, on the days of rain and on the number of users served. This result was somewhat expected because there is a high percentage of users who live in houses with gardens or external spaces, as shown in Figure 9; hence, a large amount of water consumption is necessarily used to feed the green spaces in the hot and drought period. Furthermore, the habits of users are strongly influenced by temperatures. As shown in Figure 10, during the summer, a consumption peak is observed, which is due to the concurrent high number of users and hot temperatures.

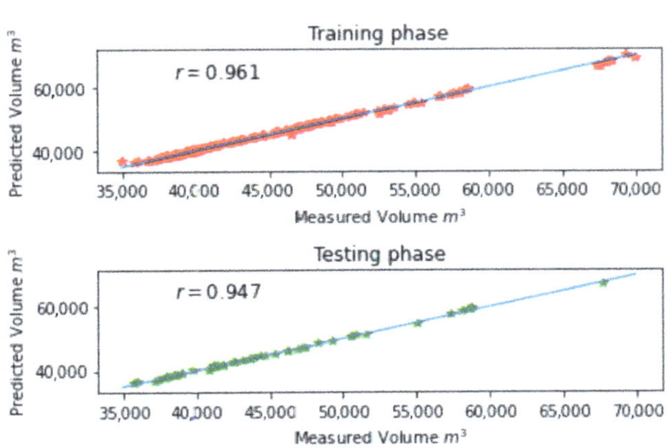

Figure 8. Comparison of the performance between the predicted and training data.

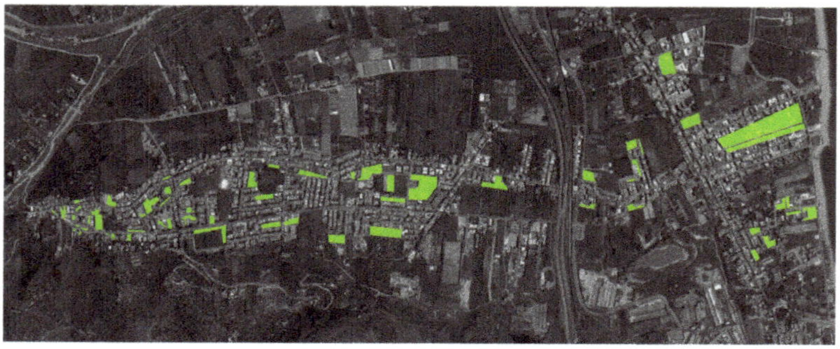

Figure 9. The total garden area supplied by the hydraulic system.

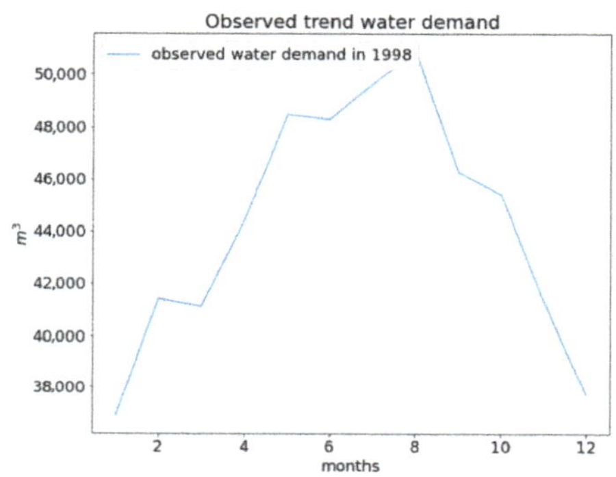

Figure 10. Typical trend of annual water consumption.

Once the ANN is trained, it can be used in order to forecast forthcoming crisis events of the reservoirs. To this aim, it is particularly important to predict the environmental parameters accordingly. For population, a fitting interpolation starting from available data was used, while the other parameters were determined through the Copernicus platform as detailed in the next section.

4. Prediction of Environmental Variables

For the prediction of future data, a toolbox made available by the Copernicus platform was used [40]. In particular, the toolbox for biodiversity-climate-suitability was developed in 2022 with the aim of identifying multiple bio-climate indicators, among which, the temperature and drought days series are also considered [40]. Essentially, the latter quantities can also be determined for the past in order to be compared with the existing data.

Considering a RCP8.5 scenario [46], such an approach was applied to both rainy days and temperature for the period 1998–2018 to detect the trends as shown in Figures 11 and 12, respectively. As well, the temperature and rainy days data are also superimposed on the same pictures. It is possible to notice from Figures 11 and 12 that there are significant differences between the observed data and the trends proposed by the RCP8.5 model for the considered area.

In particular, it is possible to note a strong difference in temperatures, while a better agreement is shown on rainy days (Figure 12). Such discrepancies can be due to the fact that the scenario modeled by the Copernicus toolbox refers to a grid with a (1 × 1 km) resolution. It was, thus, necessary to calibrate the curves proposed by Copernicus with the data observed at the rainfall station.

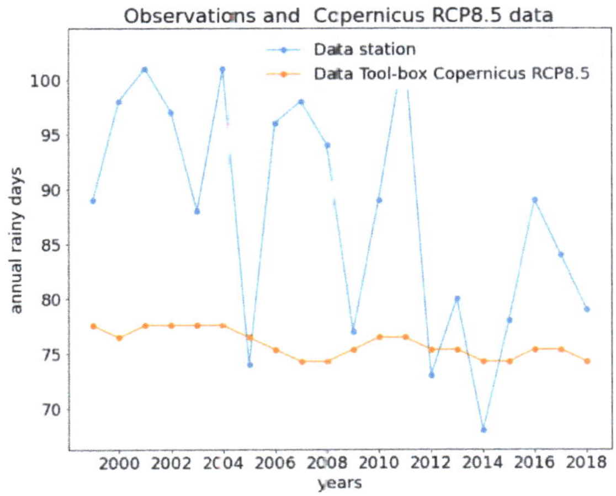

Figure 11. Comparison of rainy days provided by the hydrological annals and the estimate provided by the Copernicus toolbox.

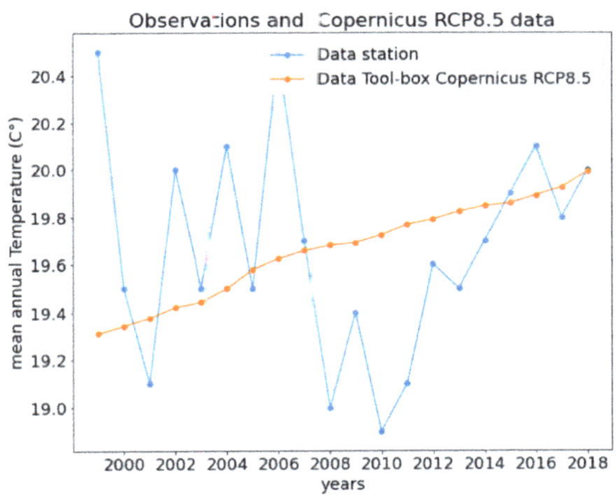

Figure 12. Comparison of temperature provided by the hydrological annals and the estimate provided by the Copernicus toolbox.

In order to estimate each quantity, the coefficient $\tilde{\pi}$, which minimizes the deviation between the observed data φ_{obse} and the model data ϕ_{mod}, was searched:

$$\tilde{\pi} \Rightarrow min\left|\left(\phi_{obse} - \tilde{\pi}\phi_{mod}\right)^2\right| \tag{6}$$

Figures 13 and 14 show the comparisons between the observed data and the Copernicus estimate with the adoption of parameter $\tilde{\pi}$. After the calibration, it is possible to note that the model well describes the observed data, at least in terms of the mean value. According to such an adaptation, Figures 15 and 16 show the future trends of both rainy days and temperatures in terms of the median value predicted by the Copernicus-RCP8.5 model, together with the 15th and 85th quantiles.

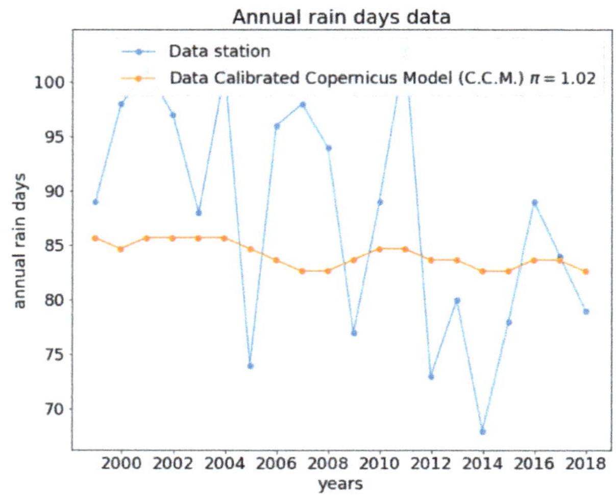

Figure 13. Comparison of rainy days provided by the hydrological annals and the calibrated estimate provided by the Copernicus toolbox.

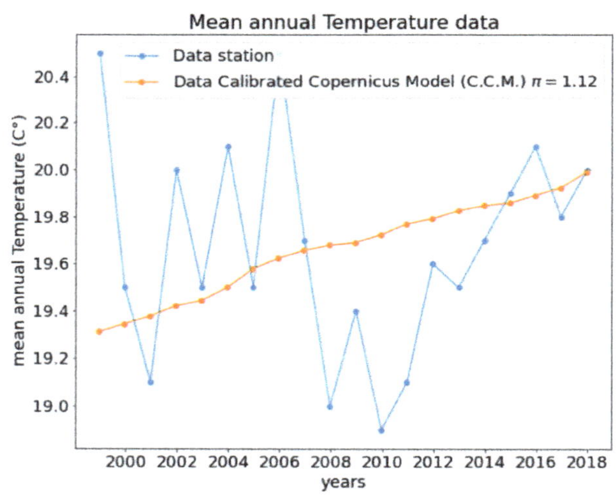

Figure 14. Comparison of temperature provided by the hydrological annals and the calibrated estimate provided by the Copernicus toolbox.

4.1. From Yearly to Monthly Water Demand

Once the average yearly trends were recovered, it was necessary to refer such values to a monthly scale through an adaptive function based on the observed quantities at the measuring station. The adaptive function was used for data on rainy days, temperatures and served users. This operation was performed through Equation (7), where $\tilde{\gamma}$ indicates the monthly trend of both such quantities.

$$\tilde{\gamma} = \tilde{c} + \sum_i \exp\left(- \frac{(t - \tilde{a})^2}{\tilde{b}} \right) \cdot \tilde{d}_i \cdot (1 + rand(\tilde{r})) \tag{7}$$

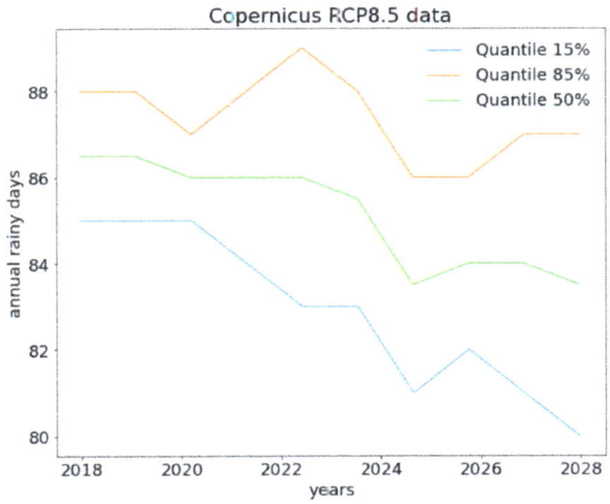

Figure 15. Future forecast—rainy days.

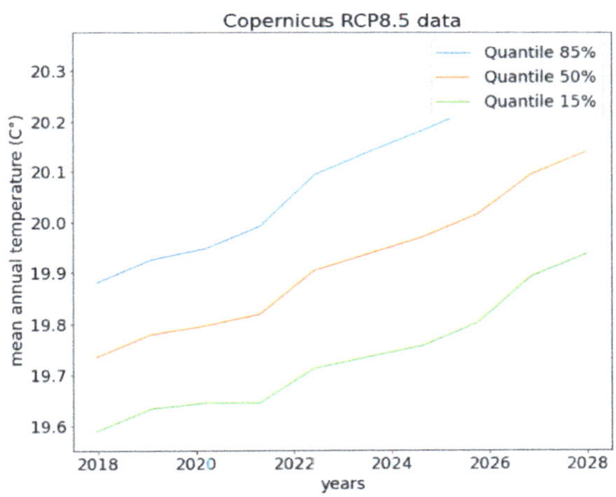

Figure 16. Future forecast—temperature.

$\tilde{a}$ and $\tilde{b}$ are calibration coefficients to be obtained starting from the observed data, while $\tilde{c}$ and $\tilde{d}$ must respect the continuity of the annual data, i.e., the integral of the adaptive function must return a value congruent with the same prediction of the calibrated Copernicus model (C.C.M.). Moreover, $rand(\tilde{r})$ is a random function that expresses the possible deviation from the average value P_{mean}. Finally, $\tilde{r}$ is a numerical interval, depending on the range of the 85th ($P_{85\%}$) and 15th ($P_{15\%}$) quantiles as predicted by the Copernicus-RCP8.5 scenario:

$$\tilde{r} = \frac{(P_{15\%} - P_{mean})}{P_{mean}} \div \frac{(P_{85\%} - P_{mean})}{P_{mean}} \tag{8}$$

Figures 17–19 show the comparisons between the function reported in Equation (7) and the data observed at the reference station. The Equation (7) model is well suited for data relating to temperature, while it shows some scatter for data on population and rainy days, due to the fact that the latter two assume a less regular monthly trend in the observations.

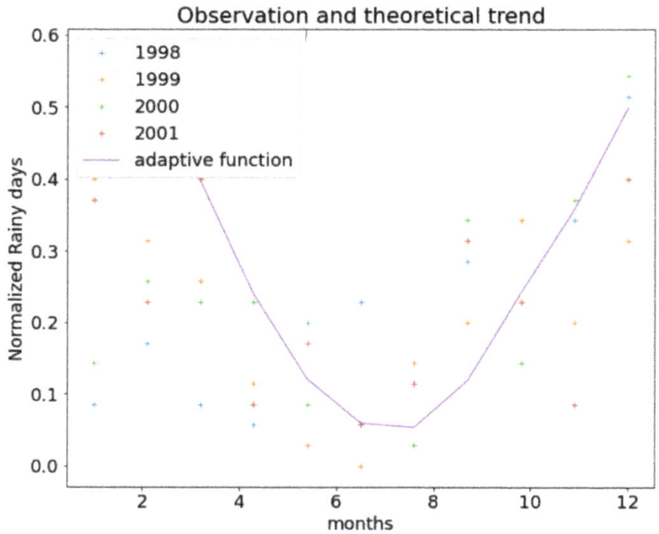

Figure 17. Comparison between the adaptive function and the observed rainy days. The values are normalized with respect to the minimum and maximum values as found in Equation (7).

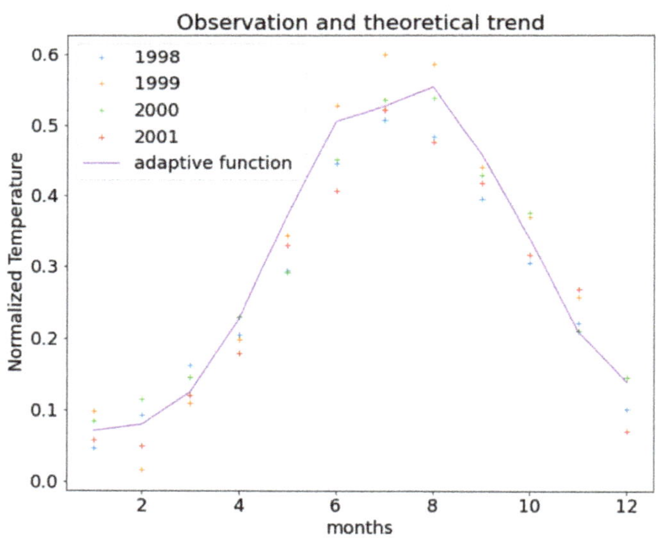

Figure 18. Comparison between the adaptive function and the observed temperature. The values are normalized with respect to the minimum and maximum values as found in Equation (7).

Regarding the population, we assumed that it will undergo a decrease of 10% in the next 10 years as stated by the global analysis conducted by Istat [47]. In addition, the suggested trend on DEMO-Istat [38] on the studied location confirms this tendency. In particular, for the town of Torregrotta, the decrease began in 2014.

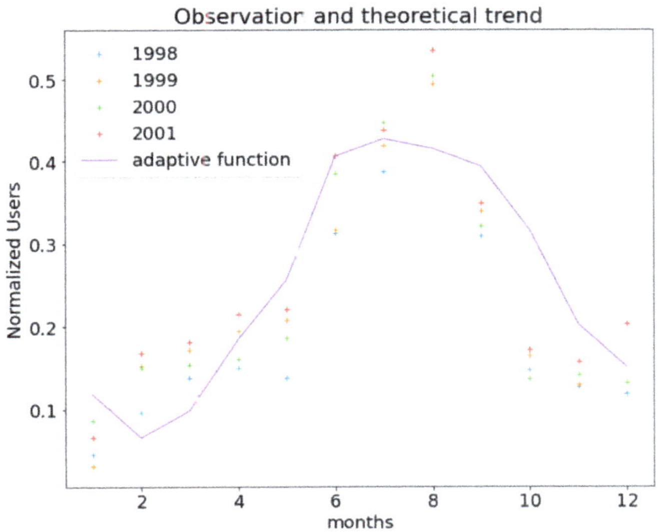

Figure 19. Comparison between the adaptive function and the observed users. The values are normalized with respect to the minimum and maximum values as found in Equation (7).

All the previously obtained data for the next 10 years were considered as the input data of the trained network. The results are shown in Figure 20, where the future prediction of consumption on a 10-year time scale is reported along with the observed data.

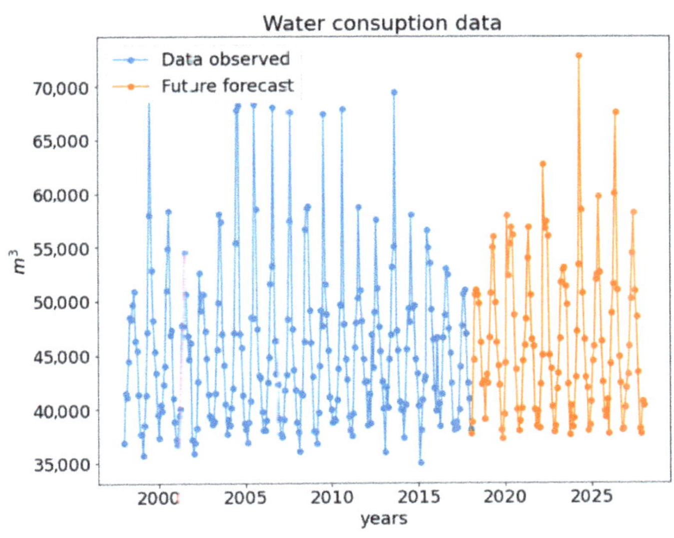

Figure 20. Observed and future water consumption up to 2028. The future trend was obtained by applying the trained ANN.

4.2. From Monthly to Hourly Water Demand

Once the monthly consumption was obtained, it is convenient to switch to hourly consumption. This procedure is entirely theoretical; therefore, it requires, as inputs, the existing observations and literature data [48]. Specifically, the first step is to obtain the daily consumption $q_{tot.day}$ in 24 h, through a peak shape function that identifies different peak consumption within a month:

$$q_{tot.day} = \tilde{q} + \phi \sum_i \exp\left(-\left(\frac{t - f_i}{\sigma_i}\right)^2\right) \tag{9}$$

where the t is expressed in days, the index i sums over 4, i.e., the number of weeks as well as the peaks per month. The maximum of $q_{tot.day}$ is reached on Saturday, when the urban district welcomes worker users coming from the neighboring towns. f_i indicates the peak position, σ_i represents the extremities at each side of the peak, and $\tilde{q}$ is a term introduced for sake of continuity.

Essentially, the sum of the monthly consumption must satisfy the prediction of C.C.M. according to Equation (10):

$$\int_0^{30\,days} q_{tot.day}\, dt = V_{tot\ month} \tag{10}$$

Based on Equation (10), it is possible to define the previously introduced coefficients by means of the following system of Equation (11):

$$\begin{cases} \phi \simeq \dfrac{V_{tot\ month}}{(30\ days) \cdot 6.5} \\ \tilde{q} \simeq \dfrac{V_{tot\ month}}{(30\ days)} \cdot \left(1 - \dfrac{\alpha}{6.5}\right) \\ \alpha = 1 \\ \sigma_i = 2 \\ f_i = 5.5 + (i-1) \cdot 5.5 \end{cases} \tag{11}$$

The results of the mentioned approach are synthesized in Figure 21, where the non-dimensional daily volume, made non-dimensional by means of the Saturday peak volume, is plotted.

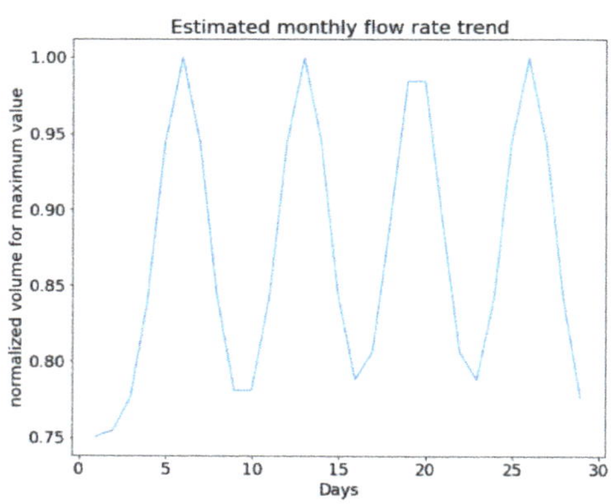

Figure 21. Graphical representation of function (9).

In the same way, the equation of the hourly consumption trend was obtained, considering that the peaks occur at 18:00 and at 8:00 [3]. Furthermore, as the inhabitants of the district total about 7000, the hourly peak coefficient was assumed equal to 3 [3]. The adopted relation is reported below:

$$q_{hour} = q + \sum_k \gamma_k \exp\left(-\left(\frac{t - f_k}{\sigma_k}\right)^2\right) \tag{12}$$

where the t is expressed in hours, and k is the number of peaks per day, equal to 2.

Similarly to Equation (9), Equation (12) must also satisfy compatibility on a daily scale with $q_{tot.day}$ volumes. Therefore, an integral is imposed that is equal to the total daily volume (Equation (13)).

$$q_{tot.day} = \int_0^{24\,hours} q_{hour}\,dt \tag{13}$$

The resolution of the previous problem leads to the following solution:

$$\begin{cases} \gamma_1 \simeq \dfrac{q_{tot.\,day} \cdot 2.2}{(24\;hours)} \\ \gamma_2 \simeq \dfrac{q_{tot.\,day} \cdot 3}{(24\;hours)} \\ \varsigma \simeq \dfrac{q_{tot.\,day}}{(24\;hours)} \cdot \left(1 - \dfrac{\alpha_1 + \alpha_2}{6}\right) \\ \alpha_1 = 2.1 \\ \alpha_2 = 3.5 \\ \sigma_1 = 5 \\ \sigma_2 = 3 \\ f_1 = 8 \\ f_2 = 18 \end{cases}$$

The results of the mentioned approach are synthesized in Figure 22, where the hourly volume, normalized by means of its maximum value, is plotted showing the two peaks as indicated before.

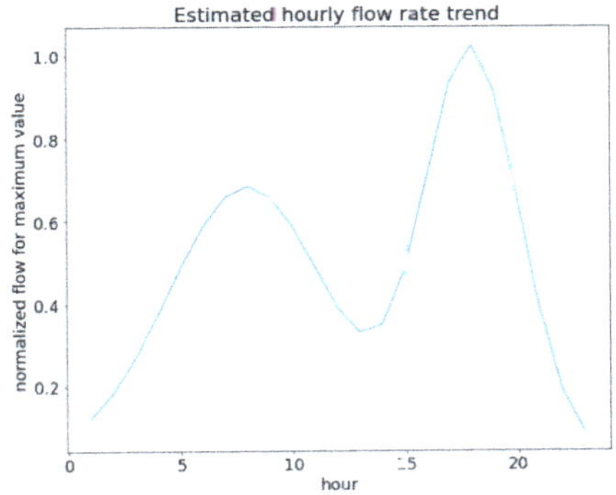

Figure 22. Application of function (12) to recover hourly volumes.

Regarding the water flowing into the reservoir to be sized, it was necessary to assume that the input volume q_{in} varied randomly around its average value $q_{in\,av.}$ [49]. To this aim, it is possible to assume that *Tank*-1 and *Tank*-2, being connected in a series, can be represented as a unique tank with an equivalent volume, equal to the sum of the two. In this way, q_{in} can be referred to as the equivalent volume. Equation (14) was, thus, applied to the entire simulated period, while Figure 23 plots the results for a one-month period.

$$q_{in} = q_{in\,av.} \cdot \left(1 + \frac{rand_1 - rand_2}{5}\right) \tag{14}$$

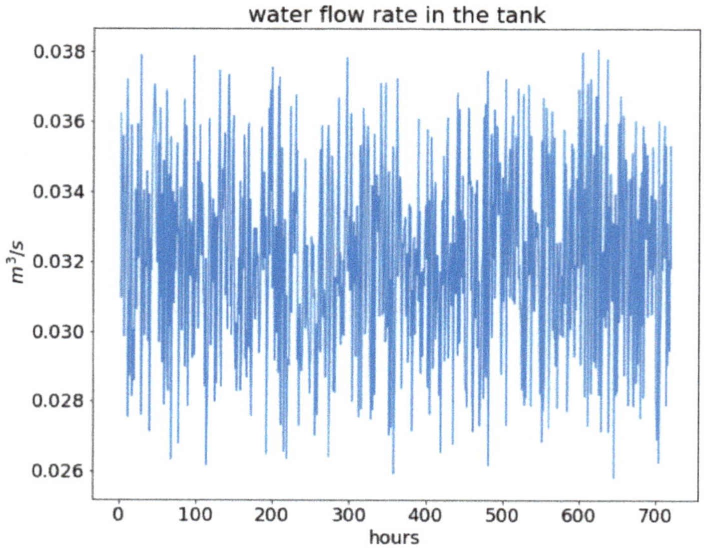

Figure 23. Random function (14) applied to the water flow rate within the tank.

5. Sizing of the Reservoir Volume

The variability of the volumes stored in the equivalent tank $V(t)$ was calculated through the numerical integration of Equation (15):

$$V(t) = \int_0^{\tau} (q_{in}(t) - q_{hour}(t))\, dt \tag{15}$$

where q_{hour} is the hourly flow determined through Equation (12), q_{in} is the hourly inlet flow Equation (14), and τ is the time-series length. The physical limit conditions impose a minimum value of 0 and a maximum equal to the total volume of the equivalent tank V_{Tot} as reported below.

$$\begin{cases} V(t_{i+1}) = V(t_i) + \int_{t_i}^{t_{i+1}} (q_{in} - q_{hour})\, dt \\ \qquad\qquad\qquad\qquad if \;\; V(t_i) < V_{Tot} \\ V(t_{i+1}) = 0 \;\; if \;\; V(t_{i+1}) < 0 \\ V(t_{i+1}) = V_{Tot} \;\; if \;\; V(t_{i+1}) > V_{Tot} \end{cases} \tag{16}$$

Equation (15) can, thus, be solved through numerical integration at finite differences, i.e.,

$$V(t_{i+1}) = V(t_i) + \left(q_{in}(t_{i+1}) - q_{hour}(t_{i+1}) \right) \cdot \Delta t \tag{17}$$

The series begins when the tank is full, i.e., $V(t_0) = V_{Tot}$; therefore, it is possible to impose V_{Tot} as the initial condition and to evaluate the total crisis hours in the time series corresponding to the simulated 10 years. The tank crisis hours will be the instants in which the tank volume equals 0. The results are plotted in Figure 24. Here, the time series of the volume is shown for a period of 20 days, and the red circle denotes the crisis event occurring during the selected period.

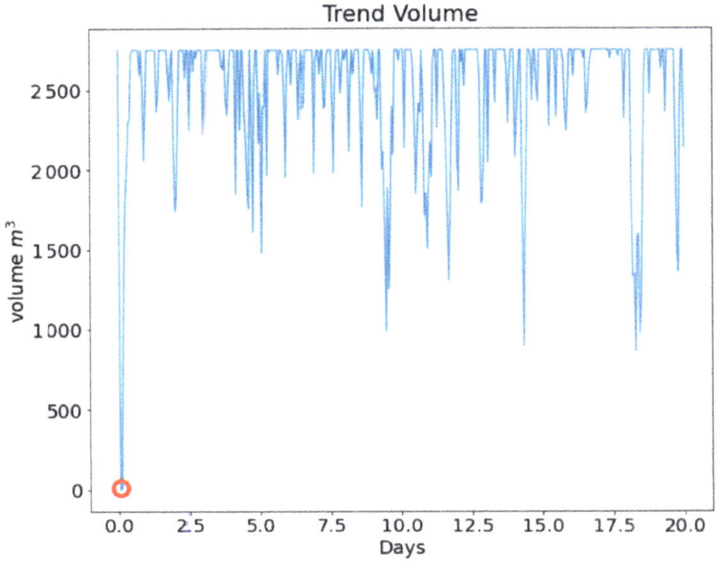

Figure 24. Volume trend in m^3 in hours. Red circle denotes crisis event.

The same operation was repeated as V_{Tot} changed in order to check the number N of crisis events during the simulated period. As is possible to observe in Figure 25, N decreases with an exponential law with the increasing of the tank volume. Hence, it is possible to identify an optimal tank volume once an acceptable number of crisis events is fixed. In particular, in the present case study, it can be observed that the number of crisis events tends to remain approximately constant for volumes higher than 2200 m^3. This fact leads us to conclude that the choice of reservoirs larger than this, while also being uneconomical, does not result in a further reduction of critical episodes.

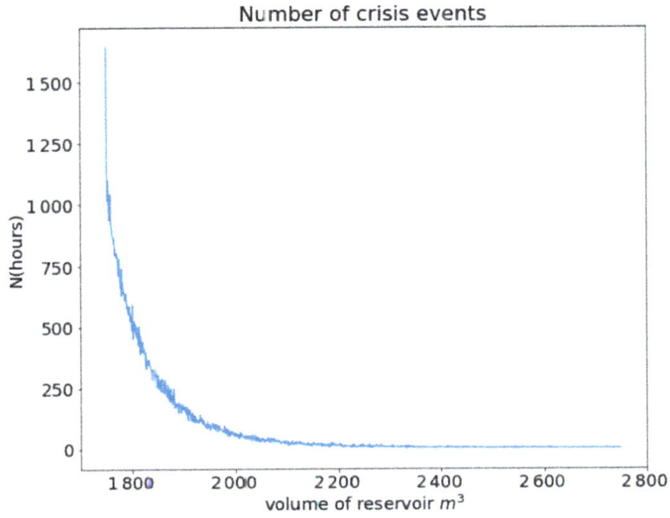

Figure 25. Reduction in the hours of crisis as a function of the adopted volume of the reservoir.

6. Conclusions

This paper proposed a new theoretical methodology to calculate the volume of an urban reservoir serving a civil network. First, data on the consumption, temperature, rainy days and number of served users were acquired. The future forecast of such environmental parameters was determined on the basis of the Copernicus service over the next 10 years. The resulting trends were applied for training an ANN with the aim of recovering the water consumption as output. Furthermore, through theoretical equations, the monthly flow rate was transformed into an hourly flow.

Numerical integration of continuity equations applied to reservoirs was then conducted to define the inlet flow. From the obtained results, it can be inferred that the number of crisis events, identified as the time steps in which the volume of the tank reaches zero, decreased exponentially toward zero as the reservoir volume increased. However, over a certain reservoir size, it became ineffective to further increase the volume to reduce the number of critical episodes. Hence, considering different total volume scenarios, this approach led to an optimization procedure that allowed the total volume of the reservoir to be determined once the number of crisis events was fixed.

Compared to other methods already used in the literature, the present approach does not explicitly consider the volumes required for different needs, such as fire protection, reserve or compensation; thus, it is suitable in applications involving small urban reservoirs.

Author Contributions: Conceptualization, B.S. and C.F.; methodology, B.S.; software, B.S.; validation, C.F.; investigation, B.S.; resources, C.F.; writing—original draft preparation, B.S.; writing—review and editing, C.F.; supervision, C.F. All authors have read and agreed to the published version of the manuscript.

Funding: This research received no external funding.

Data Availability Statement: All the data used for the present research can be found at the referenced links.

Conflicts of Interest: The authors declare no conflict of interest.

References

1. UNESCO. *The United Nations World Water Development Report 2015: Water for a Sustainable World*; Technical Report Water and Climate Change, UNESCO: Paris, France, 2015.
2. Ehsani, N.; Vörösmarty, C.J.; Fekete, B.M.; Stakhiv, E.Z. Reservoir operations under climate change: Storage capacity options to mitigate risk. *J. Hydrol.* **2017**, *555*, 435–446. [CrossRef]
3. Milano, V. *Acquedotti*; Hoepli Editor: Milano, Italy, 1996.
4. van Zyl, J.E.; Piller, O.; Le Gat, Y. Sizing municipal storage tanks based on reliability criteria. *J. Water Resour. Plan. Manag.* **2008**, *134*, 548–555. [CrossRef]
5. Assessorato Regione Sicilia. *Piano di Tutela delle Acque, Sicilia*; Assessorato Regione Sicilia: Sicily, Italy, 1999.
6. Margaritora, G.; Moriconi, P. Consumi idropotabili del comune di Roma. In Proceedings of the La Conoscenza Dei Consumi Per Una Migliore Gestione Delle Infrastrutture Acquedottistiche, Sorrento, Italy, 9–10 September 1990; pp. 9–11.
7. Stańczyk, J.; Kajewska-Szkudlarek, J.; Lipiński, P.; Rychlikowski, P. Improving short-term water demand forecasting using evolutionary algorithms. *Sci. Rep.* **2022**, *12*, 13522. [CrossRef]
8. Chen, J.; Boccelli, D. Demand Forecasting for Water Distribution Systems. *Procedia Eng.* **2014**, *70*, 339–342. [CrossRef]
9. Hemati, A.; Rippy, M.A.; Grant, S.B.; Davis, K.; Feldman, D. Deconstructing Demand: The Anthropogenic and Climatic Drivers of Urban Water Consumption. *Environ. Sci. Technol.* **2016**, *50*, 12557–12566. [CrossRef] [PubMed]
10. Ghisi, E. Parameters influencing the sizing of rainwater tanks for use in houses. *Water Resour. Manag.* **2010**, *24*, 2381–2403. [CrossRef]
11. Okeola, O.; Balogun, S. Estimating a municipal water supply reliability. *Cogent Eng.* **2015**, *2*, 1012988. [CrossRef]
12. Georgakakos, A.P.; Marks, D.H. A new method for the real-time operation of reservoir systems. *Water Resour. Res.* **1987**, *23*, 1376–1390. [CrossRef]
13. Koutsoyiannis, D. A Monte Carlo approach to water management. In Proceedings of the Geophysical Research Abstracts, European Geosciences Union General Assembly, Vienna, Austria, 22–27 April 2012; Volume 14.
14. Marton, D.; Stary, M.; Menšik, P. The influence of uncertainties in the calculation of mean monthly discharges on reservoir storage. *J. Hydrol. Hydromech.* **2011**, *59*, 228–237. [CrossRef]

15. Psarrou, E.; Tsoukalas, I.; Makropoulos, C. A Monte-Carlo-Based Method for the Optimal Placement and Operation Scheduling of Sewer Mining Units in Urban Wastewater Networks. *Water* **2018**, *10*, 200. [CrossRef]
16. Sieber, J.; Yates, D.; Huber Lee, A.; Purkey, D. WEAP: A demand priority and preference driven water planning model: Part 1, model characteristics. *Water Int.* **2005**, *30*, 487–500.
17. Sigvaldson, O.T. A simulation model for operating a multipurpose multireservoir system. *Water Resour. Res.* **1976**, *12*, 263–278. [CrossRef]
18 Klipsch, J.D.; Evans, T.A. Reservoir operations modeling with HEC-ResSim. In Proceedings of the third Federal Interagency Hydrologic Modeling Conference, Reno, NV, USA, 2–5 April 2006; Volume 3.
19. Kim, J.; Read, L.; Johnson, L.E.; Gochis, D.; Cifelli, R.; Han, H. An experiment on reservoir representation schemes to improve hydrologic prediction: Coupling the national water model with the HEC-ResSim. *Hydrol. Sci. J.* **2020**, *65*, 1652–1666. [CrossRef]
20. Lévite, H.; Sally, H.; Cour, J. Testing water demand management scenarios in a water-stressed basin in South Africa: Application of the WEAP model. *Phys. Chem. Earth Parts A/B/C* **2003**, *28*, 779–786. [CrossRef]
21. Yates, D.; Sieber, J.; Purkey, D.; Huber-Lee, A. WEAP21—A Demand-, Priority-, and Preference-Driven Water Planning Model. *Water Int.* **2009**, *30*, 487–500. [CrossRef]
22. Draper, A.J.; Munévar, A.; Arora, S.K.; Reyes, E.; Parker, N.L.; Chung, F.I.; Peterson, L.E. CalSim: Generalized model for reservoir system analysis. *J. Water Resour. Plan. Manag.* **2004**, *130*, 480–489. [CrossRef]
23. Draper, A.J.; Lund, J.R. Optimal hedging and carryover storage value. *J. Water Resour. Plan. Manag.* **2004**, *130*, 83–87. [CrossRef]
24. Zhang, D.; Lin, J.; Peng, Q.; Wang, D.; Yang, T.; Sorooshian, S.; Liu, X.; Zhuang, J. Modeling and simulating of reservoir operation using the artificial neural network, support vector regression, deep learning algorithm. *J. Hydrol.* **2018**, *565*, 720–736. [CrossRef]
25. Hejazi, M.I.; Cai, X. Input variable selection for water resources systems using a modified minimum redundancy maximum relevance (mMRMR) algorithm. *Adv. Water Resour.* **2009**, *32*, 582–593. [CrossRef]
26. Cho, K.; Kim, Y. Improving streamflow prediction in the WRF-Hydro model with LSTM networks. *J. Hydrol.* **2022**, *605*, 127297. [CrossRef]
27. Ghiassi, M.; Zimbra, D.K.; Saidane, H. Urban water demand forecasting with a dynamic artificial neural network model. *J. Water Resour. Plan. Manag.* **2008**, *134*, 138–146. [CrossRef]
28. Salloom, T.; Kaynak, O.; He, W. A novel deep neural network architecture for real-time water demand forecasting. *J. Hydrol.* **2021**, *599*, 126353. [CrossRef]
29. Zubaidi, S.L.; Abdulkareem, I.H.; Hashim, K.S.; Al-Bugharbee, H.; Ridha, H.M.; Gharghan, S.K.; Al-Qaim, F.F.; Muradov, M.; Kot, P.; Al-Khaddar, R. Hybridised artificial neural network model with slime mould algorithm: A novel methodology for prediction of urban stochastic water demand. *Water* **2020**, *12*, 2692. [CrossRef]
30. Walker, D.; Creaco, E.; Vamvakeridou-Lyroudia, L.; Farmani, R.; Kapelan, Z.; Savić, D. Forecasting domestic water consumption from smart meter readings using statistical methods and artificial neural networks. *Procedia Eng.* **2015**, *119*, 1419–1428. [CrossRef]
31. Yang, S.; Yang, D.; Chen, J.; Zhao, B. Real-time reservoir operation using recurrent neural networks and inflow forecast from a distributed hydrological model. *J. Hydrol.* **2019**, *579*, 124229. [CrossRef]
32. Özdoğan-Sarıkoç, G.; Sarıkoç, M.; Celik, M.; Dadaser-Celik, F. Reservoir volume forecasting using artificial intelligence-based models: Artificial Neural Networks, Support Vector Regression, and Long Short-Term Memory. *J. Hydrol.* **2023**, *616*, 128766. [CrossRef]
33. Oyebode, O.; Ighravwe, D.E. Urban Water Demand Forecasting: A Comparative Evaluation of Conventional and Soft Computing Techniques. *Resources* **2019**, *8*, 156. [CrossRef]
34. Merlino, G. Progetto dell'Acquedotto ACAVN, in Italian. 1972. Available online: http://www.acavn.it (accessed on 30 September 2022).
35. Risposta a Domanda Consiliare, III Area Territorio ed Ambiente, "Appunti del Fontaniere" 2019, in Italian. Available online: http://www.comune.torregrotta.me.it (accessed on 30 September 2022).
36. ACAVN. Available online: http://www.acavn.it (accessed on 30 September 2022).
37. Annali Idrologici Sicilia. Available online: https://www.regione.sicilia.it/istituzioni/regione/strutture-regionali/presidenza-regione/autorita-bacino-distretto-idrografico-sicilia/annali-idrologici (accessed on 30 September 2022).
38. DemoIstat. Available online: https://demo.istat.it/ (accessed on 30 September 2022).
39. Comune di Torregrotta. Available online: https://www.comune.torregrotta.me.it/i-area-amministrativa-e-servizi-alla-persona-ed-alle-imprese/servizio-demografico/ (accessed on 30 September 2022).
40. Copernicus. Available online: https://cds.climate.copernicus.eu/cdsapp#!/home (accessed on 30 September 2022).
41. Li, H.; Liu, Z.; Liu, K.; Zhang, Z. Predictive power of machine learning for optimizing solar water heater performance: The potential application of high-throughput screening. *Int. J. Photoenergy* **2017**, *2017*, 4194251 . [CrossRef]
42. Abadi, M.; Agarwal, A.; Barham, P.; Brevdo, E.; Chen, Z.; Citro, C.; Corrado, G.S.; Davis, A.; Dean, J.; Devin, M.; et al. Tensorflow: Large-scale machine learning on heterogeneous distributed systems. *arXiv* **2016**, arXiv:1603.04467.
43. Chollet, F.K. Available online: https://keras.io (accessed on 14 August 2019).
44. Tetko, I.V.; Livingstone, D.J.; Luik, A.I. Neural network studies. 1. Comparison of overfitting and overtraining. *J. Chem. Inf. Comput. Sci.* **1995**, *35*, 826–833. [CrossRef]
45. Li, H.; Zhang, Z.; Liu, Z. Application of artificial neural networks for catalysis: A review. *Catalysts* **2017**, *7*, 306. [CrossRef]

46. IPCC Report *Towards New Scenarios for Analysis of Emissions, Climate Change, Impacts, and Response Strategies*; Technical Report; Moss and Others: Geneva, Switzerland, 2008.
47. Istat. Available online: https://www.istat.it/it/archivio/263995 (accessed on 30 September 2022).
48. Frega, G.C. *Lezioni di Acquedotti e Fognature*; Liguori Pub.: Naples, Italy, 1984.
49. Erto, P. *Probabilità e Statistica: Per le Scienze e L'ingegneria*; McGraw-Hill: Milano, Italy, 1999.

water

MDPI

Article

Laboratory Model Test and Field In Situ Test of Distributed Optical Fiber Monitoring of Seepage in a Karst Depression Reservoir Basin

Bo Yu [1], Chunyong Shen [2,*], Hao Peng [1] and Fawang Guo [1]

1 Power China Guiyang Engineering Corporation Limited, Guiyang 550081, China
2 Power China Renewable Energy Co., Ltd., Beijing 100020, China
* Correspondence: shency_gyy@powerchina.cn; Tel.: +86-18985191345

Abstract: Karst depressions are ideal places for building reservoirs and stacking wastes, but karst depressions are mostly located in areas with strong karst development, and there is a problem of karst leakage. The study of karst depression monitoring methods can provide technical support for the construction of reservoirs in karst depressions, and leakage monitoring technology based on distributed temperature sensing (DTS) has the clear advantages of a large measurement range, high precision, continuity, wide distribution, large area, etc. In this paper, a temperature-sensing optical cable is first used to conduct leakage monitoring tests in different soil media, and the temperature change curve in different media is obtained, which verifies the feasibility of the heatable temperature-sensing optical cable to identify leakage in soil media with different moisture contents. Then, for the heatable distributed temperature-sensing optical cable, the variation law of the temperature eigenvalues of the sensing optical cable under different seepage velocities is studied using the layout method wrapped with saturated medium-coarse sand, and the relationship between the seepage rate and the temperature eigenvalues was analyzed. A regression formula for quantitative analysis of leakage velocity was established; finally, the manganese slag silo project in Songtao County in Guizhou Province, China, was used to simulate different leakage conditions on site for testing, and the data were compared and analyzed to obtain the route under different conditions. The temperature distribution law of the seepage section verifies the feasibility of the application of heatable temperature-sensing optical cables in the seepage monitoring of karst depressions.

Keywords: karst depression reservoir basin seepage; distributed temperature sensing (DTS); model test; phosphorus slag medium; temperature distribution

Citation: Yu, B.; Shen, C.; Peng, H.; Guo, F. Laboratory Model Test and Field In Situ Test of Distributed Optical Fiber Monitoring of Seepage in a Karst Depression Reservoir Basin. *Water* **2023**, *15*, 1477. https://doi.org/10.3390/w15081477

Academic Editor: Aizhong Ye

Received: 4 March 2023
Revised: 29 March 2023
Accepted: 6 April 2023
Published: 10 April 2023

1. Introduction

Karst depressions are well sealed, and are ideal places for stacking ash, tailings, waste slag, second-class industrial solids, and hazardous wastes and building reservoirs. However, karst depressions are located in the karst vertical infiltration recharge area. Once leakage occurs, such as after building a reservoir, it may cause the reservoir to fail to store water normally. Waste storage can lead to sewage entering underground rivers and causing serious environmental incidents. Therefore, research on seepage monitoring methods for anti-seepage bodies in karst depressions can provide theoretical and technical support for the resource utilization of karst depressions and has comprehensive social and economic benefits. Currently, the methods commonly used for monitoring seepage in karst depressions include pressure pipes, seepage meters [1], and flow measurement weirs [2]. Common methods for detecting leakage in karst depressions include high-density resistivity methods [3], capacitive methods, temperature tracing methods [4], transient electromagnetic methods [5], electrical methods, and ground-penetrating radar methods [6]. These monitoring and detection methods have the common characteristic of not being able to comprehensively cover the monitored objects and often cannot detect certain locations.

To improve the accuracy of monitoring and detection, a large number of sensors must be arranged at a certain density, which is extremely uneconomical. Additionally, the lifespan of sensors is limited, and it is not possible to maintain good working conditions throughout the entire lifecycle of the project. As time passes, the sensors will continue to deteriorate, leading to the inability to obtain effective data in the future.

The present study found that heat can be continuously transmitted through a medium, thus providing a new tracer quantity for monitoring the seepage flow of the structure. Among the various temperature tracking methods for seepage monitoring, the distributed fiber optic temperature sensor (DTS) has attracted the most attention. Aufleger et al. [7,8] proposed the idea of using distributed optical fibers for long-term online monitoring of earth–rock embankments and introduced the application of distributed optical fibers in water conservancy projects in a comprehensive manner, which pointed out the direction for subsequent research. Li et al. [9] analyzed the feasibility of DTS application in the leakage monitoring of the Yangtze River embankment. Li [10] developed a multiparameter optical fiber monitoring system for embankments based on DTS and verified the applicability and reliability of the monitoring system by analyzing field monitoring data. Wang et al. [11] introduced the basic principle of DTSs and their application in a concrete face rockfill dam. The comparison with traditional monitoring methods showed the advantages of this method. Wei et al. [12] and Qin et al. [13] introduced the application of DTSs in the location monitoring of leakage in the surrounding joints of concrete face rockfill dams. Tan et al. [14] verified the effectiveness of optical fiber leakage monitoring technology for core rockfill dams through dynamic simulation and model tests, which can compensate for the shortcomings of conventional monitoring methods. Harvie et al. [15] conducted an experimental study on the seepage monitoring method of tailings dams based on DTSs. Wang [16] studied the sandy soil seepage flow monitoring test based on distributed optical fiber, inferred that there is a certain linear relationship between leakage and temperature medium, carried out related research on using optical fiber to determine the location of geomembrane damage, and obtained some beneficial results. Thiele et al. [17] and Artieres et al. [18] combined distributed optical fibers with anti-seepage geotextiles commonly used in hydraulic engineering to locate and provide early warnings of structural damage through a combination of leakage detection and strain monitoring. In Vogt et al. [19], the high-precision vertical temperature distribution profile was obtained through the distributed optical fiber temperature measurement data, the temperature monitoring data were processed by the dynamic harmonic analysis method, and then the leakage velocity of the river recharge groundwater was estimated. Beck et al. [20] used two large-scale experimental studies and presented the results of the field monitoring data analysis of a dam, and DTS-based dam seepage and piping monitoring methods were introduced. Deng [21] obtained a large amount of test data by establishing a leakage test model, compiled a program to extract and organize the massive data, and determined the heating power range by preparing the test. The influence law of fiber temperature rise was identified. Liu et al. [22] applied distributed optical fiber temperature measurement technology to monitor the leakage of an underground diaphragm wall. It was proven through experiments that the use of an external, heatable optical fiber cable can monitor the temperature difference between the seepage part and the non-seepage part of the diaphragm wall well, which is the most important part of the diaphragm wall. Quantitative measurement and precise location of leakage velocity provide the basis this research. The above studies are mainly based on indoor model tests and lack in situ test verification. Moreover, most of them are aimed at dam structures and rarely involve the seepage monitoring of a karst depression reservoir basin.

This paper aims to establish a set of application systems for the leakage monitoring of karst depression reservoirs based on distributed optical fiber technology to detect leakage in time and guide repair. This paper first studies the leakage identification effect of a heatable temperature-sensing optical cable in media with different moisture contents through laboratory experiments and verifies the feasibility of the leakage monitoring application of the heated optical cable when theses temperature changes are small, which is the applicable scope of the optical cable in practical applications. The study provides guidance, then carries out a model test and quantitative analysis of the leakage rate monitoring performance of the heatable temperature-sensing optical cable. Finally, through the field simulation test of a manganese slag warehouse, it is verified that the heatable optical cable can be used in the leakage monitoring of an actual karst depression reservoir basin. The application effect provides a reference for the application of DTSs in seepage field monitoring with small temperature differences and small flow rates in the future.

2. Principle of Distributed Optical Fiber Leakage Monitoring

As shown in Figure 1, when the incident light quantum in the fiber collides with the fiber material molecules, an inelastic collision occurs, which is manifested in the generation of Stokes light with a longer wavelength and anti-Stokes light with a shorter wavelength. Since anti-Stokes light is more sensitive to the effects of temperature, the DTS system uses the Stokes light channel as the reference channel and the anti-Stokes light channel as the signal channel. The ratio of the two can eliminate the fluctuation in the light source signal As well as fiber bending and other nontemperature factors to acquire temperature information. The intensity ratio of Stokes light and anti-Stokes light and temperature satisfies the relationship of Formula (1):

$$R(T) = \frac{I_a}{I_b} = \left(\frac{\nu_a}{\nu_b}\right)^4 e^{\frac{-hc\nu}{KT}} \tag{1}$$

where R (T) is a function of the temperature to be measured; I_a and ν_a are the StOx light intensity and frequency, respectively; I_b and ν_b are the anti-Stokes light intensity and frequency, respectively; c is the speed of light; ν is the Raman frequency shift; h is Planck's constant; K is the Boltzmann constant; and T is the absolute temperature.

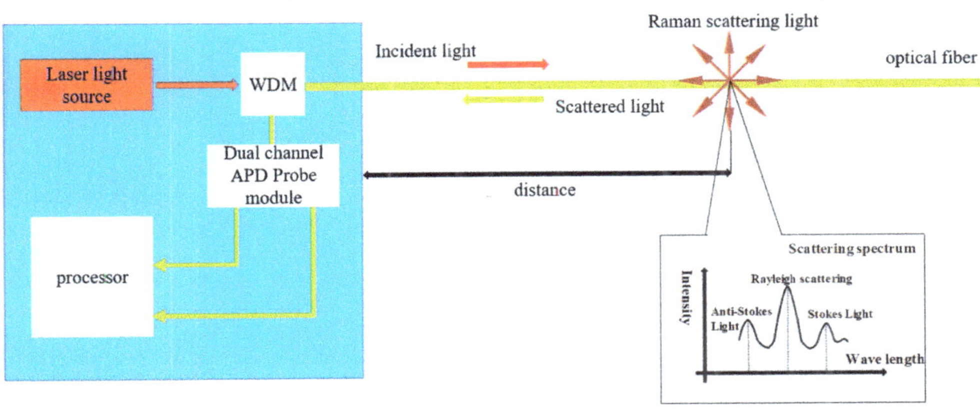

Figure 1. DTS temperature measurement principle.

The principle of optical fiber leakage monitoring is the principle of heat dissipation, as shown in Figure 2. After the copper mesh is heated, the heat will diffuse from the inside to the outside. Then, the optical fiber monitoring system is used to measure the change in temperature, and the relationship between the temperature characteristic's value and the moisture content during the heating process is used to measure the water field in the soil in a distributed manner. The thermal conductivity of soil varies with the moisture content. The higher the moisture content is, the stronger the thermal conductivity. The sensor or sensing optical cable with a self-heating function is implanted into the soil, and the fixed power is heated to establish the relationship between the degree of heat diffusion and the moisture content in the soil, thereby realizing the qualitative and quantitative monitoring of the moisture content of the surrounding soil. The degree of heat diffusion is represented by the temperature characteristic's value, which is defined as follows: After heating for a period of time, the measured temperature tends to be stable. A certain characteristic time interval is selected, and multiple temperature values measured in this time interval are arithmetically averaged to obtain the temperature average value. Then, the initial temperature is subtracted to obtain the temperature characteristic value. The higher the moisture content of the soil is, the better the thermal conductivity. After the external heat source is provided for heating, the greater the degree of heat diffusion and the greater the amount of diffusion, the less the temperature can be raised and the smaller the temperature characteristic value.

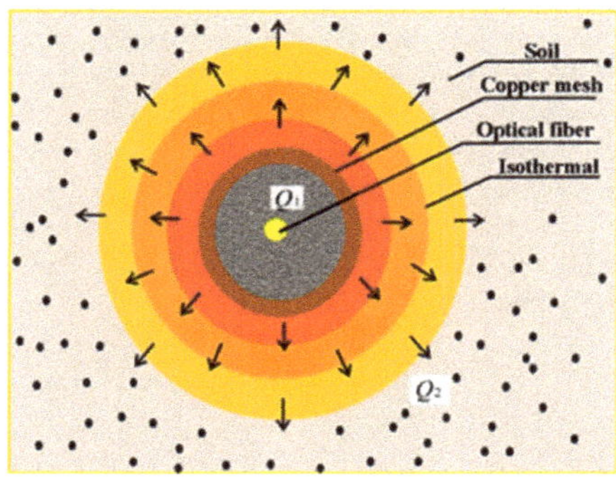

Figure 2. Schematic diagram of heat diffusion in soil after copper mesh is heated.

In the early stage of seepage development, the seepage flow was low, and the soil moisture content increased, but there was no obvious seepage phenomenon. The moisture content of the medium near the hidden leakage location will gradually increase with continuous leakage, and the temperature characteristic values of the optical cable in different moisture content environments are not the same, so the seepage can be judged by heating the optical cable and comparing the temperature characteristic value. The occurrence of leakage events, and the specific location of leakage, can be located. When the leakage has reached a certain scale, an obvious seepage phenomenon is formed at the leakage location, and the vicinity of the leakage has been saturated. By heating the optical cable and establishing the relationship between the flow velocity and the temperature characteristic value, the mathematical expression of the flow velocity and the temperature characteristic value is derived. When the temperature characteristic value changes, the corresponding flow velocity can be obtained according to the corresponding fitting formula to judge the extent of the leakage. According to the specific situation, the identification and semiquantitative analysis of the leakage location are formed.

3. Leakage Test of Heatable Temperature-Sensing Optical Cables under Different Moisture Content Media (Sand, Phosphorus Slag)

The heatable temperature-sensing optical cable selected in this test is a copper mesh heatable temperature-sensing optical cable. Its outermost layer is a black sheath, the second layer is a braided copper wire mesh wrapped with Kevlar, and the third layer is an armored tube with a sheathe and Kevlar fiber inside. The blue cable is multimode 62.5/125, the white is multimode 50/125, and the red is a single-mode cable. The average resistance of the optical cable is 0.08 Ω. When the above heatable temperature-sensing optical cable is not heated, it is regarded as a conventional temperature-sensing optical cable. The demodulation device is a DTS demodulator (suitable for the 62.5/125 fiber type), and a temperature compensation box is connected at the same time. The main test equipment used in the test is shown in Figure 3.

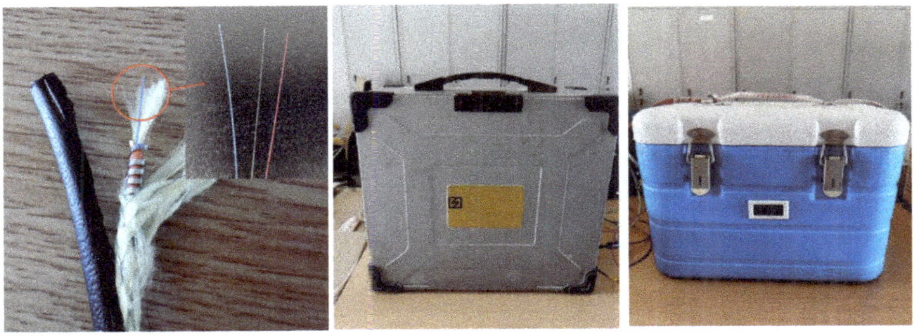

(a) (b) (c)

Figure 3. Test equipment for the leakage monitoring indoor model: (**a**) Heatable cable structure; (**b**) DTS demodulator; (**c**) Temperature compensation box.

One end of the temperature-measuring cable is connected with the jumper through a fusion splicer, and the other end of the jumper is connected with the port of the temperature compensation box, as shown in Figure 4a. Then, the other port of the temperature compensation box is connected to the channel port of the DTS demodulator. One end of the data communication line is connected to the temperature compensation box, and the other end is connected to the USB interface corresponding to the DTS host. The copper wires at both ends of the heatable temperature-sensing optical cable are connected to the DC power supply through external copper wires to form a heatable loop, as shown in Figure 4b.

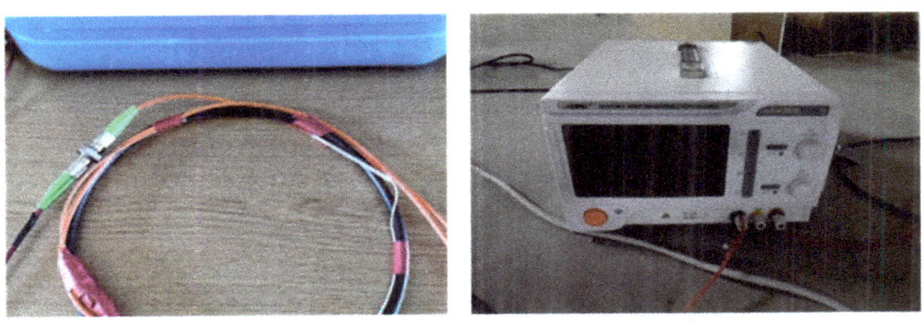

(a) (b)

Figure 4. Temperature measurement cable connection: (**a**) Jumper connection; (**b**) Copper wire connection.

A pump tube is placed above the leakage position, and the other side of the pump tube is connected to the peristaltic pump. By controlling the peristaltic pump, the water flow enters the pump tube and flows out at the leakage position, and the water diffuses downward in the medium to simulate leakage, as shown in Figure 5a,b. The size of the acrylic box is 1 m × 0.2 m × 0.2 m, and the optical cable is straightened and buried in the medium, as shown in Figure 5c. At the same time, holes are also opened at the bottom of the box, and a permeable cloth is placed to observe whether water seeps during the leakage test, as shown in Figure 5d.

The test steps are as follows:

(1) Build a test device, fill the acrylic box with the medium, and bury the optical cable in the box to straighten and fix it;
(2) Before the official start of the test, check all equipment to ensure that it is operating normally and collecting data synchronously;
(3) Use a hot air gun to heat the medium, observe the real-time temperature data of the DTS demodulator, and locate the part of the optical cable in the acrylic box After

the temperature of the heatable temperature-sensing optical cable has stabilized, a heating power of 5 w/m was applied to raise the temperature to a steady state. Then, the peristaltic pump was activated to start the leakage test. The speed of the peristaltic pump is controlled to 30 rpm, and the temperature curve is observed. After the temperature drops to a stable state, stop the test.

(4) Process the test data, analyze the overall temperature change curve of the heatable temperature-sensing optical cable in the test section and the temperature change curve at the leakage point, and evaluate the identification effect of the heatable temperature-sensing cable under leakage conditions.

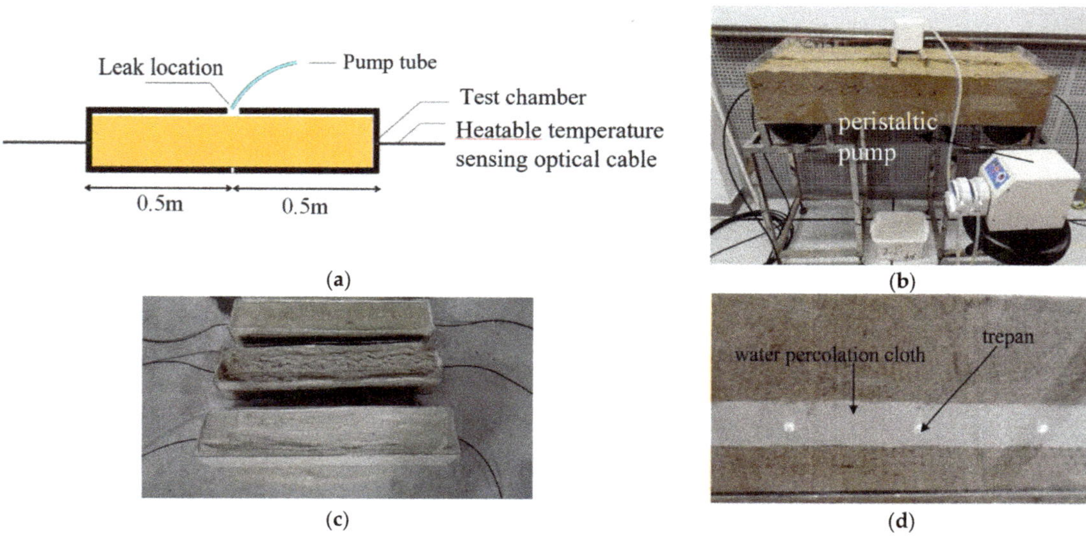

Figure 5. Leakage monitoring test system: (**a**) Schematic diagram of the leakage device; (**b**) The actual layout of the leakage device; (**c**) Arrangement of temperature-measuring optical cables; (**d**) Holes at the bottom of the device.

We tested the working conditions shown in Table 1 in turn and repeated the test steps. We analyzed and compared the temperature characteristic values of the optical cable in different media (sand and phosphorus slag) and the temperature change law of leakage, evaluated the leakage identification effect of the heatable temperature-sensing optical cable in dry medium, and analyzed and evaluated the leakage identification effect. The temperature response law of the heatable optical cable under different moisture contents was also observed.

Table 1. Model test conditions.

Number	Media Type	Moisture Content
1	Sand	dry soil
2		8%
3		14%
4		20%
5	Phosphorus slag	dry soil
6		14%
7		24%
8		34% (saturation)

3.1. Comparison of the Temperature Measurement Results of Different Media in the Dry State

The temperature change curve at the leakage position of the optical cable in the sand medium is shown in Figure 6. The figure shows that the temperature of the optical cable changes very little before and after the leakage occurs. Based on this, the heating power was increased to 10 w/m, and the test was repeated while maintaining the same leakage conditions. The results are shown in Figure 7. For the fiber optic cable in phosphorus slag medium, the temperature change curve at the leakage position is shown in Figure 8, and the heating power is the same as that in the sandy soil, which is 10 w/m.

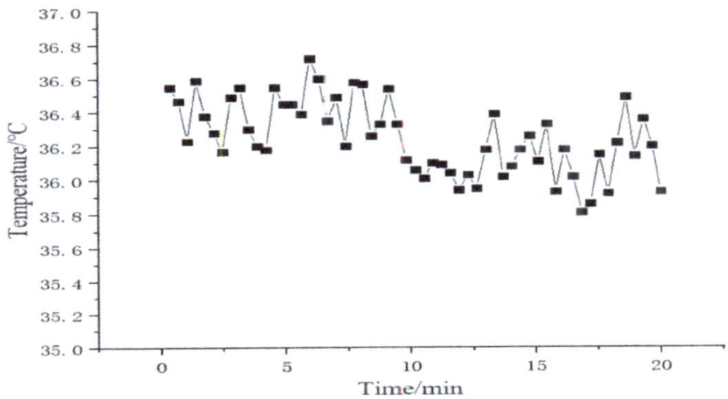

Figure 6. Temperature variation curve of the leakage position under a heating power of 5 w/m.

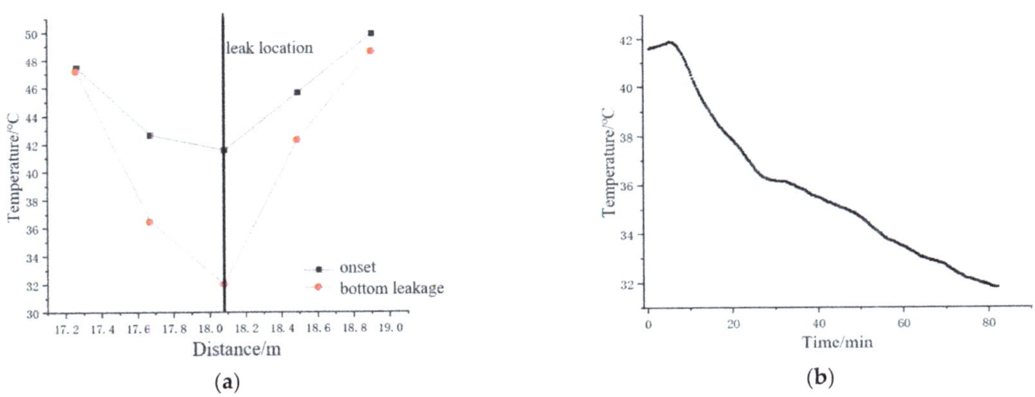

Figure 7. Temperature change curve under heating power of 10 w/m (dry sand): (**a**) Test section temperature distribution curve; (**b**) Temperature change curve at the leakage position.

It can be seen from Figure 7 that the optical cable is in the dry sand medium, from the beginning of the simulated leakage to the water seepage at the bottom, and the entire leakage process is approximately 80 min. Under a heating power of 10 w/m, when leakage occurs, the temperature at the leakage position drops by approximately 10 °C, and the temperature change gradually becomes slower as the leakage continues. It can be seen from Figure 8 that when the optical cable is in the dry phosphorus slag medium, under the heating power of 10 w/m, when leakage occurs, the temperature at the leakage position remains stable for a period of time before it begins to drop, and the drop is by approximately 12 °C. The temperature change gradually slows down as the leakage continues and finally tends to be stable. However, during the whole process, there was no water leakage at the bottom, and on the surface of the device, stratification of phosphorus slag and water appeared, as shown in Figure 9. The reason for this is that after the phosphorus slag encounters water, its cohesiveness becomes very high, resulting in poor water permeability. Eventually, the phenomenon of delamination is formed.

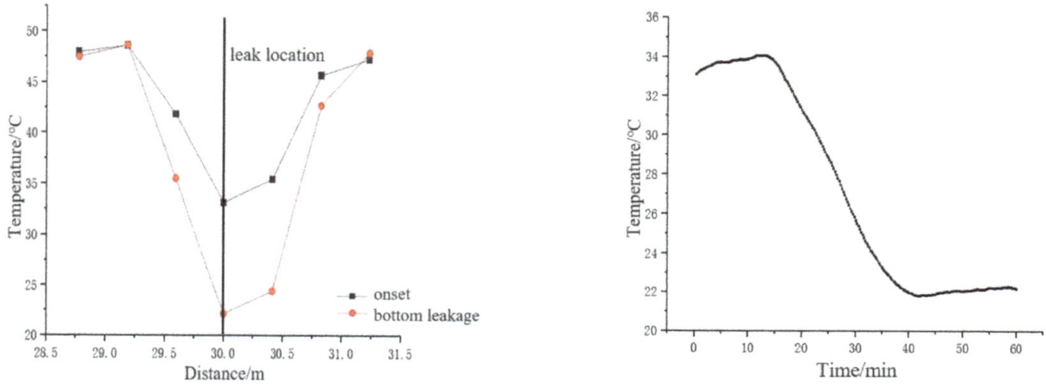

(**a**) (**b**)

Figure 8. Temperature change curve under heating power of 10 w/m (dry phosphorus slag): (**a**) Temperature distribution curve of the test section; (**b**) Temperature change curve at the leakage position.

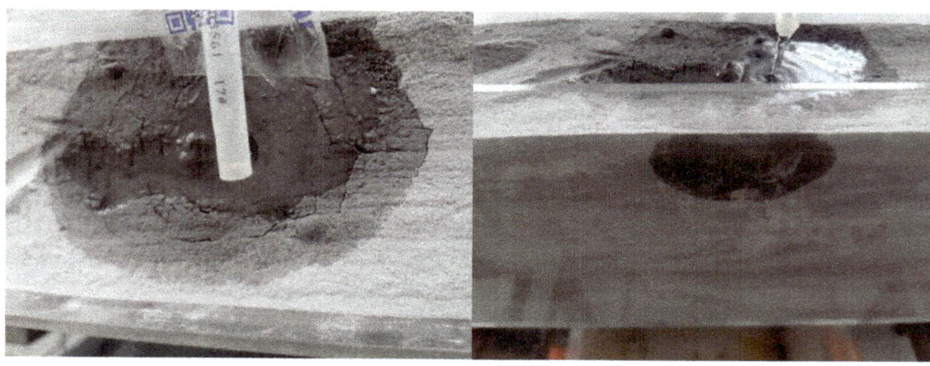

(**a**) (**b**)

(**c**) (**d**)

Figure 9. Temperature change phenomenon under a heating power of 10 w/m (dry phosphorus slag): (**a**) Early stage of leakage; (**b**) Water passing down; (**c**) The leakage spreads to the surroundings; (**d**) Phosphorus residue and water layer.

3.2. Comparison of the Temperature Measurement Results of Different Media in the Unsaturated State

In the environment of different soil moisture contents, the temperature change curves of the optical cable under different heating powers are shown in Figures 10 and 11. Figures 10 and 11 show that the temperature rise law of the optical cable in the unsaturated medium also gradually increases first and then gradually stabilizes; the temperature characteristic values of the optical cable in the unsaturated medium all increase with the increasing heating power, and the heating is stable. The time is approximately 15 min.

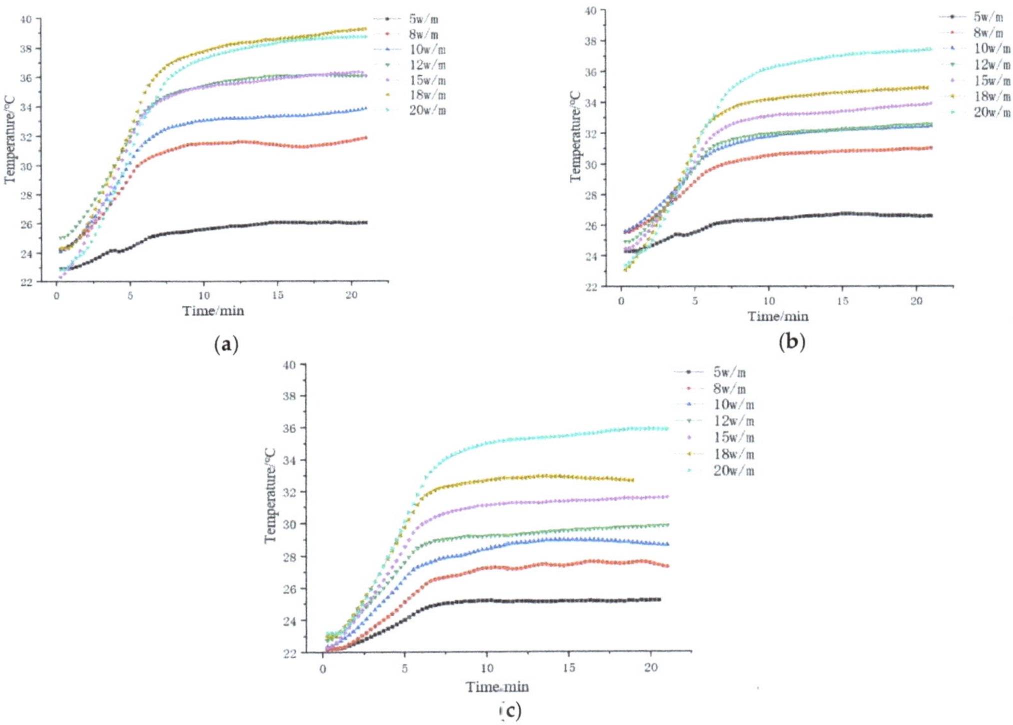

Figure 10. Temperature variation curve of sandy soil with different moisture contents under different heating powers: (**a**) Temperature curve of 8% moisture content; (**b**) Temperature curve of 14% moisture content; (**c**) Temperature curve of 20% moisture content.

The increases in the temperature of sand with different moisture contents and phosphorus slag under different heating powers are shown in Tables 2 and 3. The saturated moisture contents of sand and phosphorus slag were determined to be 24% and 34%, respectively. Table 2 shows that when the heating power is 12 w/m, the increase in temperature of the optical cable in the sand with an 8% moisture content and with a 20% moisture content differs by 4.36 °C, and there is a significant temperature difference. The minimum heating power for effectively identifying the leakage of the optical cable is 12 w/m in sandy soil with an 8% moisture content. When the heating power reaches 20 w/m, the temperature rise of the optical cable in the sand with a 14% moisture content and 20% moisture content is only 1.35 °C, and there is no obvious temperature difference. This result shows that under the sandy soil with 14% moisture content, a heating power within 20 w/m cannot effectively identify the leakage.

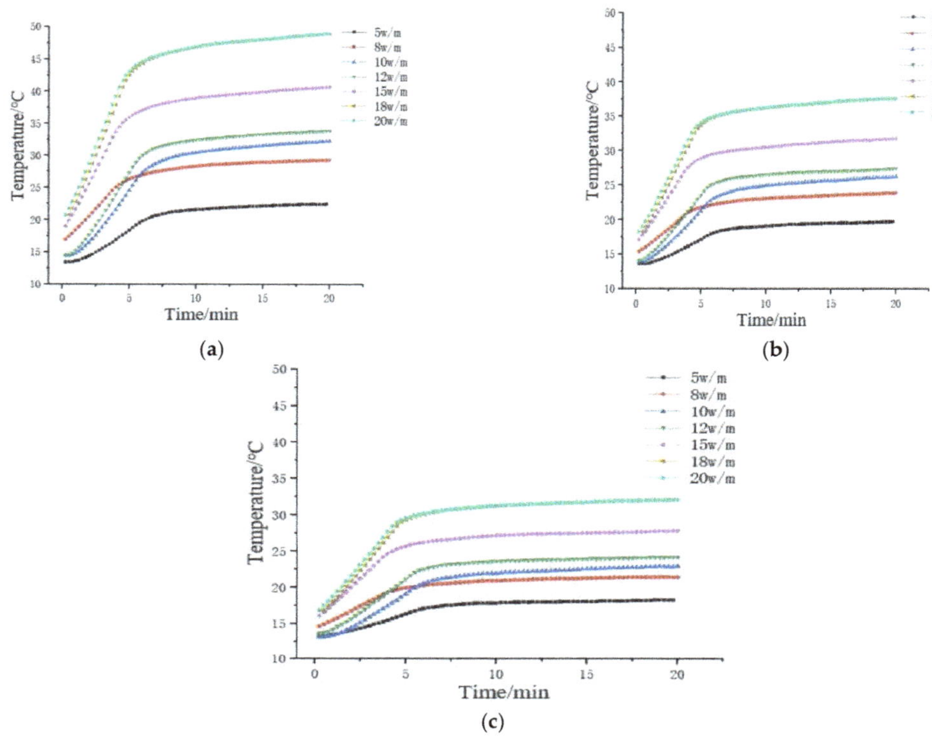

Figure 11. Temperature variation curve of phosphorus slag with different moisture content under different heating power: (**a**) Temperature curve of 14% moisture content; (**b**) Temperature curve of 24% moisture content; (**c**) Temperature curve of 34% moisture content.

It can be seen from Table 3 that when the heating power is 5 w/m, the increase in temperature of the optical cable in the phosphorus slag with a moisture content of 14% and a moisture content of 34% (saturated state) differs by approximately 4.05 °C, and there is a significant temperature difference. The minimum heating power for effectively identifying leakage is 5 w/m in phosphorus slag with a 14% moisture content. When the heating power reaches 12 w/m, the increase in temperature of the optical cable in the phosphorus slag with a 24% moisture content and a 34% moisture content (saturated state) differs by approximately 2.62 °C, and there is a significant temperature difference. The minimum heating power for effectively identifying the leakage of the optical cable is 12 w/m in phosphorus slag with a moisture content of 24%.

Table 2. Temperature increase in sandy soil with different moisture contents under different heating powers.

Heating Power (w/m)	8% Moisture Content Temperature Increase (°C)	14% Moisture Content Temperature Increase (°C)	20% Moisture Content Temperature Increase (°C)
5	3.10	2.45	2.94
8	7.39	5.29	5.27
10	9.25	6.64	6.60
12	11.03	7.31	6.80
15	13.58	9.08	8.94
18	14.29	11.54	9.90
20	15.52	13.69	12.31

Table 3. Temperature increase of phosphorus slag with different moisture contents under different heating powers.

Heating Power (w/m)	14% Moisture Content Temperature Increase (°C)	24% Moisture Content Temperature Increase (°C)	34% Moisture Content Temperature Increase (°C)
5	8.71	5.92	4.66
8	12.03	8.16	6.66
10	17.06	11.68	9.46
12	18.91	12.90	10.28
15	20.95	14.17	11.48
18	25.22	17.28	13.66
20	27.30	18.84	14.99

4. Monitoring Test and Quantitative Analysis of the Leakage Velocity of the Heatable Temperature-Sensing Optical Cable

4.1. Test Arrangement and Test Procedure

The test setup used to verify the quantitative analysis capability of leakage monitoring based on a heatable optical cable is shown in Figure 12. The seepage device is composed of a leakage box and a water valve, and a peristaltic pump is used to fill the water tank at a constant speed. The speed of water leakage is controlled by adjusting the water valve. There are three rows of holes at the bottom of the acrylic box buried in the sand, and a permeable cloth is placed in the bottom of the acrylic box. An inclined semicircular PVC pipe is placed below the box, and the measuring cylinder is placed at the end to measure the flow rate.

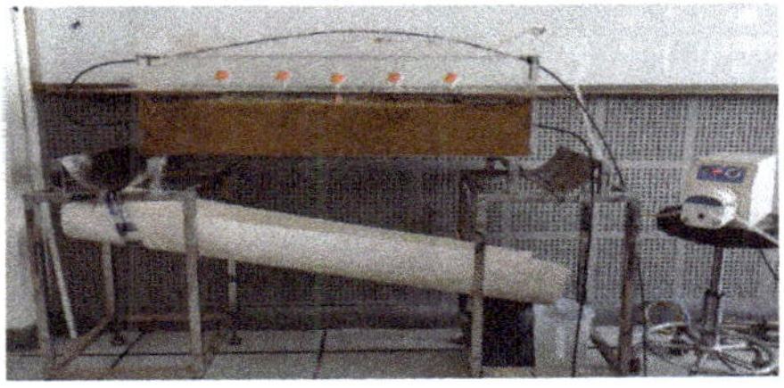

(a)

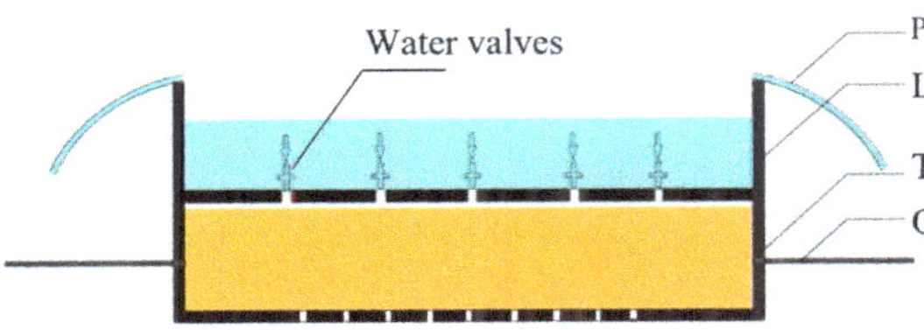

(b)

Figure 12. Leakage velocity monitoring test device: (a) The actual layout of the test device; (b) The schematic diagram of the test device.

The test steps are as follows:

(1) Fill the saturated sand medium in the acrylic box, bury the temperature-sensing optical cable in the acrylic box, and straighten and fix it;
(2) Connect one end of the optical cable and the jumper together through the fusion splicer, and connect the other end of the jumper to the port of the temperature calibrator. The other port of the temperature calibrator can be connected to the channel port of the DTS demodulator for data communication. Connect one end of the line to the temperature calibration box and the other end to the USB interface corresponding to the DTS host;
(3) Connect the copper wires at both ends of the heatable temperature-sensing optical cable to the DC power supply through external copper wires to form a heatable loop;
(4) Before the simulated leakage starts, check all equipment to ensure that it is operating normally and collecting data synchronously;
(5) Position the part of the optical cable in the acrylic box, which can be heated by a thermal drying gun; observe the real-time temperature data of the DTS demodulator and record the position data;
(6) After the temperature drops to a stable level, turn on the water valve switch and calculate the flow rate through the volume of water in the measuring cylinder at

the corresponding time. After the flow rate stabilizes, maintain the flow rate and apply different heating powers (5–10 w/m). Carry out heating, observe the real-time temperature data of the DTS demodulator, and stop heating after the temperature rises to a relatively stable level;

(7) After the temperature drops and stabilizes, change the flow rate by adjusting the water valve, and repeat step (6);

(8) Analyze the test data, compare and analyze the relationship between different flow velocities and temperature characteristic values, and establish a semiquantitative evaluation of leakage velocity.

4.2. Analysis of Test Results

The increases in temperature of the optical cable under different flow rates are shown in Table 4 and Figure 13. Figure 13 shows that there is an obvious negative correlation between the leakage velocity and the temperature characteristic value.

Table 4. Temperature characteristic values of different flow rates under different powers.

Heating Power (w/m)	Flow Velocity v (mm/s)					
	0.00	2.90	4.60	7.00	10.10	11.78
5	2.36	2.11	2.04	1.89	1.82	1.79
8	3.45	3.30	2.84	2.71	2.66	2.63
10	4.49	4.05	3.93	3.68	3.54	3.49

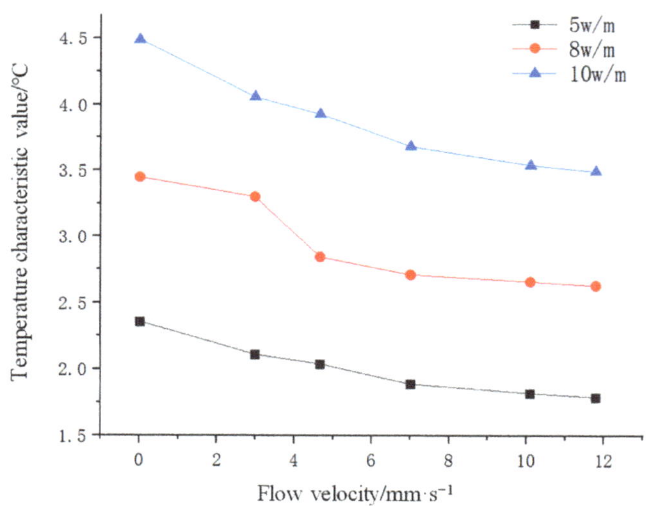

Figure 13. Temperature characteristic values at different flow rates.

To further analyze the correlation between the flow rate and the temperature eigenvalues, the data under three heating powers were fitted, as shown in Figures 14–16. Figures 14–16 show that polynomial fitting is more in line with the laws of the experimental data. The comparison shows that the quadratic polynomial fitting curve has a high degree of fitting to the test data and is more reliable.

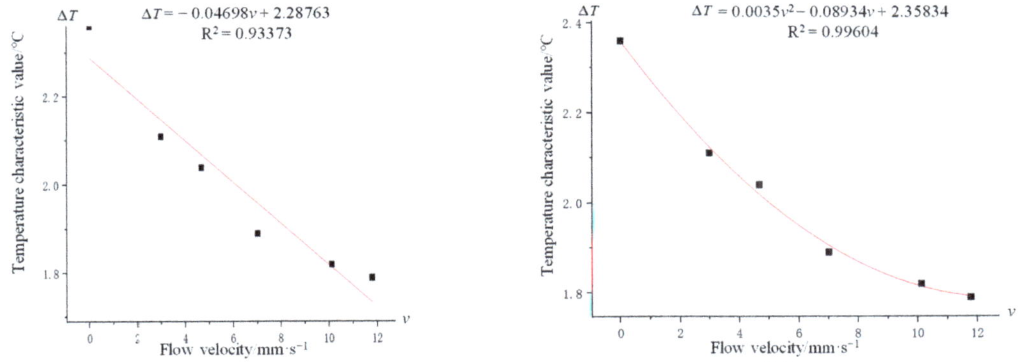

(a) (b)

Figure 14. The data of heating power 5 w/m corresponding to the results of different fitting methods: (**a**) Linear fit; (**b**) Polynomial fit.

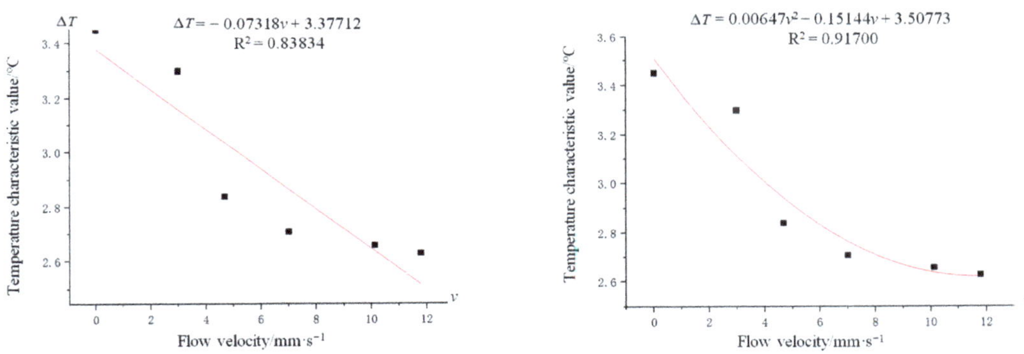

(a) (b)

Figure 15. The data for the heating power of 8 w/m correspond to the results of different fitting methods: (**a**) Linear fit; (**b**) Polynomial fit.

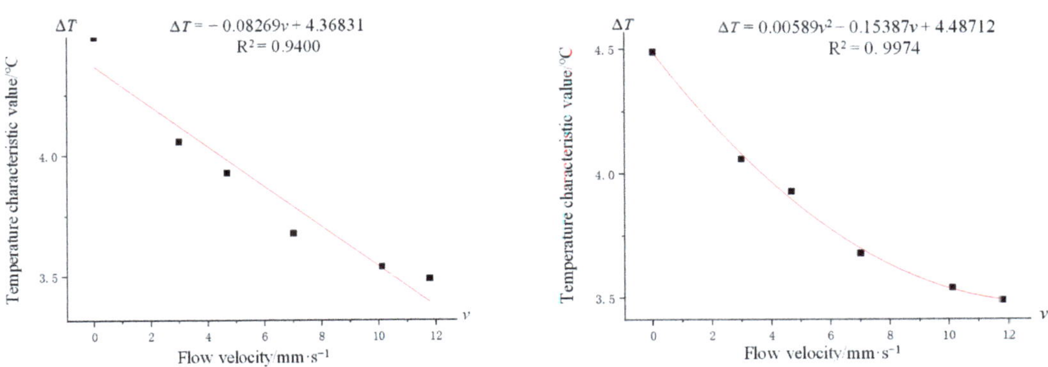

(a) (b)

Figure 16. The data of heating power 10 w/m corresponding to the results of different fitting methods: (**a**) Linear fit; (**b**) Polynomial fit.

5. On-Site In Situ Test

The manganese slag silo project in Songtao County, Guizhou Province, China, was selected for an on-site simulated leakage test to test the effectiveness of leakage monitoring using the heatable optical cable in an actual project. As shown in Figure 17, the test area is a square area of 40 m × 40 m, and the optical cables are arranged along the area in an S-shape. The test simulates leakage by using a water pump to inject water into the reserved PE (polyethylene) pipe. The PE pipe is 40 m long and has 6 holes. The length of the heated optical cable in the test area is approximately 840 m, and the average resistance value is 15 Ω/km.

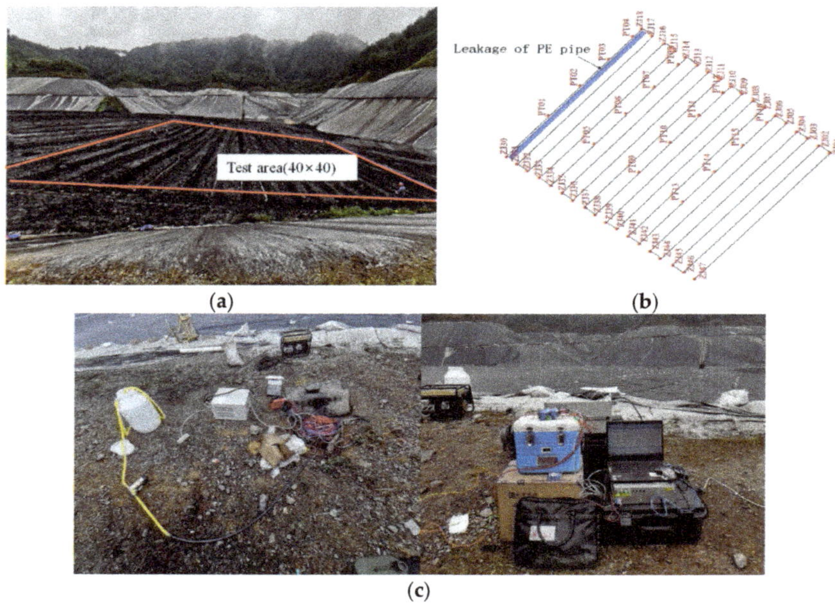

Figure 17. On-site in situ test layout: (**a**) Test area; (**b**) Distribution of optical cables and leakage pipes; (**c**) Test device.

During the test, the two types of optical cables (normal type and heating type) are connected to the equipment, and their normal operation is debugged and checked. As shown in Table 5, the test of working conditions I~IV is carried out in turn to observe the variation law of the temperature measurement value and the leakage identification effect of the two types of temperature-sensing optical cables under the same leakage conditions.

Table 5. Field test conditions.

Condition Number	Content
I	Heat the cable to a stable temperature with 2 w/m heating power (the maximum power that the field device can provide)
II	Heat the cable at 2 w/m until the temperature stabilizes to simulate leakage
III	Stop heating and wait for the temperature to drop and stabilize, then reheat the optical cable with 2 w/m heating power to stabilize
IV	Simulate leakage again after heating stabilizes

The test results of working Condition I are shown in Figure 18. Due to the power limitation of the equipment, the overall temperature of the heatable temperature-sensing optical cable after heating is only approximately 1 °C higher. The test results of working condition II are shown in Figures 19 and 20. The temperature of the heatable cable drops significantly at the position of 880–920 m, with a drop of approximately 2–3.22 °C, while the temperature of the ordinary cable also drops at the position of 880–920 m, with a drop of 1.49–2.35 °C, which is 1.49–2.35 °C compared to the heatable cable. The temperature gradient of the heatable cable before and after the leakage is not obvious, mainly because the heating power is low, resulting in an insignificant heating effect, indicating an insignificant temperature gradient after leakage occurs, and the reason for the temperature change in the two optical cables is mainly due to the water temperature. The main reason for the temperature changes in the two types of optical cables is the temperature difference between the water and the soil, but it can be seen that the temperature gradient has expanded by heating.

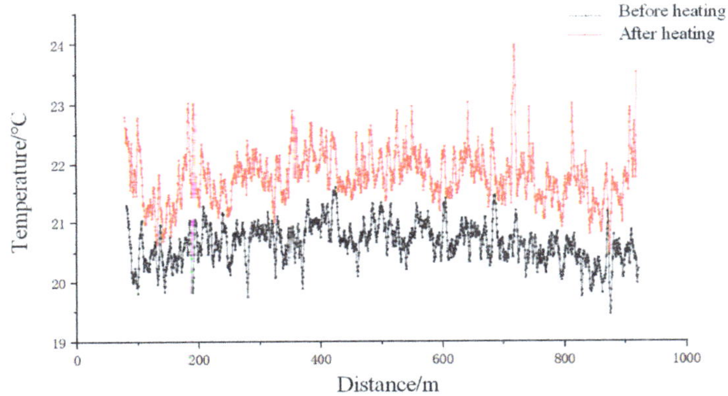

Figure 18. Temperature change in the optical cable before and after heating in working Condition I.

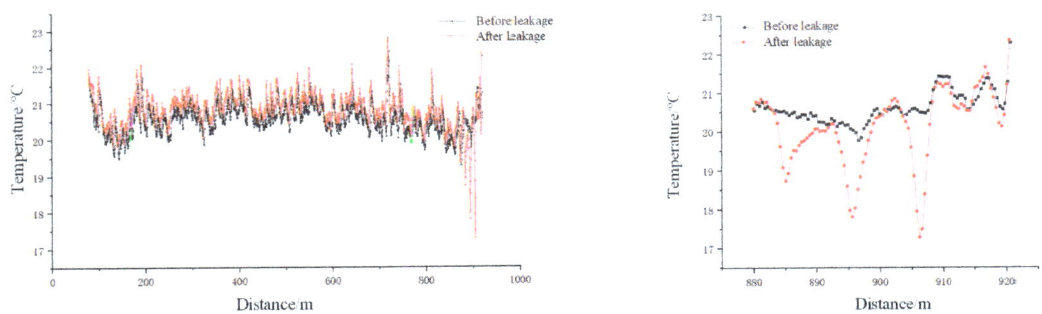

(a) (b)

Figure 19. Temperature change in the heatable optical cable under working condition II: (**a**) Temperature distribution along the route; (**b**) Temperature distribution in the leakage section.

The results of working condition III are shown in Figure 21. The temperature change in the leaking section is larger when the optical cable is heated twice, before and after leaking, but the temperature change is smaller because the heating effect is not obvious. The results of Case IV are shown in Figures 22 and 23. Consistent with the results of working condition II, the temperature dropped significantly at the leakage position. The temperature drop of the ordinary optical cable was 0.59–1.81 °C, while the temperature drop of the heatable cable reached 1.89–3.82 °C. The higher the value is, the better the recognition effect.

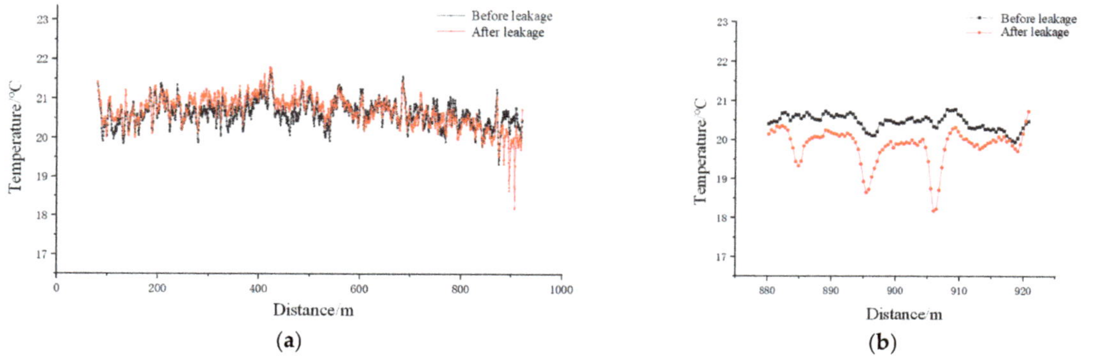

Figure 20. Temperature change in the ordinary optical cable in working condition II: (**a**) Temperature distribution along the route; (**b**) The temperature distribution of the leakage section.

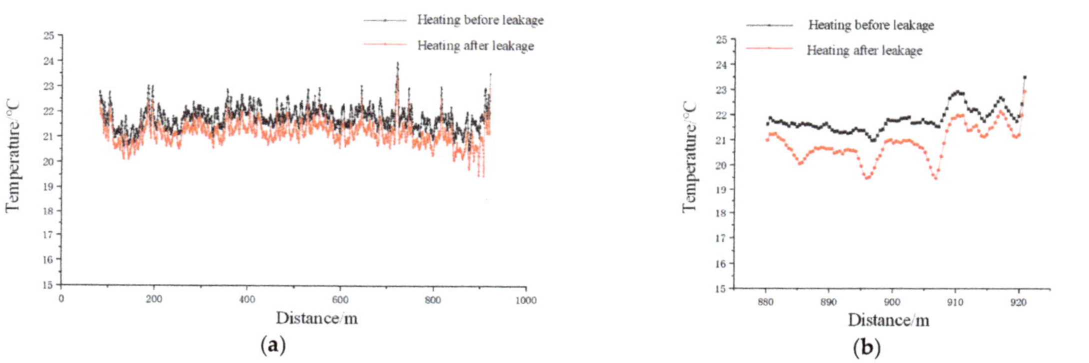

Figure 21. Temperature comparison of the two heating times before and after the leakage of the heatable optical cable under working condition III: (**a**) Temperature distribution along the route; (**b**) Temperature distribution of leakage section.

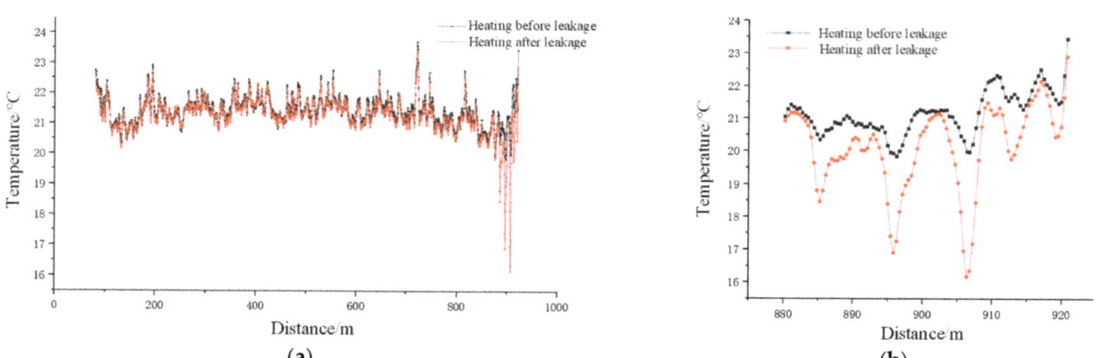

Figure 22. Temperature change in heatable optical cable under working condition IV: (**a**) Temperature distribution along the route; (**b**) Temperature distribution of leakage section.

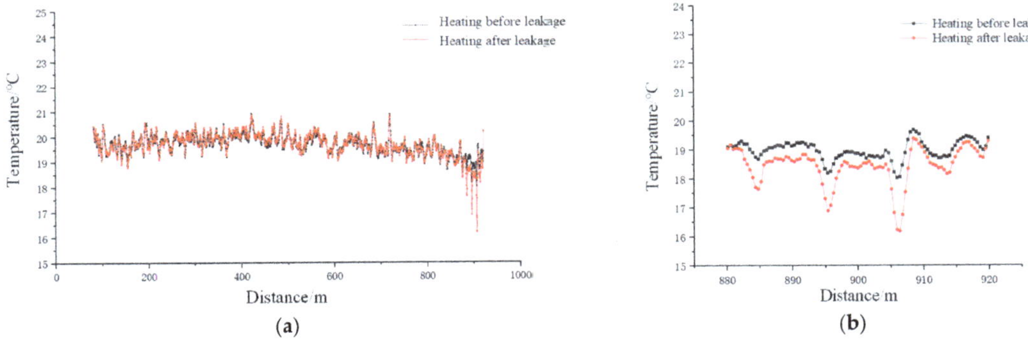

Figure 23. Temperature change in ordinary optical cable under working condition IV: (**a**) Temperature distribution along the route; (**b**) Temperature distribution of leakage section.

6. Discussion

(1) In the experiment described in this paper, the permeability of phosphorus slag was worse than that of sand, and its response to leakage events was slower. Therefore, when laying temperature measurement cables, placing the cables in soil layers with good permeability or setting permeable soil layers around the cables when burying them should be considered to enhance their detection accuracy and response speed to leakage events.

(2) Through the experiments conducted in this study, it was found that the temperature characteristics of optical cables in sand and slag with different moisture contents increased with the increase in heating power. Moreover, the temperature increase in the optical cable at the same power level basically decreased with the increasing moisture content. Specifically, when the moisture content of the sand medium itself is less than 8%, leakage occurs, and the minimum heating power for effectively identifying leakage using the heated optical cable is 12 w/m. When the moisture content of the sand medium itself is greater than 14%, leakage occurs, and a heating power within 20 w/m cannot effectively identify the leakage. When the moisture content of the slag medium itself is less than 14%, leakage occurs, and the minimum heating power for effectively identifying leakage using the heated optical cable is 5 w/m. When the moisture content of the slag medium itself is between 14% and 24%, leakage occurs, and the minimum heating power for effectively identifying leakage using the heated optical cable is 12 w/m. The heating power of the heated optical cable is directly related to its heating effect. If the heating power is too low, it may not meet practical needs, and if the heating power is too high, it will result in energy waste and overheating, which may cause damage to the facility. Currently, scholars have paid little attention to this issue, and in practical engineering, reasonable heating power can be selected through indoor model experiments to provide a basis for actual engineering measurement.

(3) The establishment of a quantitative relationship between seepage velocity and temperature characteristic values is the basis for the quantitative monitoring of seepage. Currently, research in this field involves conducting measurement experiments in indoor models and then applying the analysis results of the indoor experimental data to actual engineering. This paper also adopts a similar approach, and the analysis of experimental data shows that the quadratic polynomial fitting curve has high reliability.

7. Conclusions

Through the experimental study of the feasibility of the heatable temperature-sensing optical cable in karst depression engineering applications, the following conclusions were drawn:

(1) The leakage test results of the heatable temperature-sensing optical cable in different dry and unsaturated media show that the leakage identification effect of the heatable temperature-sensing optical cable is very obvious, which verifies the feasibility and precise positioning function of leakage identification of optical cables in dry media. Because phosphorus slag has worse water permeability than sand, the response speed to leakage events is slower.

(2) In the unsaturated medium test, the difference in moisture content of the medium can be identified by heating the optical cable, and increasing the heating power can increase the identification effect in the range of moisture content identified.

(3) In the semiquantitative identification test of leakage velocity, the temperature characteristic value under different conditions was measured by controlling the flow rate and heating power. The results showed that there was a significant negative correlation between leakage velocity and temperature characteristic value. The greater the flow rate is, the higher the temperature. The smaller the eigenvalue is, the more stable it is. By comparing the data, the quadratic polynomial fitting curve has high reliability.

(4) The temperature behavior of two types of temperature-sensing optical cables (conventional type and heating type) under the same leakage conditions was studied through in situ tests. The degree of response of the heatable fiber optic cable to leakage is very significant, and the result is accurately identifying the location of the leakage.

Author Contributions: B.Y.: Conceptualization, Methodology, Writing—Original Draft; C.S.: Visualization, Writing—Review and Editing; H.P.: Data Curation, Investigation; F.G.: Resources, Formal analysis. All authors have read and agreed to the published version of the manuscript.

Funding: This research is supported by the major research project of Power China Guiyang Engineering Corporation Limited (YJZD2020-02).

Data Availability Statement: The data is confidential (provided upon request).

Acknowledgments: The authors are grateful to all participants for their efforts.

Conflicts of Interest: The authors declare no conflict of interest.

References

1. Zhao, Z.R.; Zhang, L.Z. Research on seepage safety monitoring technology for dams. *Water Resour. Hydropower Eng. Des.* **2001**, *3*, 44–48.
2. Yan, W.J. Technical improvement of dam seepage observation by measuring weir method in Yantan Hydropower Station. *Hongshui River* **2016**, *35*, 76–79.
3. Wang, H.B. Application of high density resistivity method in the detection of embankment hidden danger. *J. Zhejiang Univ. Water Resour. Electr. Power* **2010**, *22*, 31–34. [CrossRef]
4. Duan, X. Research on dam leakage detection based on temperature tracer method. *Sci. Technol. Eng.* **2014**, *14*, 140–143.
5. Fang, C.G.; Ge, H.G.; Jia, Y.M. Technical problems of transient electromagnetic method used in the detection of hidden dangers of dike leakage. *Dam Obs. Geotech. Tests* **2001**, *5*, 21–24.
6. Ren, A.W.; Ke, B.R.; Cheng, J.S. Cause analysis and non-destructive testing verification of reservoir leakage in karst area. *J. Hydraul. Eng.* **2014**, *45*, 119–124.
7. Aufleger, M.; Goltz, M.; Dornstadter, J. Distributed fiber optic temperature measurements in embankment dams with central core new benchmark for seepage monitoring. In Proceedings of the International Symposium on Dams and Reservoirs under Changing Challenges during the 79th Annual Meeting of the International Commission on Large Dams, Lucerne, Switzerland, 1–3 June 2011.
8. Aufleger, M.; Conrad, M.; Goltz, M. Innovative dam monitoring tools based on distributed temperature measurement. *Jordan J. Civ. Eng.* **2007**, *1*, 29–37.
9. Li, R.Y.; Chen, X.P.; Wang, Z.W. Application of temperature indication method in seepage monitoring of Yangtze River levee. *J. Yangtze River Sci. Res. Inst.* **2000**, *17*, 48–57.

10. Li, Y.X. Study on Multiparameter Monitoring of Embankment Based on Distributed Fiber-Optic Sensing Technology. Master's Thesis, Nanjing University, Nanjing, China, 2021. [CrossRef]
11. Wang, Y.H.; Cai, D.S.; Xiao, H.L. Application of DTS in leakage monitoring of concrete face rockfill dam. *China Rural Water Hydropower* **2006**, *8*, 101–102. [CrossRef]
12. Wei, D.R.; Zhao, H.C.; Qin, Y.T. Leakage monitoring technology based on optical fiber temperature measurement. *Zhejiang Hydrotech.* **2004**, *2*, 19–21. [CrossRef]
13. Qin, L.T.; Liu, J.M.; Xia, X.P. Distributed optical fiber temperature monitoring system and its application. *Dam Saf.* **2004**, *01*, 45–48. [CrossRef]
14. Tan, Z.W.; Hu, L.Z. Study on Optical Fiber Seepage Monitoring Technology of the Vertical Core Rockfill Dam. *Water Power* **2011**, *37*, 83–85. [CrossRef]
15. Harvie, W.B.; Peppler, R.A. Sensor-Enabled Geosynthetics Monitoring Stability and Leak Detection in Earthen Construction Technical Paper #159. In Proceedings of the World of Coal Ash Conference, Lexington, KY, USA, 22–25 April 2013.
16. Wang, Z.H. Study on Seepage Monitoring Method of Soil and Stone Dam Based on Distributed of Optical Fiber Temperature Measurement Technology. Ph.D. Thesis, Xi'an University of Technology, Xi'an, China, 2019.
17. Thiele, E.; Helbig, R.; Krebber, K. Dike Monitoring. In Proceedings of the 4th International Symposium on Flood Defense, Managing flood Risk Reliability and Vulnerability, Toronto, ON, Canada, 6–8 May 2008.
18. Artieres, O.; Galian, M.; Royet, P. Fiber optics monitoring solution for canal dykes. In Proceedings of the Planc Mmx Congress, Liverpool, UK, 10–14 May 2010.
19. Vogt, T.; Schneider, P.; Woernle, L.H. Estimation of seepage rates in a losing stream by means of fiber-optic high-resolution vertical temperature profiling. *J. Hydrol.* **2010**, *380*, 154–164. [CrossRef]
20. Beck, Y.L.; Cunat, P.; Reboud, M. Improvement of leakage monitoring in dikes by the use of distributed fiber optics sensors. In Proceedings of the 6th International Conference on Scour and Erosion, Paris, France, 27–31 August 2012.
21. Deng, X.W. Model Test of Leakage Monitoring Based on Distributed Optical Fiber Temperature Sensing Technology. Master's Thesis, Hubei University of Technology, Wuhan, China, 2011.
22. Liu, T.; Sun, W.J.; Kou, H.L. Experimental Study of Leakage Monitoring of Diaphragm Walls Based on Distributed Optical Fiber Temperature Measurement Technology. *Sensors* **2019**, *19*, 2269. [CrossRef] [PubMed]

water

MDPI

Article

A Prediction Model and Factor Importance Analysis of Multiple Measuring Points for Concrete Face Rockfill Dam during the Operation Period

Lei Shao [1], Ting Wang [2,*], Youde Wang [3], Zilong Wang [2] and Kaiyi Min [2]

1 National Engineering Research Center for Complex System Simulation Technology Application, (Aerospace Science and Industry Simulation Technology Co., Ltd.), Beijing 100000, China
2 State Key Laboratory of Eco-hydraulics in Northwest Arid Region, Xi'an University of Technology, Xi'an 710048, China
3 Sinohydro Engineering Bureau 4 Co., Ltd., Xi'ning 810000, China
* Correspondence: b22009@xaut.edu.cn

Abstract: Dam settlement monitoring is a crucial project in the safety management of concrete face rockfill dams (CFRD) over their whole life cycle. With the development of an automatic monitoring system, a large amount of settlement data was collected. To precisely predict the structural health of dams, a combined multiple monitoring points (MMP) model and a machine learning model has been developed. In this paper, based on the physical factors of the CFRD, we comprehensively analyzed the influence of water level load transfer, rockfill rheology and soil properties on the settlement during the impoundment operation period. Then, we established a space-time distribution model of the CFRD during its operation period under multiple factors. An extreme gradient boosting (XGBoost) model was used for fitting prediction, and the model was evaluated using various performance indicators. The results show that spatial parameters such as the upper filling height, rockfill thickness, panel-point distance and soil material correlate to the deformation characteristics of the rockfill dam. Taking the monitoring data of the settlement of the Liyuan CFRD as an example, the new MMP model was evaluated and used to predict the settlement of the full-section points with higher accuracy, which has certain application and popularization value for related projects. Then, to evaluate the contribution of the components of the new MMP model, the SHapley Additive explanation (SHAP) methods are used to evaluate the importance of the selected factors, and the reasonability of these factors is verified.

Keywords: multiple measuring points; concrete face rockfill dam; XGBoost; dam settlement monitoring; factor importance analysis

Citation: Shao, L.; Wang, T.; Wang, Y.; Wang, Z.; Min, K. A Prediction Model and Factor Importance Analysis of Multiple Measuring Points for Concrete Face Rockfill Dam during the Operation Period. *Water* **2023**, *15*, 1081. https://doi.org/10.3390/w15061081

Academic Editor: Chin H Wu

Received: 29 January 2023
Revised: 3 March 2023
Accepted: 9 March 2023
Published: 11 March 2023

1. Introduction

Settlement deformation monitoring is a crucial project in the safety management of concrete face rockfill dams (CFRD) over their whole life cycle [1]. Performing systematic monitoring and data prediction of the structural state of the dam is beneficial to dam managers in that it helps them understand and accurately evaluate the running state of the dam in a timely manner, provide early warnings and implement maintenance measures [2,3]. With the development of automatic monitoring systems, the number of dam deformation monitoring sensors has increased greatly. The method of using deformation data to evaluate dam structure health has gradually changed from a single monitoring point model to a combination of a multiple monitoring points (MMP) model and a machine learning (ML) model [4–7]. Therefore, it is of great scientific significance to research a new model of the settlement of CFRDs to gain control of the overall deformation trends of these dams.

The traditional MMP model has been used to evaluate the deformation of CFRDs. Sun et al. [8] applied the MMP model to the prediction of settlement displacement during the operation period of a CFRD and verified the applicability of the model. However, the function expression was not improved according to the deformation characteristics of CFRD, and the continuous influence of rheology and soil compaction on the settlement was ignored. This is because the traditional MMP model is derived from a simplified model of the concrete dam, where the concrete dam's body is assumed to be a rigid structure. A simple model is not perfectly suitable to represent a CFRD, for a CFRD is a complex of elasticity, plasticity and viscosity. Given the complicated model factors and modeling, Huang et al. [9] and Chai et al. [10] used the Back Propagation (BP) algorithm, which is a learning algorithm based on the gradient descent method and suitable for multilayer neuronal networks, to replace the combination of a statistical model and MMP model. They also expanded the sample space using the monitoring data of multiple measuring points to build the corresponding displacement monitoring model and reflect the overall deformation of the dam. Li et al. [11] and Lu et al. [12] took into account the factors influencing rockfill settlement and derived the spatiotemporal function expression for the instantaneous and continuous soil deformation caused by filling during the construction period. However, during the storage period of the rockfill dam, the settlement displacement is mainly composed of a water level component and a time component, so it is necessary to further analyze the influence of hydraulic pressure on the settlement during the operation period based on the original model [13,14]. Therefore, the expressions of multiple measuring points conforming to the deformation law of a rockfill dam need to be derived according to the deformation characteristics of the CFRD.

In the past few decades, machine learning has made great advances to solve problems in the field of civil engineering, such as predicting the material properties of concrete and monitoring the health of structures [15]. The artificial neural network algorithm is the earliest machine learning algorithm [16]. Artificial neural networks are used to predict the shear strength of concrete beams reinforced by composite materials. Kim and Kim [17] adopted the artificial neural network method to establish an artificial intelligence prediction model to predict the relative settlement of the dam top based on the measured settlement data of 30 faced rockfill dams. Support vector regression was then developed to effectively solve nonlinear regression estimation problems. Su et al. [18] proposed a method of optimizing algorithm parameters and input vectors and a new method for updating model monitoring in real-time based on the dam safety monitoring model based on support vector machines (SVM), which improved the accuracy of prediction. Furthermore, the Support Vector Regression (SVR) algorithm is often combined with other algorithms, such as Fuzzy Logic algorithms (FL) and Genetic Algorithms (GA), etc., to improve the training speed and prediction accuracy. Marandi et al. [19] predicted the settlement of faced rockfill dams using a genetic programming algorithm. Moreover, the decision tree algorithm benefits from its straightforwardness, observability and interpretability, providing a convenient and reliable solution to the regression problem. Decision tree generation algorithms include the ID3 algorithm, the C4.5 algorithm and the Classification and Regression Tree (CART) algorithm. The Gradient Boosting Decision Tree (GBDT) is the most representative decision tree algorithm and is combined with ensemble learning. XGBoost (extreme gradient boosting) is an ensemble learning method based on the GBDT model optimization proposed by Chen and Guestrin [20], which can be widely used in machine learning challenges and data science applications [21]. Lim and Chi [22] adopted the XGBoost algorithm to evaluate the degree of damage done to a bridge and used the Shapley value to evaluate the influence of each parameter in the input variable on the predicted results. Shi et al. [23] employed an XGBoost model to predict the stability of a landslide dam considering missing data and show that the XGBoost model can improve model accuracy compared with the rapid evaluation methods widely used at present. Wakjira and his team [24,25] presented a total of seven ML models, namely kernel ridge regression, K-nearest neighbors, support vector regression, classification and regression trees, random forest, gradient boosted trees and XGBoost, to choose the best predictive model for fabric-reinforced, cementitious matrix-strengthened beams. They show that the XGBoost model is the most accurate model with the highest coefficient of determination. Nguyen et al. [26] applied the XGBoost model, an artificial neural network (ANN) and random forest (RF) to the prediction of the punching shear resistance of reinforced concrete (R/C) interior slabs without shear reinforcement. Their results show that the XGBoost model presented the most accurate prediction among all models, with the coefficient of determination (R^2) for the testing dataset being equal to 0.9578.

XGBoost is composed of the integration of several weak learners. The basic idea is to fit the residual of the previous weak learner's training by adding new weak learners and obtaining the predicted score of each sample at the end of the training. Finally, the predicted scores of all weak learners are added together, namely the predicted value of the sample. XGBoost has undergone a lot of optimization based on the Boosting model, which includes: (1) performing second-order Taylor expansion on the objective function, which improved the model accuracy; (2) adding the regularization term to the objective function to reduce the complexity and effectively prevent overfitting; and (3) the sample automatically learning the splitting direction of the missing value to process the missing value. XGBoost further upgrades Bagging and Boost's initial integrated algorithm. XGBoost also has certain improvements in sample selection, parallel computation and missing value handling, etc. [27]. Together, these improvements make XGBoost show significant advantages in computational efficiency and prediction results, making it one of the most popular algorithms nowadays [26]. Compared with other decision tree algorithms, the XGBoost algorithm has unique advantages, such as the introduction of penalty functions and random variables for decision tree structure to reduce the overfitting phenomenon of the model [24,25,28–31]. At the same time, each decision tree of the XGBoost model is completely independent, which enables the XGBoost algorithm to have superparallel computing capabilities [20]. Since the multi-factor MMP model contains discrete characteristic variables and requires precision analysis of the model, we adopt XGBoost to construct the MMP model.

Factor importance analysis can reflect the reasonability of components of the MMP model and assist in screening these factors to make the model more robust [18]. Considering that XGBoost cannot determine what kind of correlations (positive or negative) exist between these factors and the settlement, we further use the SHAP method to conduct the factor importance analysis. SHAP is a model interpretation package developed based on Python that can interpret the output of machine learning models. The basic idea is to calculate the marginal contribution the factors bring into the model. Then, taking into account the different marginal contributions of the factor in all cases, the SHAP value is the contribution value assigned to each factor in the sample. The SHAP method can realize a visual analysis and explain the value of the model from the characteristic factors [32]. As an explainable machine learning model, the use of a unified SHAP method was investigated to explain the predicted response and rank the input factors and their interactions [24,25].

In this paper, the influence of the load transfer of the water level, the flow of the rockfill body and the characteristics of the soil on the settlement during the operation period are comprehensively analyzed. The spatial parameters which are more consistent with the deformation characteristics, such as the upper fill elevation, the thickness of the rockfill and the distance between the measuring point and the face plate, are used to replace the original position coordinates. Based on this, the MMP model of the rockfill dam during the operation period under the action of multiple factors is established. It is proven that the combined model of the new MMP model and the XGBoost algorithm has a higher prediction accuracy regarding the CFRD using the actual engineering data. Then, the factor importance analysis was conducted to figure out the orders of importance of these selected factors. This work has certain reference values for the safety monitoring of dams.

2. Materials and Methods

2.1. The Multi-Factor and Multi-Monitoring Point Statistical Model

At present, the main statistical model for dam deformation monitoring is hydrostatic-seasonal-time (HST), which uses the function expression of time, reservoir water level and periodic temperature to fit the dam deformation rule. The HST model is presented as the sum of the water pressure component (δ_H), temperature component (δ_S) and time effect component (δ_T) [3,11]:

$$\delta(h,t,s) = \delta_H(h,t) + \delta_S(s) + \delta_T(t) \tag{1}$$

Based on the original HST model, the mathematical expressions of the water pressure component δ_H, the temperature component δ_S and the time effect component δ_T are improved. The rheological component $\delta\varepsilon$ and the material component δ_m are added.

2.1.1. The Water Level Component δ_H

The water level component δ_H is the sum of the water level component of settlement measurement points δ'_h and the pre-reservoir water level component $f(h)$:

$$\delta_h = \delta'_h + f(\overline{h}) = b_0 + b_1 \cdot Ah^{1-n} \cdot \frac{d_1}{d_2} + b_3\overline{h} \tag{2}$$

where b_0, b_1, b_2 and b_3 are the coefficients of regression; h is the elevation difference between the water level and the measuring point; d_1 is the rockfill thickness; d_2 is the distance from the measuring point to the face panel (Figure 1); $\overline{h}$ represents the average water levels in the 3 days before the observation date; n is the modulus of elasticity index; and A is a constant.

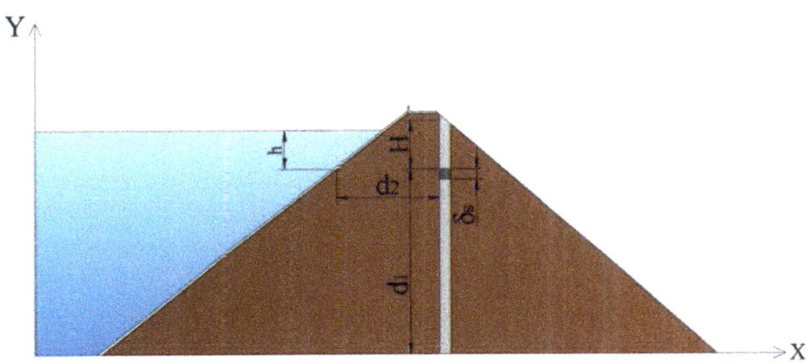

Figure 1. Schematic diagram of monitoring point affected by rheology and soil weight.

2.1.2. The Temperature Component δ_S

The relationship between the deformation of rockfill body δ_S and temperature with the annual periodicity of the freezing period can be expressed by using the periodic function:

$$\delta_S = \sum_{i=1}^{2} \left(\cos\frac{2\pi it}{365} + \sin\frac{2\pi it}{365}\right) \tag{3}$$

where t is the cumulative number of days from the monitoring date to the starting date.

2.1.3. The Time Effect Component δ_T

According to the monitoring data of the CFRD, the settlement in the late settlement period tends to converge gradually and eventually becomes stable. The settlement vs. per unit dam height (the proportion of settlement amount to dam height) has a linear relationship with the logarithm of time, as follows:

$$s = \frac{\alpha H'}{100}(\ln\theta - \ln\theta_0) \tag{4}$$

where α is the regression coefficient; H' is the height of the dam; $\theta = t/100$; θ_0 = the cumulative number of days since the selected monitoring date/100; and n' is the soil porosity at the measurement point.

2.1.4. The Rheological Component δ_ε

The expression of the rheological component of the dam body should be combined with the rewritten formula of the vertical compression modulus [33] and the rheological curve, as shown below:

$$\delta_\varepsilon = a_0 + a_1(1 - e^{-Dt})\frac{\gamma H d_1}{E_{rc}} \quad (5)$$

where a_0 and a_1 are the coefficients of regression; D is the initial relative deformation rate; γ is the bulk density of the filled rockfill; H is height of filled rockfill above the measuring point; E_{rc} is the tangential modulus of the rockfill and according to the Duncan–Chang model; and E_{rc} can be presented as:

$$E_{rc} = KP_a(\frac{\sigma_3}{P_a})^n[1 - \frac{R_f(\sigma_1 - \sigma_3)(1 - \sin\varphi)}{2c\cos\varphi + 2\sigma_3\sin\varphi}]^2 \quad (6)$$

where R_f is the failure ratio; σ_1, σ_3 are the large and small principal stresses, respectively; c is the cohesion force; φ is the friction angle; K is the tangent elastic modulus; and P_a is the atmospheric pressure.

Then, substituting Equation (6) into Equation (5), we can obtain the expression δ_ε:

$$\delta_\varepsilon = a_0 + a_1(1 - e^{-Dt})A_1 H^{1-n} \cdot d_1 \quad (7)$$

where n is the elastic modulus index; ξ is the lateral pressure coefficient, and

$$A_1 = \frac{\gamma^{1-n}}{[1 - \frac{R_f(1-\xi)(1-\sin\varphi)}{2\xi\sin\varphi}]^2 \cdot K \cdot \xi^n} \quad (8)$$

2.1.5. The Material Component δ_m

In this paper, three material parameters, namely, the soil porosity at the measurement point n', the coefficient of compact permeability k, and the dry density ρ are brought into the regression analysis according to discrete variables. It is assumed that the functional relationship between a certain point settlement and its material parameters is as follows:

$$\delta_m = f(n', k, \rho) \quad (9)$$

where f is the regression formula.

To sum up, the statistical model expression of the settlement of the rockfill during the operation period is established, which is the sum of the above components:

$$\begin{aligned}\delta = \delta_h + \delta_S + \delta_T + \delta_\varepsilon + \delta_m = a_1(1 - e^{-Dt})A_1 H^{1-n} \cdot d_1 + b_1 h^{1-n} \cdot \tfrac{d_1}{d_2} \\ + b_2(1 - e^{-Dt})h + b_3\overline{h} + c_1(\ln\theta - \ln\theta_0) + d_1 f(n', k, \rho) + \alpha\end{aligned} \quad (10)$$

It should be noted that the temperature component is very small for the settlement of the mearing points, so it was not included in the final statistical model. According to the multi-factor MMP model, it can be found that the settlement deformation of the rockfill body is related to the three position factors of the upper fill elevation (H), the rockfill thickness (d_1) and the distance from the measuring point to the face panel (d_2). These three parameters can be used to represent the position coordinates of any point in the dam, so they can be used as variables to represent the spatial position. When the reservoir water level elevation is fixed, the deformation at different points is also related to the filling material at the location. These parameters are used to represent the comprehensive influence of the rockfill crushing characteristics, compression deformation properties or other factors in different rockfill areas, to explain the reasons for the differences in settlement values at different measuring points under the same external environmental conditions.

2.2. XGBoost Model for Multiple Monitoring Points Model

When the XGBoost prediction model of the MMP model for the settlement of CFRD is established, the input variables are 13 influence factors affecting the settlement, which are obtained from the statistical model (Equation (10)) of the running period of the rockfill. For the specific input variables, see Table 1. The output variable is the settlement value of 12 measuring points in the whole section. To show the superiority of the XGBoost model, we also perform the prediction using the base learner, which is a classification and regression tree (CART). A comparison between the performance of the XGBoost model and the CART model is shown to indicate the necessity of using the XGBoost model over simpler white box models.

Table 1. Input variable set.

	Water Level Component (δ_H)	Rheology–Soil Weight Component ($\delta\varepsilon$)	Time Effect Component (δ_T)	Material Component (δ_m)
Factors	$h^{1-n}(d_1/d_2), \bar{h}(1-e^{-Dt})h$	$(1-e^{-Dt})H^{1-n}\cdot d_1 d_1, d_2, H, Hd_1$	$\ln\theta, \ln\theta_0$	n', k, ρ

2.3. Hyperparameter Optimization and Performance Measures

During model training, it is necessary to optimize the hyperparameters in the machine learning algorithm. Common hyperparameter optimization methods include grid search [29,34], random search, Bayesian optimization [35], etc. Grid search requires traversing all hyperparameter combinations, which consumes a lot of computational resources, which are already limited, and model training is slow. The random search method randomly selects different combinations of hyperparameters with great randomness. Bayesian optimization is a method of automatic adjustment of hyperparameters, which will track the evaluation results of every combination of hyperparameters tried in the past and form a probabilistic model. This model is called the "proxy" of the objective function. Before trying the next set of hyperparameters, the Bayesian optimization method will refer to this proxy model and select the hyperparameter with the best performance on the proxy function to evaluate the actual objective function. Compared with grid search and random search, Bayesian optimization has better performance and can reduce the computing time. In this paper, the method of Bayesian optimization combined with K-fold cross-validation is used to optimize the hyperparameters of the model. Using at the hyperparameter values selected by the Bayesian optimization algorithm, the K-fold cross-validation method was used to evaluate the performance of the model under the selected combination of hyperparameters. When the hyperparameters are optimized by Bayesian optimization, the domain space is the value range of the hyperparameter to be searched, and the objective function is the evaluation index value of the model's prediction performance on the verification set using the current combination of hyperparameters. The specific steps are as follows:

(1) Establish the alternative probability model of the objective function: some hyperparameters are randomly generated in the domain space, and K-fold cross-validation is used to train and evaluate the model. The evaluation results are used to describe the ability b of these models and the prior data set $\mathbf{O} = \{(a_1, b_1) \ldots, (a_k, b_k)\}$. A Gaussian model, GM, is fitted based on O fitting. (2) Select the hyperparameter with the best performance on the agent function: find the maximum hyperparameter a' under GM through the collection function. (3) Apply the selected optimal hyperparameter to the objective function: the model is trained and evaluated based on the hyperparameter a' and K-fold cross-validation, and the evaluation results are used to describe the ability of the model b'. (4) Update the proxy model and add (a', b') to set $\mathbf{O}$. (5) Repeat steps (2)~(4) until the maximum number of iterations or running time is reached.

The specific steps of K-fold cross-validation taken to train and evaluate the model are as follows: (1) Divide the training set into K parts. (2) i ($i = 1, 2, \ldots, K$) is the test set, and the remaining $K-1$ is the training set. K data sets are constructed. (3) For the current combination of hyperparameter values, K different models are trained based on the K data sets in step (2) and test index values are calculated. (4) Calculate the mean value of the K test index values as the evaluation value of the model's prediction performance using the combination of corresponding hyperparameter values.

The XGBoost algorithm has many parameters, and different selections of parameter will affect the model's prediction performance. The hyperparameters of the XGBoost model selected in this paper to be optimized and their implications are shown in Table 2, along with the range settings of each hyperparameter and the optimal value results under 5-fold cross-validation. The eta parameter is the shrinkage step size, and the model robustness can be improved by reducing the weight of each step. The value range is [0.01, 0.3]. Max_depth is the maximum depth of the decision tree. As the depth of the tree increases, the model will have a higher grasp of the local factors of the sample. The value range is [3, 10]. Learning_rate is the learning rate, which affects the speed at which the parameter is updated to the optimal value. The value range is [0.05, 0.3]. N is the maximum number of iterations or the maximum number of weak learners, which mainly affects the fitting degree of the model.

Table 2. Parameters of XGBoost Model.

Parameters	Value	Range	Note
eta	0.2	[0.01, 0.3]	the shrinkage step size
max_depth	5	[3, 10]	the maximum depth of the decision tree
learning_rate	0.1	[0.05, 0.3]	the learning rate
N	160	/	the maximum number of iterations

The mean absolute error (MAE), mean absolute percentage error (MAPE), root mean square error (RMSE) and coefficient of determination (R^2) of different measurement points are used as quantitative indicators to evaluate the prediction ability of the model. The calculation of MAE, MAPE and RMSE and R^2 are presented mathematically by Equations (12)–(14):

$$MAE = \frac{1}{N}\sum_{i=1}^{N}|y_i - \hat{y}_i| \tag{11}$$

$$MAPE = \frac{1}{N}\sum_{i=1}^{N}\left|\frac{y_i - \hat{y}_i}{y_i}\right| \tag{12}$$

$$RMSE = \sqrt{\frac{1}{N}\sum_{i=1}^{N}(y_i - \hat{y}_i)^2} \tag{13}$$

$$R^2 = 1 - \frac{\sum_{i=1}^{N}(y_i - \hat{y}_i)^2}{\sum_{i=1}^{N}(y_i - \overline{y})^2} \tag{14}$$

where y and $\hat{y}$ the target and predicted values respectively, $\overline{y}$ is the average of the target values and N is the number of data points.

2.4. Factor Importance Analysis Based on SHAP

The SHAP functions interpret the impact of each factor on the predicted value of each individual and provide the visible factor importance analysis [36,37]. It is an additive explanation model constructed by SHAP inspired by cooperative game theory with all the characteristics treated as "contributors" [38]. For each predicted sample, the model generates a predicted value, and the SHAP value is the value assigned to each factor in the sample [39]. The process of this analysis is as follows: Suppose that the ith sample is x_i, the factor j of the ith sample is x_{ij}, m is the number of factors in the model, the predicted value of the sample is y_i, the baseline of the whole model (usually the mean value of the target variable of all samples) is y_{base}, and the SHAP value is shown as follows [36]:

$$y_i = y_{base} + f_1(x_{i1}) + f_2(x_{i2}) + \cdots + f_{i*j}(x_{ij}) + \cdots + f_{i*j*m}(x_{mj}) \tag{15}$$

where $f(x_{ij})$ is the SHAP value of x_i, e.g., $f(x_{i1})$ is the contribution of the first factor in the ith sample to the final predicted value y_i. When $f(x_{i1}) > 0$, it indicates that this factor increases the predicted value, showing a positive correlation; otherwise, it indicates that this factor reduces the predicted value, which is a negative correlation. The new MMP model and the traditional model have 13 + 11 influencing factors, which are chosen as influencing factors (see Table 3). The factors of the new MMP model are represented by X series, while the factors of the traditional model are represented by Y series. It should be noted that all the factors are independent and have distinct meanings.

Table 3. Influence variables set.

Components from the MMP Model	NO.	Factors	Components from the Traditional Model	NO.	Factors
Water level component	X1	$h^{1-n}(d_1/d_2)$	Water level component	Y1	h
	X2	$\bar{h}$		Y2	h^2
	X3	$(1-e^{-Ct})h$		Y3	h^3
Rheology–soil weight component	X4	$(1-e^{-Dt})H^{1-n} \cdot d_1$	Temperaturecomponent	Y4	$\sin\frac{2\pi t}{365}$
	X5	d_2		Y5	$\cos\frac{4\pi t}{365}$
	X6	d_1		Y6	$\cos\frac{2\pi t}{365}$
	X7	H		Y7	$\sin\frac{4\pi t}{365}$
	X8	Hd_1	Location component	Y8	x
Time effect component	X9	$\ln\theta$		Y9	y
	X10	$\ln\theta_0$	Time effect component	Y10	$\ln\theta$
Material component	X11	n'		Y11	$\ln\theta_0$
	X12	k	Note: Time effect components are same in the two models; $X5(d_2)$ and $Y9(y)$ are the same.		
	X13	r			

3. Case Study

We use the actual monitoring data of the Liyuan CFRD to evaluate the feasibility of settlement prediction using the newly derived MMP model. The Liyuan CFRD is located in the middle reach of the Jinsha River, Lijiang City, Yunnan Province. The dam began to be filled in the middle of August 2011, and the main part of the dam was completed at the end of July 2013. The construction period of the concrete panel was from 28 November 2013 to 28 May 2014, and the water storage began on 10 November 2014. The project's reservoir has the capacity for cycle regulation, the normal water level is 1618 m, the dead water level is 1605 m, and the flood control limit water level is 1605 m (early July to early August). The maximum dam height of the Liyuan CFRD is 155 m, and the normal water storage capacity corresponds to 727 million m^3.

We use the water level settlement gauge to monitor the settlement of the dam. The gauges were arranged along the river at a distance of 35 m to obtain the vertical displacement of the dam during the construction period and the operation period (see Figure 2). There are three monitoring points on the settlement's maximum cross-section which are arranged horizontally along a straight line at the right side of the dam at 0 + 223 m. As shown in Figure 2, most of the measuring points are located in the main rockfill area, and some of them are located in the secondary rockfill area and the downstream rockfill area The arrangement and range of 12 measuring points on the cross-section covers the whole section and can reflect the settlement of the whole rockfill dam under external load well, which is helpful in studying the prediction performance of the new MMP model. The monitoring data selected for modeling in this paper are from November 2014 to March 2016, that is, the period from the initial storage to the normal water level. The data were recorded once a week. A total of 924 effective data points can fully reflect the variation trend of the settlement at the initial stage of the storage operation.

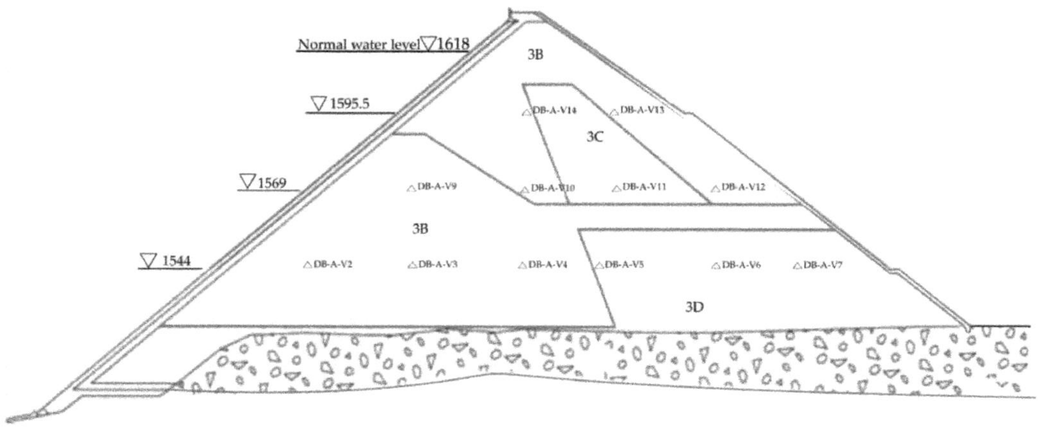

Figure 2. The layout of monitoring points of Liyuan dam.

4. Results

4.1. Prediction Accuracy of the New MMP Model

To verify the prediction accuracy and effectiveness of the new MMP model proposed in this paper, a full-section deformation prediction model is established for the settlement of the Liyuan CFRD measured at 12 monitoring points. We select 77 sets of settlement data from the initial impounding in November 2014 to the normal water level in March 2016 as samples for the analysis. A total of 924 valid data from 12 measurement points can fully reflect the variation trend of the initial impounding settlement. The first 67 groups are used as training data, and the last 10 groups are used as test data. The prediction period is from July 2015 to March 2016. The period is 8 months. The long-period test set is helpful to evaluate the prediction performance of the new MMP model.

The predicted residual distribution of the new MMP model is shown in Figure 3. As can be seen from Figure 3 and Table 4, the normal distribution of the fitting residuals of the 804 training groups used in the XGBoost model is mainly between −1 mm and 2 mm; the mean and median of the residuals in the training set are about 1 mm, while the MAE and RMSE values of each measurement point in the training set are mostly in the fitting error range of 0–3 mm. The results show that the new MMP model can fit the deformation of measuring points in the same section well. In the test set, the mean and median of the residuals of 120 groups were slightly higher than 2 mm, and the prediction errors of each measurement point were also around 2–3 mm. The difference between the fitting and predicted residual values were small, indicating that the overall performance of the XGBoost model was precise and had a good prediction performance.

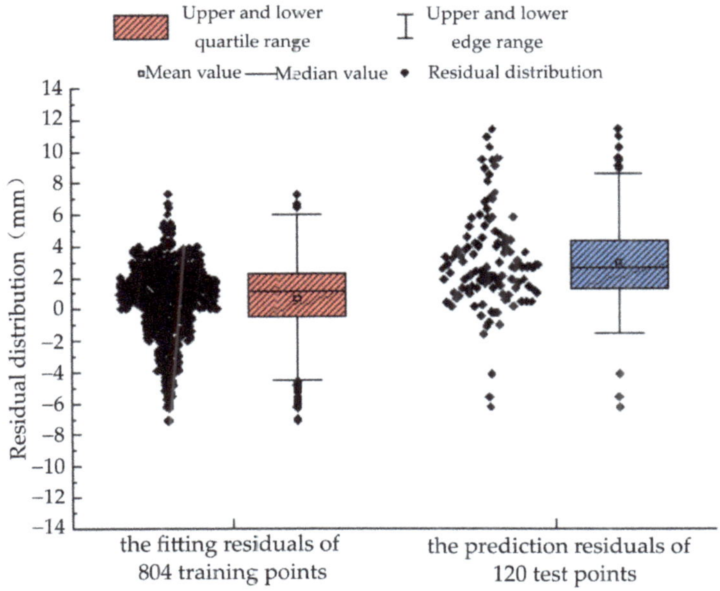

Figure 3. Box chart of model fitting and prediction residuals.

Table 4. Settlement fitting and prediction performance analysis of new multiple points monitoring model.

Measuring Point	Training Set				Test Set by XGboost				Test Set by CART			
	MAE /mm	RMSE /mm	MAPE/%	R^2	MAE /mm	RMSE /mm	MAPE/%	R^2	MAE /mm	RMSE /mm	MAPE/%	R^2
V2	3.41	4.18	1.54	0.88	7.31	7.74	2.65	0.85	9.54	8.65	5.88	0.49
V3	1.34	1.57	0.34	0.97	1.95	2.18	0.47	0.94	6.87	8.64	3.76	0.56
V4	2.24	2.58	0.54	0.95	4.00	4.19	0.88	0.93	5.88	6.00	1.55	0.58
V5	1.49	1.92	0.24	0.98	3.17	3.54	0.50	0.92	4.67	4.99	1.09	0.58
V6	0.93	1.37	0.17	0.99	1.35	2.05	0.24	0.96	4.56	3.77	1.88	0.64
V7	1.27	2.02	0.39	0.94	1.97	2.49	0.55	0.92	3.64	5.85	1.55	0.56
V9	2.39	2.86	0.62	0.95	1.77	2.09	0.41	0.93	5.47	6.16	1.2	0.59
V10	1.80	2.23	0.26	0.98	1.67	2.23	0.24	0.95	2.67	4.73	0.39	0.65
V11	2.44	2.68	0.49	0.96	2.14	2.48	0.42	0.93	4.34	4.53	1.14	0.67
V12	0.68	2.04	0.22	0.97	1.22	3.15	0.34	0.94	1.54	3.86	0.39	0.73
V14	1.23	2.22	0.34	0.94	4.12	4.79	1.00	0.90	8.12	8.42	3.55	0.62
V15	0.61	2.46	0.23	0.95	7.08	7.55	2.03	0.85	10.57	10.05	7.89	0.50

We then compared the subsidence obtained from XGBoost model and the CART model. The prediction accuracy of the two algorithms (XGBoost and CART) can be quantitatively compared using the data in Table 4. R^2 is used to measure the correlation between the actual value and the predicted value. The larger R^2 is, the more accurate the prediction of the algorithm is. In the test set, only the R^2 of XGBoost algorithm is above 0.90, larger than that of the CART model (around 0.6). This phenomenon shows that the XGBoost algorithm has a higher prediction accuracy than its base algorithm (CART). The RMSE, MAE and MAPE are all used to measure the difference between the actual value and the predicted value. It can be seen that the model's MAE and RMSE values at each measurement point in the training set and the test set are smaller than 3 mm, while the RMSE and MAPE of the test set from the CART algorithms are above 3 mm. The MAPE values of the training set and test set from the XGBoost algorithm are less than 1%, much smaller than those from the CART algorithm, which are mostly higher than 1%. It can be seen that for predicting the settlement of the CFRD, the predicted value of XGBoost shows the best correlation with the actual value and the smallest error. This finding demonstrates the superiority of the XGBoost model and indicates the importance of the use of the XGBoost algorithm.

We also provide a scatter plot of the predicted and actual response values of one typical monitoring point, V9, in Figure 4. The figure shows that the predicted values of the settlement of the measuring point continuously increase. This trend is the same for the measuring value. Furthermore, the prediction accuracy of the XGBoost model is higher than the CART model. This result is consistent with the MAE, RMSE and MAPE values.

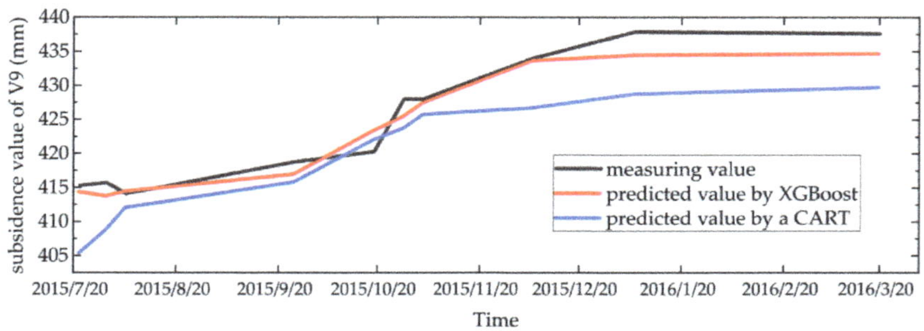

Figure 4. Predicted values and monitoring values of monitoring point V9 using different algorithms.

All in all, the new MMP model has the best prediction accuracy for measuring points V9-V10 in the maximum settlement area, and the MAE and RMSE values of the three measuring points are all about 2 mm. The prediction errors of most measuring points are within 1.5–3 mm, and the MAPE is also lower than 1%, indicating that the new MMP model had a good performance in predicting global measuring points. Moreover, the stability of the model also needs to be assessed in future works.

4.2. Orders of Importance of Factors by SHAP

Figure 5 shows the evaluation results of characteristic values of factors obtained by the SHAP method. Each row in Figure 5 represents a factor, each point represents a monitoring sample, and the x-axis represents the SHAP value. A redder color represents a larger value of the impact factor, and a bluer color represents the smaller the value of the impact factor itself. It can be seen that the orders of importance of factors of the new MMP point model are generally higher than those of the traditional model, except for the coordinate Y8. This finding indicates that the new MMP model is more suitable for predicting the settlement of CFRDs. It can be intuitively seen from the evaluation results that the newly added spatial location components X7, X5 and X4 are important contribution factors of the model, and they are all coupling factors related to the upper fill elevation. In addition, the upper fill elevation (H) is the most important characteristic factor. The rheological component X4, which includes the upper fill elevation, makes a great contribution to the model. The new spatial location also has higher contribution and adaptability than x and y in the traditional model, indicating that the position component derived from the new MMP model is more advantageous. Furthermore, points X4 and X8 gradually turn red with the increase in the SHAP value, indicating that these factors have a positive relationship with the settlement. In another word, the increase in the upper fill elevation value leads to an increase in settlement value. The upper fill elevation (X7) and the distance from the measuring point to the panel (X5) are also positively correlated with the settlement. Interestingly, median values of features X7 and X5 (represented by a purple color in Figure 5) can have the maximum contribution value to the settlement. Combined with the characteristic distribution of the spatial location parameters, the training characteristics of the MMP model can be roughly inferred, that is, the settlement at the top of the dam and at its foundation is small, the settlement at downstream of the dam body is larger, the maximum settlement is about half of the dam body's elevation, the settlement at the downstream rockfill area is slightly larger than that at the upstream side and the predicted value is basically consistent with the actual deformation.

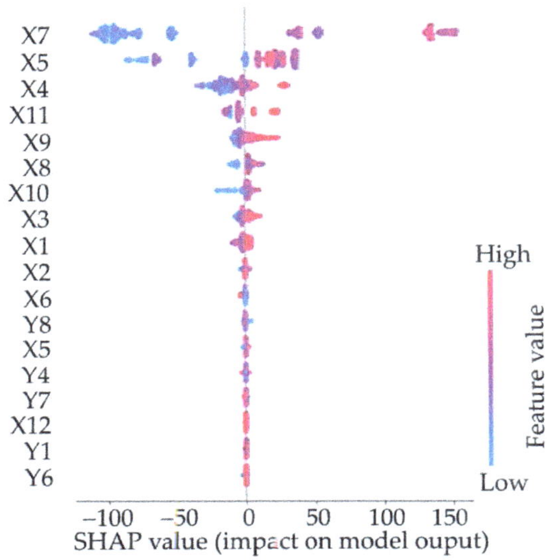

Figure 5. Evaluation results of factor values based on SHAP. A redder color represents a larger impact factor, and a bluer color is a smaller impact factor.

From Figure 5, we can further analyze the orders of importance of the water level, time and temperature components. In the water level component, the significance of X1 (which represents the transfer effect of water pressure) is higher than that of traditional water pressure Y1. This indicates that the combination of water pressure and spatial position can effectively improve the settlement. At the same time, the importance of X1 is also higher than the mean water level pressure X2 and the rheology–water pressure component X3, which proves that the hydrostatic load is the main factor of the settlement. Both X9 and X10 rank high because the time component is used to characterize the irreversible sedimentation over time. The least important component is the temperature component, as it is shown that Y4–Y7 have little influence on settlement, which basically conforms to the actual situation that the deformation of the CFRD has little correlation with periodic changes in temperature.

We further present the mean of the absolute SHAP values for each factor across the whole dataset in Figure 6. We only present several important factors; the other factors that have very low SHAP values are not present. The results show that in the MMP model, the contribution of the factors related to the spatial location (X7 and X5) to the settlement is the largest, while the contributions of the three traditional BP components of water level (X3), time and temperature (X9, X10 and other factors not shown in Figure 6), which represent temporal changes, are somewhat weaker. As a sum of a spatial model and a temporal model, the settlement of the measuring point consists of the completed settlement reference value in the filling period and the temporal value in the operating period. The spatial location component (X7 and X5) and material component (X11) determine the basic settlement, while the water level (X3) and time component (X9 and X10) affect the temporal variation in the settlement. Compared with the time series value with a smaller variation range, the basic settlement at the measuring point accounts for the main part of the settlement value, often reaching more than 80% of the total settlement. Therefore, in the monitoring data set of the full-section measuring points during the study of the water storage operation period, the spatial component (X7 and X5) and the material component have the largest average SHAP values contributing to the settlement. Moreover, their contribution is greater than that of the temporal component (X9 and X10), which also confirms the prediction accuracy of the MMP model. Furthermore, it is validated that it is necessary to pay attention to the spatially related variables in the study of the overall dam deformation.

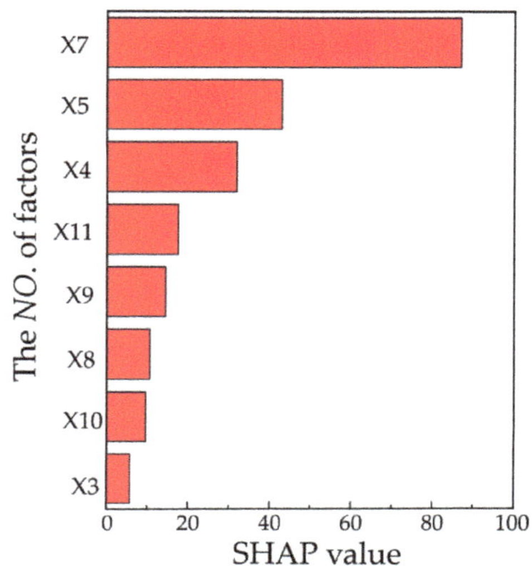

Figure 6. The mean of the absolute SHAP values for each factor across the whole dataset.

Two typical single prediction plots using the XGBoost model are shown in Figure 7. The length and color of the bars in Figure 7 show the degree of significance and direction (negative or positive) of the effect of each factor, respectively. The base value denotes the average of the observed response values. The prediction of point V9 on 30 March 2016 is shown in Figure 7a. The upper fill elevation H (X7) showed the highest effect, followed by the distance from the measuring point to the panel d (X5). Their values are lower than the median, and they both have a negative effect on the settlement. This phenomenon is same with the global SHAP value shown in Figure 5, where the points of X7 and X5 are in blue and often have a negative SHAP value and thus have a negative effect on the subsidence. What is more, X9 has the largest positive effect on the subsidence for V9. This is because X9 is related to the monitoring days and has the largest value. The prediction of point V10 on 30 March 2016 is shown in Figure 7b. All factors in Figure 7b show positive influences. V10 is located in the middle part of the CFRD, and has the largest settlement. Among these factors, the upper fill elevation H (X7) showed the highest effect. The distance from the measuring point to the panel d_1 (X5) ranks second. This phenomenon is the same with V9. However, although X5 has a median value, it has the highest SHAP value; this corresponds to the purple dots of X5 in the large SHAP value areas in Figure 5. Based on the individual analysis of the MMP model, the prediction process and prediction basis of the model for each specific sample can be understood. According to the data, the point position has the largest impact on the subsidence, and the monitoring time also has significant effects.

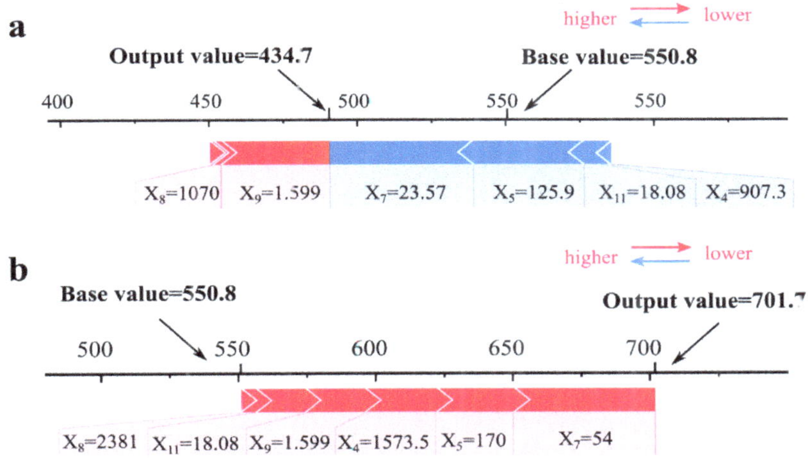

Figure 7. Evaluation results of factors based on SHAP values. (**a**) The prediction of point V9 on 30 March 2016; (**b**) The prediction of point V10 on 30 March 2016. A redder color represents a larger impact factor, and a bluer color represents a smaller impact factor.

To sum up, from the feature importance analysis, it can be seen that the contribution of the influence factors of the newly derived MMP model is better than that of the traditional variables. The average SHAP value of the spatial component and the material component to the output settlement is the largest, and the factor contribution is greater than that of the temporal correlation component, which confirms that the prediction accuracy of the MMP model is closely related to the spatial location component. In the study of the overall deformation behavior of a CFRD, it is necessary to pay attention to the spatially related variables.

5. Conclusions

The MMP prediction model is based on the physical cause analysis of a CFRD settlement and the expansion of spatial components. The influence of water level load transfer, rockfill rheology and soil properties on the settlement during the operation period of impoundment is comprehensively analyzed, and a space-time distribution model of the CFRD during the operation period under the action of multiple factors is established. The XGBoost model was used for fitting prediction, and the model was evaluated by various performance indicators. Taking the settlement monitoring data of the Liyuan CFRD as an example, the new MMP model, under the action of multiple factors, can predict the settlement of full section points with higher accuracy, which has certain application and popularization value for related projects. From the factor importance analysis, it can be confirmed that the contribution of the influencing variables of the MMP model to the model is better than that of the traditional variables. The SHAP value of the spatial component and the material component to the output settlement value is the largest, and the factor contribution is greater than that of the time component, which confirms that the prediction accuracy of the MMP model is closely related to the spatial location component. In the study of the overall deformation behavior of CFRD, it is necessary to pay attention to the spatially related variables. Our work elucidates the high prediction accuracy of the newly established MMP model and provides a benchmark for the investigation of the safety management of a CFRD over its whole life cycle. Practitioners can predict future dam deformation based on dam deformation data, obtain the dominant factors and repair them in advance. However, this MMP and XGBoost model has only been validated on one CFRD, more data from other CFRDs are needed to test the MMP and XGBoost model. In addition, this new model was developed from the physical analysis of a CFRD, and whether or not it is suited for other dams also needs to be investigated in future work.

Author Contributions: Conceptualization, T.W.; Methodology, Z.W.; Software, Y.W.; Validation, Z.W.; Formal analysis, K.M.; Investigation, Y.W.; Writing—original draft, L.S.; Writing—review and editing, T.W.; Supervision, T.W. All authors have read and agreed to the published version of the manuscript.

Funding: This research was funded by the National Natural Science Foundation of China [52079109].

Data Availability Statement: Not applicable.

Conflicts of Interest: The authors declare no conflict of interest.

References

1. Tatin, M.; Briffaut, M.; Dufour, F.; Simon, A.; Fabre, J.-P. Thermal displacements of concrete dams: Accounting for water temperature in statistical models. *Eng. Struct.* **2015**, *91*, 26–39. [CrossRef]
2. Hu, Y.; Shao, C.; Gu, C.; Meng, Z. Concrete Dam Displacement Prediction Based on an ISODATA-GMM Clustering and Random Coefficient Model. *Water* **2019**, *11*, 714. [CrossRef]
3. Min, K.; Li, Y.; Yin, Q.; Wen, L. Research on prediction performance of multiple monitoring points model based on support vector machine. *IOP Conf. Ser. Mater. Sci. Eng.* **2020**, *794*, 012038. [CrossRef]
4. He, J.-P.; Tu, Y.-Y.; Shi, Y.-Q. Fusion Model of Multi Monitoring Points on Dam Based on Bayes Theory. *Procedia Eng.* **2011**, *15*, 2133–2138. [CrossRef]
5. Cheng, L.; Zheng, D. Two online dam safety monitoring models based on the process of extracting environmental effect. *Adv. Eng. Softw.* **2013**, *57*, 48–56. [CrossRef]
6. Salazar, F.; Morán, R.; Toledo, M.Á.; Oñate, E. Data-Based Models for the Prediction of Dam Behaviour: A Review and Some Methodological Considerations. *Arch. Comput. Methods Eng.* **2017**, *24*, 1–21. [CrossRef]
7. Arlot, S.; Celisse, A. A survey of cross-validation procedures for model selection. *Statist. Surv.* **2010**, *4*, 40–79. [CrossRef]
8. Sun, X.; Zheng, D.; Zhou, M. Spatiotemporal Prediction Model for Settlement Value of Face Rockfill Dam During Operation Period. *J. China Three Gorges Univ. Nat. Sci.* **2019**, *41*, 5–8.
9. Huang, D.; Liu, B.; Dai, W. Building multiple-point displacement model of Wuqiangxi dam based on BP neural network. *Geotech. Investig. Surv.* **2017**, *45*, 62–64.
10. Chai, L.; Qi, D.; Wu, H. Application of multi-point and multidirectional BP Network Model in Dam deformation monitoring. *Water Resour. Power* **2014**, *32*, 94–97.
11. Li, Y.; Min, K.; Zhang, Y.; Wen, L. Prediction of the failure point settlement in rockfill dams based on spatial-temporal data and multiple-monitoring-point models. *Eng. Struct.* **2021**, *243*, 112658. [CrossRef]
12. Lu, X.; Wu, Z.; Zhou, Z.; Chen, J. Research on the Prediction Model of Deformation of High Core Rockfill Dam During Construction Period. *Adv. Eng. Sci.* **2017**, *49*, 61–69. [CrossRef]
13. Kang, F.; Liu, J.; Li, J.; Li, S. Concrete dam deformation prediction model for health monitoring based on extreme learning machine. *Struct. Control Health Monit.* **2017**, *24*, e1997. [CrossRef]
14. Wei, B.; Yuan, D.; Xu, Z.; Li, L. Modified hybrid forecast model considering chaotic residual errors for dam deformation. *Struct. Control Health Monit.* **2018**, *25*, e2188. [CrossRef]
15. Taffese, W.Z.; Sistonen, E. Machine learning for durability and service-life assessment of reinforced concrete structures: Recent advances and future directions. *Autom. Constr.* **2017**, *77*, 1–14. [CrossRef]
16. McCulloch, W.S.; Pitts, W. A logical calculus of the ideas immanent in nervous activity. *Bull. Math. Biophys.* **1943**, *5*, 115–133. [CrossRef]
17. Kim, Y.-S.; Kim, B.-T. Prediction of relative crest settlement of concrete-faced rockfill dams analyzed using an artificial neural network model. *Comput. Geotech.* **2008**, *35*, 313–322. [CrossRef]
18. Su, H.; Chen, Z.; Wen, Z. Performance improvement method of support vector machine-based model monitoring dam safety: Performance Improvement Method of Monitoring Model of Dam Safety. *Struct. Control Health Monit.* **2016**, *23*, 252–266. [CrossRef]
19. Marandi, S.M.; VaeziNejad, S.M.; Khavari, E. Prediction of Concrete Faced Rock Fill Dams Settlements Using Genetic Programming Algorithm. *IJG* **2012**, *3*, 601–609. [CrossRef]
20. Chen, T.; Guestrin, C. Xgboost: A scalable tree boosting system. In Proceedings of the 22nd ACM SIGKDD International Conference on Knowledge Discovery and Data Mining, San Francisco, CA, USA, 13–17 August 2016; pp. 785–794.
21. Nguyen, L.T.K.; Chung, H.-H.; Tuliao, K.V.; Lin, T.M.Y. Using XGBoost and Skip-Gram Model to Predict Online Review Popularity. *SAGE Open* **2020**, *10*, 215824402098331. [CrossRef]
22. Lim, S.; Chi, S. Xgboost application on bridge management systems for proactive damage estimation. *Adv. Eng. Inform.* **2019**, *41*, 100922. [CrossRef]
23. Shi, N.; Li, Y.; Wen, L.; Zhang, Y. Rapid prediction of landslide dam stability considering the missing data using XGBoost algorithm. *Landslides* **2022**, *19*, 2951–2963. [CrossRef]
24. Wakjira, T.G.; Rahmzadeh, A.; Alam, M.S.; Tremblay, R. Explainable machine learning based efficient prediction tool for lateral cyclic response of post-tensioned base rocking steel bridge piers. *Structures* **2022**, *44*, 947–964. [CrossRef]
25. Wakjira, T.G.; Ibrahim, M.; Ebead, U.; Alam, M.S. Explainable machine learning model and reliability analysis for flexural capacity prediction of RC beams strengthened in flexure with FRCM. *Eng. Struct.* **2022**, *255*, 113903. [CrossRef]

26. Nguyen, H.D.; Truong, G.T.; Shin, M. Development of extreme gradient boosting model for prediction of punching shear resistance of r/c interior slabs. *Eng. Struct.* **2021**, *235*, 112067. [CrossRef]
27. Wang, L.; Wu, C.; Tang, L.; Zhang, W.; Lacasse, S.; Liu, H.; Gao, L. Efficient reliability analysis of earth dam slope stability using extreme gradient boosting method. *Acta Geotech.* **2020**, *15*, 3135–3150. [CrossRef]
28. Wakjira, T.G.; Ebead, U.; Alam, M.S. Machine learning-based shear capacity prediction and reliability analysis of shear-critical RC beams strengthened with inorganic composites. *Case Stud. Constr. Mater.* **2022**, *16*, e01008. [CrossRef]
29. Wakjira, T.G.; Abushanab, A.; Ebead, U.; Alnahhal, W. FAI: Fast, accurate, and intelligent approach and prediction tool for flexural capacity of FRP-RC beams based on super-learner machine learning model. *Mater. Today Commun.* **2022**, *33*, 104461. [CrossRef]
30. AlKhereibi, A.H.; Wakjira, T.G.; Kucukvar, M.; Onat, N.C. Predictive Machine Learning Algorithms for Metro Ridership Based on Urban Land Use Policies in Support of Transit-Oriented Development. *Sustainability* **2023**, *15*, 1718. [CrossRef]
31. Al-Hamrani, A.; Wakjira, T.G.; Alnahhal, W.; Ebead, U. Sensitivity analysis and genetic algorithm-based shear capacity model for basalt FRC one-way slabs reinforced with BFRP bars. *Compos. Struct.* **2023**, *305*, 116473. [CrossRef]
32. Ishfaque, M.; Salman, S.; Jadoon, K.Z.; Danish, A.A.K.; Bangash, K.U.; Qianwei, D. Understanding the Effect of Hydro-Climatological Parameters on Dam Seepage Using Shapley Additive Explanation (SHAP): A Case Study of Earth-Fill Tarbela Dam, Pakistan. *Water* **2022**, *14*, 2598. [CrossRef]
33. Sigtryggsdóttir, F.G.; Snæbjörnsson, J.T.; Grande, L. Statistical Model for Dam-Settlement Prediction and Structural-Health Assessment. *J. Geotech. Geoenviron. Eng.* **2018**, *144*, 04018059. [CrossRef]
34. Liashchynskyi, P.; Liashchynskyi, P. Grid search, random search, genetic algorithm: A big comparison for NAS. *arXiv* **2019**, arXiv:1912.06059.
35. Snoek, J.; Larochelle, H.; Adams, R.P. Practical bayesian optimization of machine learning algorithms. *Adv. Neural Inf. Process. Syst.* **2012**, *25*, 2960–2968.
36. Dong, W.; Huang, Y.; Lehane, B.; Ma, G. An artificial intelligence-based conductivity prediction and feature analysis of carbon fiber reinforced cementitious composite for non-destructive structural health monitoring. *Eng. Struct.* **2022**, *266*, 114578. [CrossRef]
37. Panda, C.; Mishra, A.K.; Dash, A.K.; Nawab, H. Predicting and explaining severity of road accident using artificial intelligence techniques, SHAP and feature analysis. *Int. J. Crashworthiness* **2022**. [CrossRef]
38. Štrumbelj, E.; Kononenko, I. Explaining prediction models and individual predictions with feature contributions. *Knowl. Inf. Syst.* **2014**, *41*, 647–665. [CrossRef]
39. Shapley, L.S. A Value for N-Person Games. In *Classics in Game Theory*; Princeton University Press: Princeton, NJ, USA, 1997; Volume 69.

water

MDPI

Article

Hydrometer Design Based on Thin-Film Resistive Sensor for Water Measurement in Residential Buildings

Laís dos S. Gonçalves [1], Khrissy A. R. Medeiros [2] and Carlos R. Hall Barbosa [3,*]

[1] Postgraduate in Chemical and Materials Engineering Department (DEQM), Pontifical Catholic University of Rio de Janeiro, Rio de Janeiro 22451-900, Brazil
[2] Optical Fiber Sensors Laboratory (LSFO), Mechanical Engineering Department (DEM), Pontifical Catholic University of Rio de Janeiro, Rio de Janeiro 22451-900, Brazil
[3] Postgraduate Program in Metrology, Pontifical Catholic University of Rio de Janeiro, Rio de Janeiro 22451-900, Brazil
* Correspondence: hall@puc-rio.br

Abstract: Because of economic, population, and consumption patterns changes, the use of freshwater has increased significantly in the last 100 years. Notably, measurement is essential to encourage water conservation. Thus, the present study aims to evaluate the applicability of a thin-film resistive sensor (bend sensor) with different coatings for implementation in individualized water measurement systems. The motivation of this work is to propose a volumetric meter using flow control valves that ordinarily are already present in a building's hydraulic installations. Methodologically, the following are presented: the system developed for the electromechanical and thermal characterization of the sensor, the sensor computational simulation performed using Ansys® software, and for the electronic circuit designed in LTSpice® software, the artificial neural network used to estimate the flow and the volume estimates from the trapezoidal pulses. The results obtained allowed us to assess that, taking into account the type of coating, the sensor coated with polyester has better behavior for the proposed hydrometer. In addition, this evaluation allowed us to conclude that the bend sensor demonstrated its feasibility to be used as a transducer of this novel type of volumetric meter and can be easily inserted inside a hydraulic component, such as a flow control valve, for example.

Keywords: bend sensor; computational simulation; individual measurement; resistive sensor; volume measurement; water consumption

Citation: Gonçalves, L.d.S.; Medeiros, K.A.R.; Barbosa, C.R.H. Hydrometer Design Based on Thin-Film Resistive Sensor for Water Measurement in Residential Buildings. *Water* **2023**, *15*, 1045. https://doi.org/10.3390/w15061045

Academic Editors: Jie Yang, Chunhui Ma and Lin Cheng

Received: 7 February 2023
Revised: 3 March 2023
Accepted: 7 March 2023
Published: 9 March 2023

1. Introduction

The importance of water is indisputable, directly affecting the social, political, and economic development of society. Therefore, interest in hydraulic knowledge dates back to the Egyptian and Mesopotamian civilizations—located near the Tigris and Euphrates rivers—which used the waters and fertile lands on the banks of these rivers to survive and were called hydraulic civilizations [1,2]. In turn, in the Indus Valley (currently located in Afghanistan and part of India and Afghanistan), the Harappa civilization had the most advanced sanitary systems of the time where all houses had water wells and drains for sewage disposal [3]. Regarding the Roman Empire hydraulic works, the Proserpina dam (Mérida) and the Nîmes aqueduct [4] stand out. Posteriorly, with the advancement of knowledge of hydraulic systems, in the 19th century, hydraulic machines—such as turbines—and hydroelectric plants began to appear [5]. Finally, with 20th- and 21st-century computer technology, it is possible to perform computer simulations to predict and analyze projects in the water sector. With this, we emphasize that the interest in the study and advancement of technologies in the hydraulic sector is timeless.

In the past, public baths were usual, corresponding to an average daily consumption per inhabitant of around 750 L [6]. Despite seeming to be an obvious waste, part of this consumption was used for other purposes—for example, cleaning the villages. Even so, it indicates a high consumption compared to current figures. According to the National

Sanitation Information System (SNIS) [7], in Brazil, the average per capita consumption of water is 152.1 L/inhabitant/day. This difference is related to changes in habits whether due to the advancement of technologies in the water sector or to greater waste awareness.

In this sense, the importance of monitoring water consumption in urban environments, especially in large buildings, is clearly observed [8]. This need for control and monitoring can be evidenced through numerous studies in different areas that aim at more responsible consumption. Ali et al. [9] presented a study in three manufacturing plants of a leather chemical industry with strategies to minimize water consumption. Ozturk and Cinperi [10] also reported a decrease in water consumption in a wool fabric factory. In addition, Gabarda-Mallorquí et al. [11] showed a study of the hotel industry in Lloret de Mar, a tourist destination on the Mediterranean coast, which presents results that contribute to achieving efficient use of water and dealing with changes in water availability.

In general, when comparing the indicators of water losses in the national scene with the standards of developed countries, which have an average of 15%, it is possible to observe a supply that still denotes a distance from the technological frontier in terms of efficiency. According to Oliveira et al. [12], in Brazil in 2018, the rate of losses in distribution was 38.45%, 0.16% more than in 2017, demonstrating a worsening and pointing to the urgency of greater efforts to reduce losses.

Therefore, it is necessary to emphasize that potable water losses happen in different ways, the most common being leaks, thefts, and reading errors or inaccurate readings due to water meters being very old [12].

Being so, it is essential to use effective procedures, such as monitoring and measuring water, to encourage its conservation. This can be performed with the aid of hydrometers that, conceptually, are instruments of measurement and a constant indication of the volume of water that passes through them. In addition, the building sector in Brazil is increasingly showing interest and progress in water consumption individual measurements.

Commonly, the measurement of both collective and individual water consumption is conducted with conventional hydrometers, but their installation for individualization requires adaptation work in most multifamily buildings. In order for the measurement to be individualized, it is necessary to install water meters in each branch of the existing columns (Figure 1a,b). Alternatively, it is necessary to have a place in the common area of the corridors to position the meters and the installation of specific columns with individualized water meters for each autonomous unit (Figure 1c). Noting that in the case of buildings with many floors, the renovation to be carried out is large-scale and very expensive.

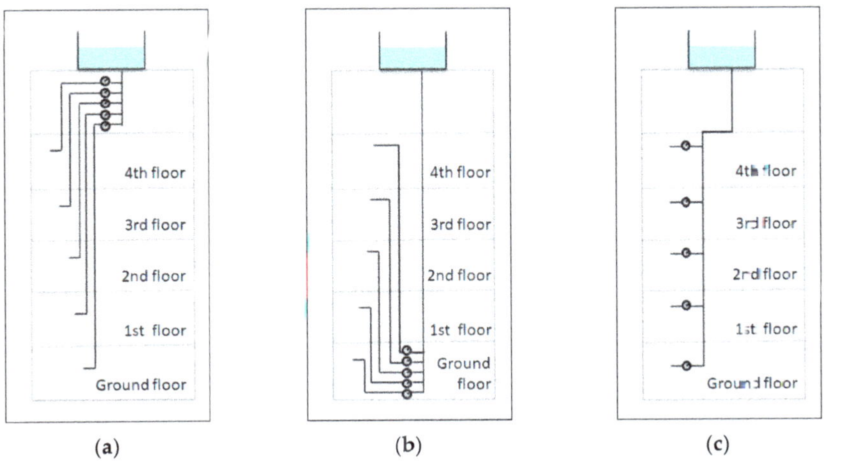

Figure 1. Possible locations for hydrometers: (a) on the top floor; (b) on the ground floor; and (c) on the building floors.

Thus, the availability of a flow and/or volume meter that could be more easily integrated into existing hydraulic installations would contribute greatly to the greater dissemination of the individual measurement systems (IMSs) in buildings with collective metering given that individualized measurement provides numerous benefits, both for the owner and for the condominium, as well as for the development of a sustainable society, as it helps in saving water, reduces waste, and financially promotes a reduction in the bill, as pointed out by some case studies identified in the literature [13–15]. For example, the study carried out by Souza and Kalbusch, 2017 [14] presents a 34% reduction in per capita consumption of buildings that changes collective measurement to individual measurement.

In this sense, the present research sought to propose and evaluate a new type of hydrometer to promote advances in the study of these water consumption meters, more specifically for measuring the volume of water consumed in autonomous units of residential buildings.

Thereby, this article aims to research and develop a new methodology for measuring water consumption using a thin-film resistive sensor. In addition, the proposal is to use valves (gate valves) commonly present in residential hydraulic systems, consequently dispensing the need for large-scale work to transform the collective measurement of water consumption into individual measurement in autonomous unit buildings; that is, there will be savings in the installation time of individual meters and a faster return on investment due to the installation of the proposed water meters being on a small scale compared to conventional water meters.

Structurally, this work is organized as follows: Section 2 describes the thin-film resistive sensor, the proposed hydrometer, and the setup used in the electromechanical and electrothermal characterizations; the project made in the Ansys® (2021R2) software of the proposed volumetric meter using the bend sensor; the computer simulation via Ansys® software to obtain the relationship between the resistance of the sensor and the water flow in a building pipe; and, still, the electronic circuit simulation via LTSpice XVII® (v. 17.0.30.0) software to obtain a signal with adequate levels and a good signal-to-noise ratio, that is, a less noisy signal. In addition, the theoretical foundations of artificial neural networks are also shown. Sections 3 and 4 present the results obtained and the discussions and recommendations for future work, respectively.

2. Materials and Methods

2.1. Thin-Film Resistive Sensor

The thin-film resistive sensor used in this work, herein called 'bend sensor', is a flexible sensor that varies its electrical resistance, as it is bent. This happens because the sensor is made by combining a single thin layer of a plastic substrate with a resistive material. This type of sensor separates this resistive material into several microcracks when it is bent, which, as a consequence of the bending movement, defines an increase or a decrease in the material's electrical conductivity [16], and the radius of curvature or sensor angular deflection establishes the sensor electrical resistance.

It is worth mentioning that the bend sensor's basic electrical characteristics are established according to the resistive material, the substrate, and the type of coating used in its construction. With respect to the resistive material, typically carbon or polymer elements are used. The substrate properties, together with the conductive component, define sensor flexibility. Finally, as for the coating, the bend sensor may not have any type but also might include some type of coating, for example, silicone rubber, adhesive rubber, polyester, or polyimide. It is important to highlight that the coating provides chemical and mechanical protection to the sensor [17].

Because the bend sensor has relevant electromechanical properties, the interest of several researchers is growing. In this way, its application is spread in several areas. Most of the studies related to the bend sensor are in the health sector, in which this type of sensor is used for numerous purposes [18–24]. In addition, the thin-film resistive sensor is used for other functionalities, such as soil monitoring [25,26]. In turn, unlike these other areas, the water sector does not present many studies using the bend sensor for flow measurement. However, some studies demonstrate the feasibility and applicability of this type of sensor for this purpose. For example, Fan et al. [27] employ the sensor to monitor the airflow velocity in real-time, helping to save energy consumption in a wastewater treatment process; Xu et al. [28] report a way to monitor water flow in pipelines using a flexible resistive film; Srinivasan et al. [29] use the sensor as a cantilever beam to measure the flow for moderate flow applications; and Stewart et al. [30] developed a velocity sensor using the bend sensor for application in streams.

In summary, these thin-film resistive sensors have various applications in the most diverse areas, such as medical, aeronautical, automotive, robotics, and even soil studies. Over the last few years, this type of sensor has attracted a lot of attention from researchers because it can vary its resistance when flexed, as well as being light, robust, and low cost. These sensors are easily adaptable in hydraulic systems of autonomous units (residential, commercial, or mixed) in buildings. In addition, it is possible to design and build interfaces that result in the effective coupling of electronic components with the bend sensor.

For the development of this study, we used two thin-film resistive sensors from the manufacturer Flexpoint (Figure 2), each one is 1 inch (24.5 mm) long, 0.28 inches (7.11 mm) wide, and 0.005 inches (0.127 mm) thick. Of these two sensors, one is coated with polyester and the other with polyimide.

(a) (b)

Figure 2. 1-inch bend resistive sensors: (**a**) polyester bend sensor and (**b**) polyimide bend sensor.

The new hydrometer proposal—based on a thin-film resistive sensor for measuring water consumption in autonomous units—consists mainly of using flow control valves that are already present in the building's hydraulic installation.

Thereby, the proposal is to insert the bend sensor into the valve so that it can be adapted to work as a volumetric meter using the valve's mechanical structure, especially the wedge—the component responsible for the release (valve open) and blockage (valve closed) of the fluid passage. This adaptation is possible because it is feasible to remove the core from these flow control valve types, enabling the use of these valves that are already installed in the piping of buildings.

Figure 3 shows a gate-type flow control valve: In Figure 3a, the valve is open (retracted wedge) with the sensor positioned inside it, allowing the passage and flow measurement, and in Figure 3b, the valve is closed (wedge obstructing the passage), preventing flow.

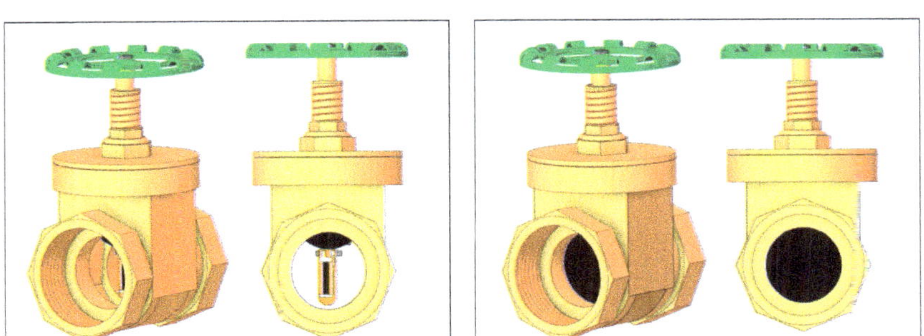

(a) (b)

Figure 3. Gate valve 3D drawing: (**a**) open and (**b**) closed.

To adapt the valve as a volumetric meter, it is proposed to insert an additional fixed part (gray part) inside the cover (Figure 4a) to work as a support for the bend sensor to be positioned inside the valve. In a meter initial proposal and laboratory evaluation, it is suggested to use epoxy resin for fixing the sensor in this additional part, as this type of resin is highly resistant and even hardens in water, as was performed in the study by Xu et al. [28]. However, due to the epoxy resin toxicity, it is necessary to investigate other material options to be used to fix the sensor within the proposed volumetric meter idea. Furthermore, in practical terms, to standardize the proposed gauge model, it would be necessary to manufacture a wedge with a hole on the side for the gray part to pass through and an opening in the center to make space for the sensor, as shown Figure 4b.

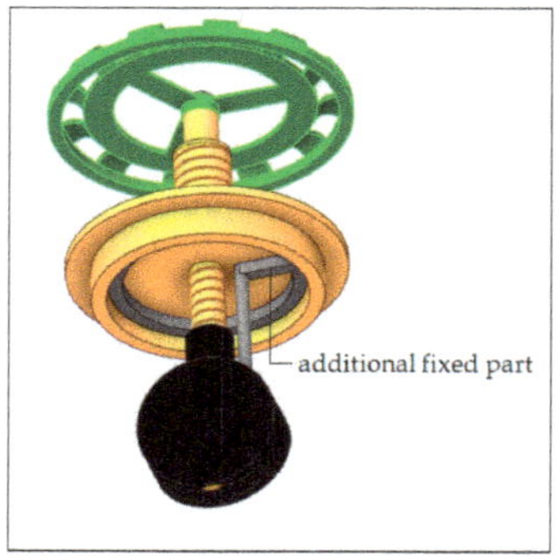

(a)

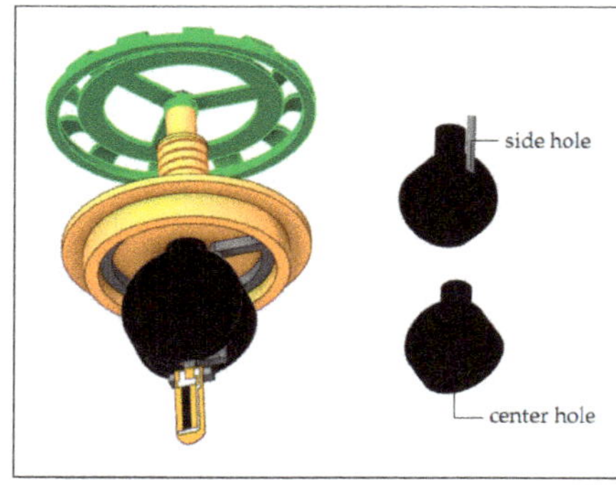

(b)

Figure 4. Gate valve cover with rubber wedge fitted with the sensor fixed to the gray part: (**a**) valve closed and (**b**) valve open.

Initially, it was necessary to better understand some properties of the bend sensor and its behavior when flexed at different angles within an exploratory study, which sought to determine the feasibility of using it to measure water consumption. Thus, two types of characterizations of thin-film resistive sensors through different methods were made. The first characterization method sought to evaluate the sensor's electromechanical behavior, and the second analyzed its electrothermal characteristics.

For this, a setup was developed and used in both characterizations, containing a Flexpoint bend sensor, a Keysight U1233A portable digital multimeter, a screw to flex the bend sensor, a fixed support with a welded nut to guide the screw, a protoboard as a base for the bend sensor, and a protractor to measure the sensor bending angle. This setup is shown in Figure 5. Additionally, the same characterizations—electromechanical and thermal—of article [31] were made but with smaller sensors (1 inch). It should be mentioned that this configuration requires the operator to manually flex the bend sensor to the required angle and visually position it. As a result, these measurements are susceptible to parallax errors, so differences in the values measured in the two characterizations may occur.

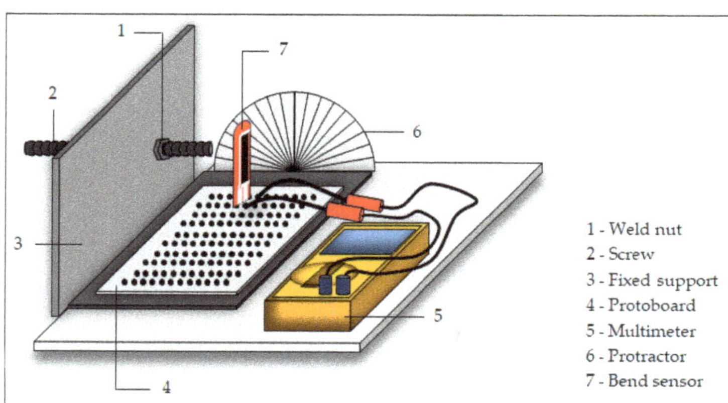

Figure 5. Test setup schematic representation.

2.2. Ansys Simulation

To computationally analyze the sensor bend flexing caused by the flow of water in a pipe, it was necessary to carry out a study through a simulation using computer-aided engineering (CAE) via Ansys® software through a decoupled system. In other words, first, a simulation was carried out in Fluent Ansys® of the water flow inside a pipe to obtain the pressure value (Figure 6a) and, after that, another simulation was carried out in Mechanical Ansys® to obtain the maximum sensor deformation when the pressure calculated in the previous step is applied (Figure 6b).

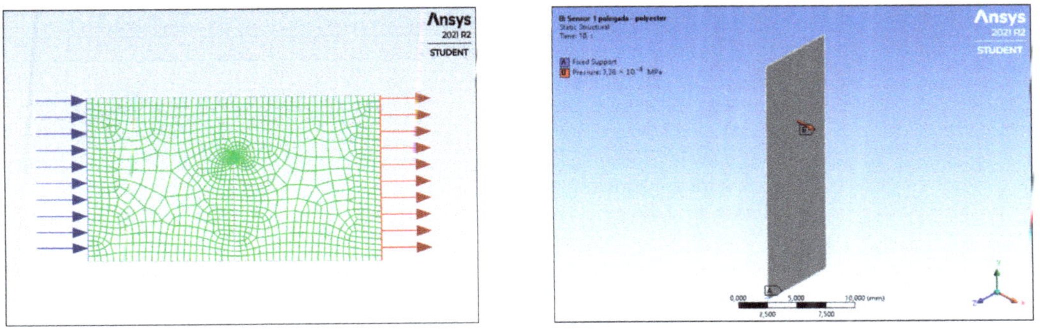

(a) (b)

Figure 6. Ansys simulation: (**a**) Fluent Ansys for obtaining the pressure value and (**b**) Mechanical Ansys for estimating the sensor maximum deformation.

Methodologically, to simulate the water flow through a pipe, a numerical simulation of computational fluid dynamics (CFD) was performed. The geometry and mesh generation were completed in SpaceClaim, the flow parameters configuration and the numerical solutions were performed in Fluent, and the visualization and analysis were conducted in CFD-Post.

With the aim to configure the simulation of water flow, the Reynolds number (Re) was used according to Equation (1), which is used to determine the flow regime of a given fluid. This regime can be classified as laminar or turbulent depending on the Reynolds number (Re). Typically, in internal flows, Re values lower than 2300 are considered laminar flows, and values higher than 2300 are considered turbulent flows.

$$Re = \frac{\rho U D_h}{\mu}, \quad (1)$$

where ρ and μ are the fluid and dynamic viscosity, respectively, U is the average fluid velocity, and D_h is the hydraulic diameter defined by

$$D_h = \frac{4\,A}{P}, \quad (2)$$

where A and P are, respectively, the cross-sectional area and the wetted perimeter in which the flow occurs.

The velocities used for the flow simulation came from historical data obtained from measurements performed with an ultrasonic flow meter and a 1 $\frac{1}{2}$ inch diameter pipe (typically used in hydraulic systems of autonomous units in residential buildings). These velocities vary between 0.0502 m/s and 0.5307 m/s. Table 1 presents the data used in the one-inch-long resistive sensor simulation and data for calculating the Reynolds number. The fluid used in the simulation was water (μ = 1.003 × 10^{-3} Pa·s and ρ = 998.2 kg/m^3 at 20 °C) without considering the air that is commonly present in the pipes.

Table 1. Data used in the CFD simulation of the bend sensor.

Parameters			Value	Units
Inner diameter D			$1\frac{1}{2}$	inch
			0.0381	m
Area A			0.00114	m
Wetted perimeter P			0.1197	m
Hydraulic diameter D_h			0.0381	m
Reynolds number	velocity	max.	21,992	-
		min.	5688	

To generate the velocity field and obtain the pressure value, it was necessary to choose the appropriate flow model for the simulation. For all velocities, the Reynolds number was higher than 2300, thus configuring a turbulent flow, so the turbulence model used was the shear stress transport.

Regarding the simulation via Mechanical Ansys®, the sensor was considered analogous to a cantilever beam subjected to a uniformly distributed load q in which the maximum deflection and the maximum rotation angle occur at the free end of the beam.

Therefore, for the purpose of this study, the method used was finite element analysis. The geometry generation and the sensor parameters configuration, as well as its dimensions and the type of coating (polyester with a density equal to 1380 kg/m^3, modulus of elasticity equal to 3.65 GPa, and Poisson's ratio equal to 0.48 [32]), were performed by the Ansys Workbench.

On the other hand, to configure the structural simulation, the pressure applied to the sensor was obtained with the flow simulation. With this, it was possible to know the maximum deformation corresponding to each speed and, consequently, the sensor's maximum deflection angle.

Finally, after obtaining these results and using the regression equation obtained in the electromechanical characterization (resistance as a function of deflection angle), it was possible to estimate the resistances for each angle from the simulation (R_S).

2.3. LTSpice Simulation

Subsequently, to compose and evaluate the functioning of the proposed water meter, a computer simulation was carried out in LTSpice® of an electronic circuit responsible for powering and conditioning the electrical signals, generating an output voltage related to the flow in the gate valve.

Aiming to perform a simple conversion of deflection to voltage, the bend sensor was connected to a resistor in a voltage divider configuration. The simplest configuration uses just two resistors in series (R_1 e R_2) and an input voltage (V_{in}), thus generating an output voltage (V_{out}) given by

$$V_{out} = R_2.\left(\frac{V_{in}}{R_1 + R_2}\right), \tag{3}$$

In general, for the purpose of the given electrical voltage signal to have adequate levels and a good signal-to-noise ratio, in other words, the lesser effect of background noise on the signal measurement, it must go through some processes, such as amplification, filtering, and linearization.

The resistance values ($R_S = R_2$) obtained from the simulation performed in the Ansys® software were used to properly configure the bend sensor circuit. In addition, using the Solver function of the Excel® (2007) software, another exponential regression (Equation (4)) was performed to obtain a relationship between the simulation resistance (R_S) and the flow, which was calculated by multiplying the flow velocity and the pipe cross-section area.

$$R(\dot{q}) = R_0 + Ae^{\alpha(\dot{q}-\dot{q}_0)}, \tag{4}$$

where R_0 is the initial resistance, $\dot{q}$ is the instantaneous flow corresponding to the sensor deflection, and $\dot{q}_0$ is the initial volumetric flow rate.

In the simulation in LTSpice®, aiming to obtain a voltage signal as a flow function, a component was created that simulates the bend sensor's resistive behavior. This component was used in the electronic circuit with processes such as offset, linearization, and amplification. Figure 7 shows how the electronic circuit simulation was performed in this work: (i) The component that simulates the bend sensor (X1) composes a voltage divider with R1, thus generating an output voltage signal (in addition, the resistance value R1 from Figure 7 was chosen to maximize the desired deflection sensitivity range), and (ii) after that, this signal went through the offset, amplification, and linearization processes. Thus, the simulation generated an electrical voltage signal as a flow function with adequate levels and a good signal-to-noise ratio, in other words, with a reduced effect of background noise on the signal measurement.

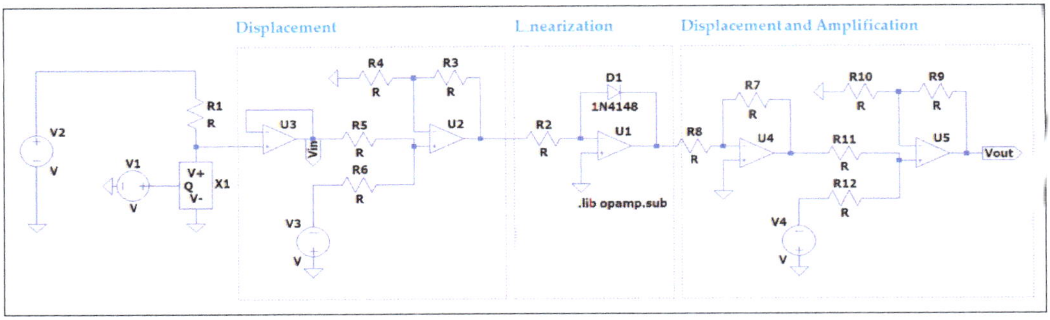

Figure 7. Electronic circuit configuration in LTSpice®.

In this way, the complete system provided a circuit output voltage (output variable *Vout*) which corresponds to a certain instantaneous flow (input variable controlled by voltage source V1 in Figure 7).

However, for this research, it is necessary to estimate the flow corresponding to this electrical voltage, that is, to obtain the relationship between flow and electrical voltage. For this purpose, the computational intelligence tools called artificial neural networks (ANNs) were used.

2.4. Inverse Problem Solution

In 1943, the first concepts about artificial neural networks were introduced by Warren S. Mcculloch and Walter Pitts [33] who proposed a model of artificial intelligence analogous to the form, behavior, and functions of a biological neuron. In other words, the input signals and the weights from the artificial neuron correspond, respectively, to the dendrites and the synapses—which are the connection of a dendrite with the cell body—from the biological neuron. Furthermore, the stimuli processing is equivalent to the soma function that occurs in the cell body and the activation function and represents the triggering threshold of the biological neuron [34].

Basically, it is enough to combine numerous artificial neurons to obtain the so-called artificial neural networks (ANNs). One of their most relevant purposes is the relationship between independent (input) and dependent (output) variables, which is determined using a learning process in which a set of data is provided to the network. Also noteworthy is the fact that an ANN is a model that does not need to be guided by physical laws; in other words, its parameters do not need to have any physical meaning [35].

Additionally, ANNs have reduced computational time, high precision, and the ability to generate nonlinear relationships between the independent and dependent system variables. Thus, in addition to performing the approximation of functions, ANNs are capable of performing pattern recognition, predictions, and image processing, enabling their application in numerous areas [35], for example, in solar energy systems [36–39]. ANNs are also used for classification and facial recognition [40–42] to create models to predict the condition of airport pavements [43] and for stock market studies [44–47].

ANNs are also present in research related to water systems, such as in the detection and recognition of leaks in pipes [48,49]; the assessment of the water quality index in the Red Sea, Sudan [50] and the Godavari River, India [51]; and in predicting the efficiency of heavy metal removal from aqueous solutions of biochar systems [52].

In short, these multiple applications are consequences of the artificial neural networks' various configuration possibilities that can be characterized by the pattern of connections between the units (structure), the learning method, and the activation function [53].

Regarding structure, ANNs can be classified as single-layer or multilayer feedforward networks and recurrent networks. Feedforward networks differ from recurrent networks because they do not have a recurrent loop.

In relation to the learning method, the training model is established by how the parameters are modified, being able to be supervised, unsupervised, and by reinforcement. In supervised mode, the input and output data set are known, but the environment is not known; in other words, the network parameters adjustment is performed by associating the input signal with an error signal (difference between the output signal desired and the one provided by the network). In contrast, the unsupervised one does not require the desired output value from the network—the system collects properties from the patterns set, grouping them into classes inherent to the data. In contrast, the reinforcement training system learns to perform a certain task only based on the results of its experience with an interaction with the environment [54]. Furthermore, it is possible to validate the training model using several algorithms, for example, cross-validation, which is a standard statistical tool that uses a different data set from those used to adjust the network parameters; in other words, the data set is randomly divided into two sets: training—which is subdivided into estimation (used to select the model) and validation (used to test or validate the model)—and test [34].

As for the activation function, it can be from several types, the most used being binary step function, linear function, sigmoid, tanh, ReLU, and radial basis function (RBF).

Thus, in this research, we chose to use a feedforward network with supervised learning and the radial basis function as the activation function (two layers) because other adjustments were tested (for example, exponential and polynomial) but had large errors.

In light of what was presented previously about LTSpice® and ANNs, the following describes what is conventionally called 'Inverse Problem Solution': The simulated system generated a voltage signal from the bend sensor as an output having a flow signal as an input; however, for this study, the flow must be the output variable. Therefore, the ANN was used to obtain the volumetric flow rate signal (output) as a voltage function (input). This training was carried out through Matlab's (R2019b) 'newrbe'; function—in which the input parameters were the electrical voltages simulated by the LTSpice®, and the outputs were the desired flow rates. With this, it was possible to obtain a trained network that, when receiving the voltage values, estimates the volumetric flow rate values.

Finally, a volume estimate was made by different means—analytical and numerical calculation—aiming at a simulated validation of the proposed measurement concept with the thin-film resistive sensor.

2.5. Simulation of Meter Operation

Assuming that the hydrometer must measure the total volume of water consumed by an autonomous unit, it is understood that the meter proposed in this work needs to estimate the volume from the instantaneous flow.

Because of this, it is necessary to consider the reality of residential building pipes given that water only flows at certain moments, such as when opening a faucet, for example. Therefore, to validate the simulations performed on the proposed hydrometer based on a resistive sensor, two different cases of trapezoidal flow pulses were generated starting from zero (no water flow in the pipe) and reaching a certain flow value, which depends on which hydraulic system was activated.

In this research, a water tap—flow curve from the faucet (Figure 8a)—and a conventional shower—flow curve traditional shower (Figure 8b)—both from the manufacturer Docol, were considered to extract the flow values to generate the trapezoidal pulse. To approach a more realistic scenario, small fluctuations in the trapezoidal pulse maximum flow were added, and 15 s faucet activation and 15 min shower using the conventional shower were considered.

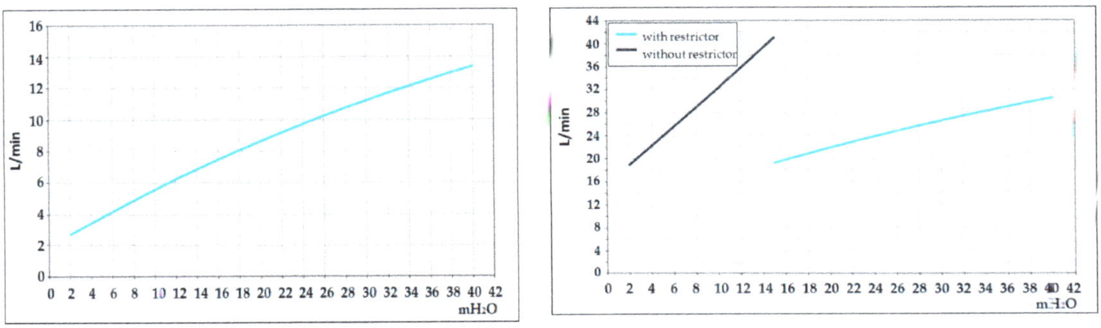

(a) (b)

Figure 8. Flow curve from the manufacturer Docol: (**a**) water tap and (**b**) traditional shower.

The configuration of the trapezoidal pulse in the two cases applied considered a twelve-floor building, and two autonomous units, one on the fourth floor and another on the eleventh floor. It is worth mentioning that, as each floor is 3 m high, the unit on the fourth floor has 24 m of water column (mH_2O), and the one on the eleventh has 3 mH_2O.

For the fourth floor unit, the faucet has a flow of 9.5 L/min (0.57 m^3/h), so the trapezoidal pulse was generated with fluctuations around this flow (Figure 9a), and for the shower—considering the use of a pressure restrictor as indicated by the manufacturer—the pulse was generated around the flow of 24 L/min (1.44 m^3/h), as shown in Figure 9b.

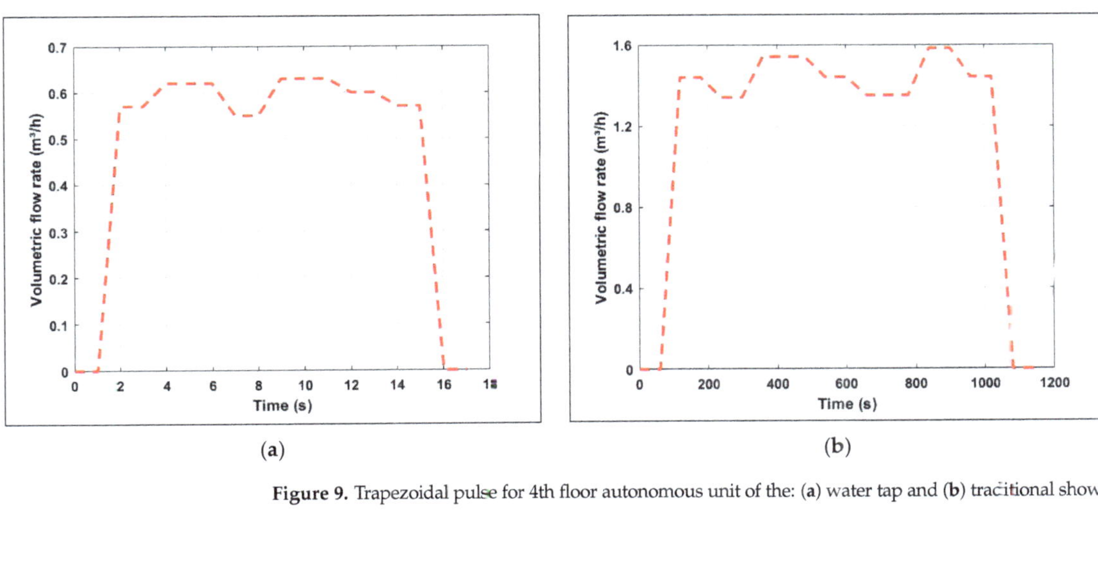

(a) (b)

Figure 9. Trapezoidal pulse for 4th floor autonomous unit of the: (**a**) water tap and (**b**) traditional shower.

Regarding the eleventh floor unit, the faucet trapezoidal pulse is shown in Figure 10a and has fluctuations around the flow rate of 3 L/min (0.18 m^3/h). Whereas, for the same unit, the traditional shower without restrictor trapezoidal pulse is shown in Figure 10b and varies around the flow rate of 20.5 L/min (1.23 m^3/h).

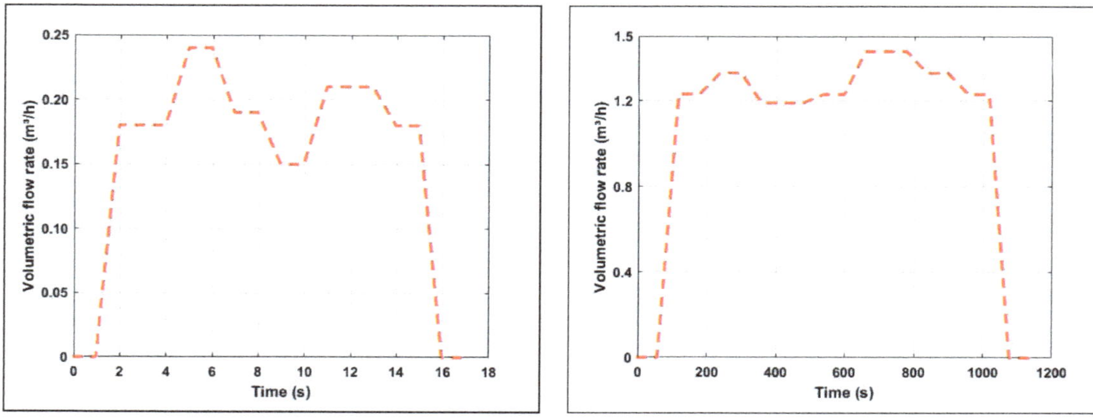

(**a**) (**b**)

Figure 10. Trapezoidal pulse for 11th floor autonomous unit of the: (**a**) faucet and (**b**) traditional shower.

Subsequently, these pulses were used in the LTSpice® software (defining the values of voltage source V1 in Figure 7) to obtain the sensor output voltage signal. With this, the simulation response of the bend sensor (voltage) towards the hydraulic system activation (flow) was obtained as a function of time. These voltage values were provided to the trained artificial neural network to obtain the corresponding volumetric flow rate output values.

Finally, to estimate the volume from these volumetric flow rates found, two calculations were performed by different means, the first by analytical calculations and the second by numerical calculations.

The analytical estimate was made from the trapezoidal pulse generated using

$$\partial vol = Q_v \, \partial t \rightarrow \int_0^v (dvol) = \int_0^t Q_v \, dt, \tag{5}$$

where Q_v is the volumetric flow that represents the volume (*vol*) of a fluid that flows through a section of a pipe in a time interval (t).

The second estimate, in turn, was performed by numerically integrating the flow values from the neural network using the Matlab software trapezoidal method, resulting in the volume calculation. Figure 11 presents, in a summarized and schematic way, the steps that were performed to obtain the volume by the two methods.

It is important to point out that the pressure value from the hydraulic system flow curve is different from the one from the computer simulation via Fluent Ansys. The first pressure is called static, which corresponds to the pressure of the water when it is stopped in the pipe, and, according to the ABNT NBR 5626 standard, in a building installation, a pressure of 40 mH_2O must not be exceeded [55]. On the other hand, the pressure of the computer simulation is called dynamic, which is when there is the movement of water in the pipe at the moment when some hydraulic system is activated. Moreover, dynamic pressure is static pressure minus distributed and localized head losses.

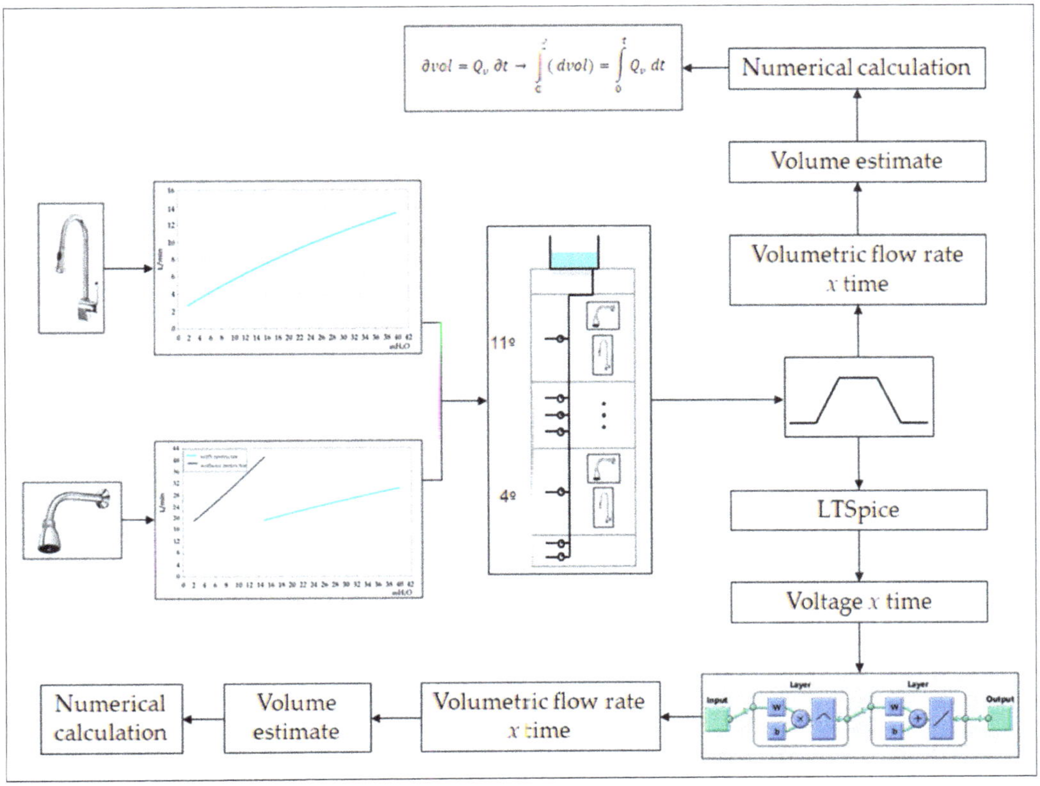

Figure 11. Schematic representation of the steps taken to validate the measurement concept of the proposed meter.

3. Results

This section presents the resistance behavior of thin film resistive sensors when subjected to angular (electromechanical characterization) and temperature (electrothermal characterization) variations; simulations in Ansys and LTSpice® software to find a correlation between the sensor resistance and the flow velocity; and finally, a mathematical formulation for estimating the volume, aiming at the bend sensor application to measure the volume of water drained in a pipe of autonomous units.

Initially, intending to facilitate the understanding of the organization of the content of the results, Figure 12 shows a section mapping.

3.1. Bend Sensor Behavior Evaluation

3.1.1. Electromechanical Characterization

This characterization had as its principal objective, to know preliminarily the characteristics of the sensor resistance variation in relation to its angular deflection and, consequently, to analyze its behavior. As previously mentioned, two one-inch-long bend sensors with different coatings—one coated with polyester and the other with polyimide-coated—were analyzed.

As this characterization method made it possible to measure the resistances in the increment and decrement of the angle, it was possible to obtain the hysteresis curve and calculate the maximum hysteresis for each sensor bend. The regression equations and the coefficients of determination (R^2) were obtained via Excel® software from the average resistances (between the values in the increment and decrement of the angle).

The results for the sensor with polyester coating are shown in Figure 13a. This sensor presented a maximum hysteresis equal to 3.5% of full scale, an average resistance variation between 1.40 kΩ and 5.28 kΩ, and the equation $R(\theta) = 1.306 + 0.03986\ e^{0.0768\theta}$ as an exponential regression with a coefficient of determination (R^2) of 0.9988. Figure 13b shows the results of the 1-inch-long sensor with a polyimide coating. This sensor showed a maximum hysteresis equal to 13% of the full scale and an average resistance variation between 3.51 kΩ and 10.65 kΩ, and the equation $R(\theta) = 3.492 + 0.08506\ e^{0.07374\theta}$ as an exponential regression with the coefficient of determination (R^2) of 0.9964.

When considering the change in velocity of the water flow inside a pipe, it becomes imperative that the proposed hydrometer based on a resistive sensor has the lowest possible hysteresis value, in other words, the highest agreement between the 'loading' and 'unloading' curves. Therefore, comparing these results from the electromechanical characterization, the sensor coated with polyester showed better applicability for measuring water consumption due to having a lower hysteresis value.

On the other hand, given that the proposal is to use the resistive sensor in a pipe to measure the volume of water, this sensor, in turn, must have characteristics that support this type of application. Therefore, the sensors were subjected to electrothermal characterization.

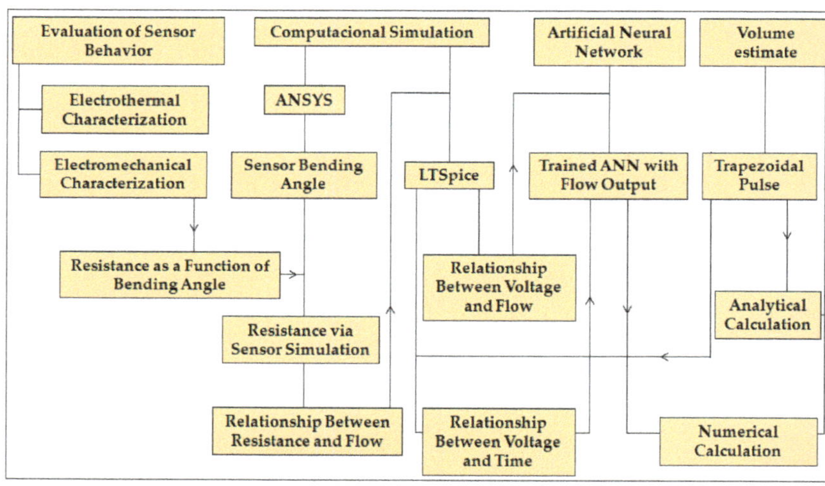

Figure 12. Content mapping of Section 3.

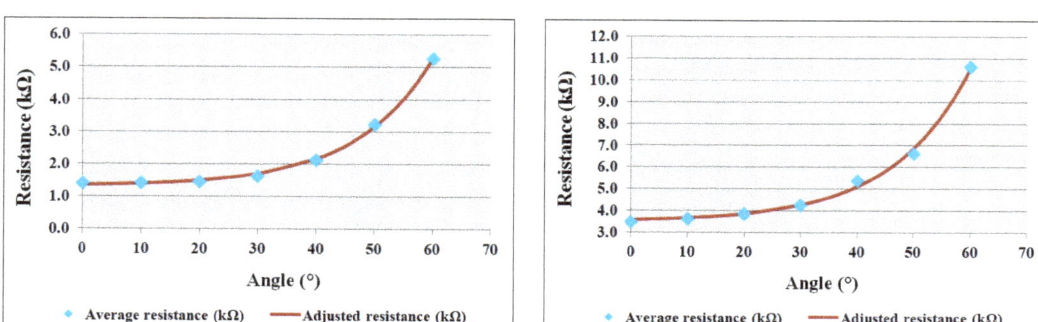

(a) (b)

Figure 13. Average and adjusted resistance as a function of the bend sensor deflection angle with (**a**) polyester coating and (**b**) polyimide coating.

3.1.2. Electrothermal Characterization

The main objective of this characterization was to know and analyze the sensor resistance variation characteristics when subjected to temperature changes. In the same way as the electromechanical characterization, two one-inch-long thin film resistive sensors were evaluated: one with a polyester coating and the other with a polyimide coating. The sensors characterized were the same for both characterizations.

In this methodology, the curvature angle was fixed, and the sensor resistances were measured in the temperature increment and decrement. Consequently, it was feasible to calculate and evaluate the hysteresis concerning the temperature.

First, the polyester-coated bend sensor was characterized. This sensor showed the same behavior for all angles tested when subjected to climatic chamber heating and cooling. As the temperature increased, the sensor resistance decreased, and when the temperature reduced, the resistance increased (Figure 14), noting, thus, a good agreement between the 'loading' and 'unloading' curves of the sensor.

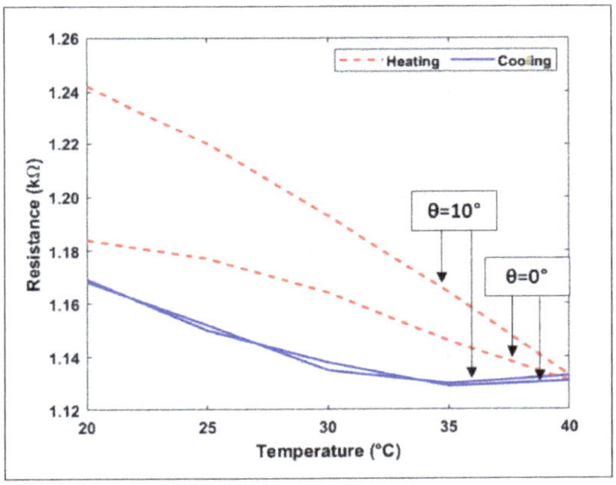

(a)

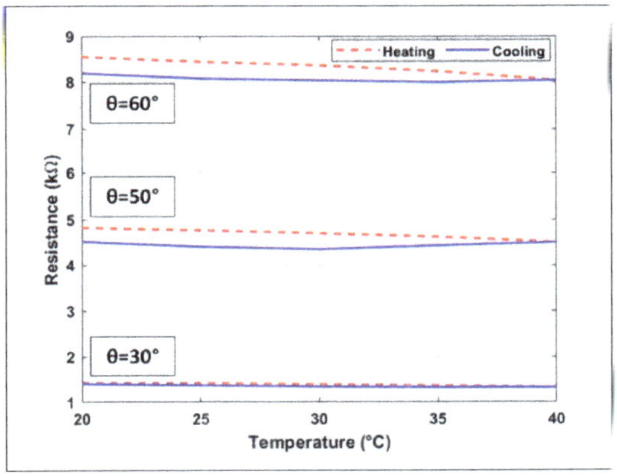

(b)

Figure 14. Resistance of the polyester-coated bend sensor to temperature variation for deflection angles: (**a**) 0° and 10° and (**b**) 30°, 50°, and 60°.

In the sequence, the other bend sensor characterized was polyimide coated. This sensor, when submitted to the climatic chamber cycle, presented an irregular behavior between the tested angles (Figure 15). At a 0° angle, when heating, the resistance values decreased, and during cooling, these values also decreased (Figure 15a). In turn, for angles of 10°, 30°, 50°, and 60°, this behavior did not remain. At these angles, as the temperature increased, the sensor resistance also increased, and during cooling, as the temperature reduced, the sensor resistance continued to rise (Figure 15b)—thus denoting an irregular behavior between the sensor 'loading' and 'unloading' curves.

In light of this result, the same characterization was repeated with an identical sensor, having obtained the same results as the first, thus confirming the anomalous behavior of that type of sensor when subjected to the same evaluation conditions.

Thus, the bend sensor with polyester coating showed a more regular and satisfactory behavior demonstrated by a better agreement between the 'loading' and 'unloading' curves when compared to the sensor with polyimide coating. For this reason, the results obtained from this sensor were plotted, indicating the quantities measured in different ways: from the three-dimensional point of view (Figure 16a) and the bend sensor resistance responses concerning the angle for each temperature (Figure 16b).

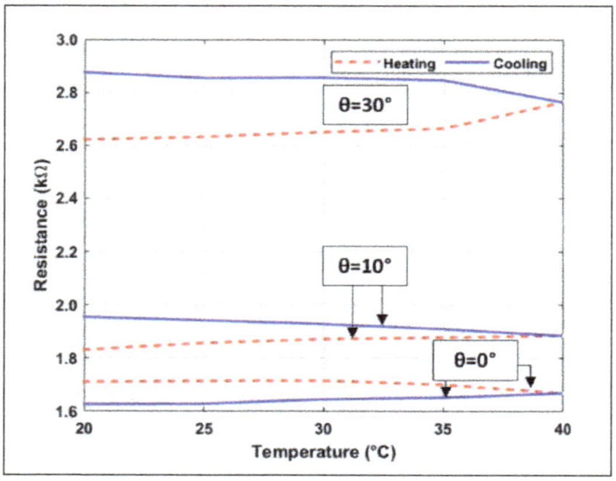

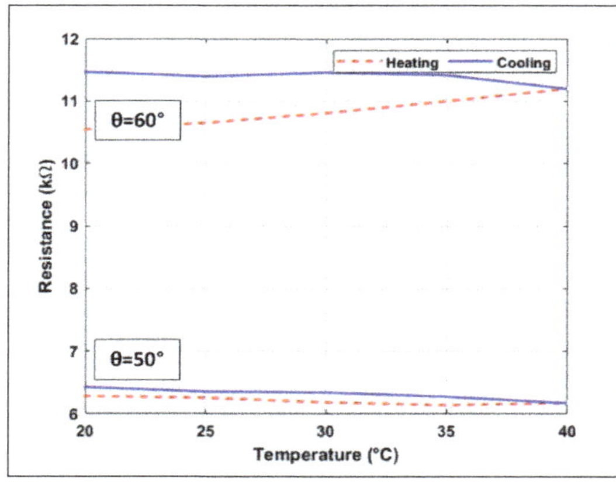

(a) (b)

Figure 15. Resistance of the polyimide-coated bend sensor to temperature variation for deflection angles: (**a**) 0°, 10°, and 30° and (**b**) 50° and 60°.

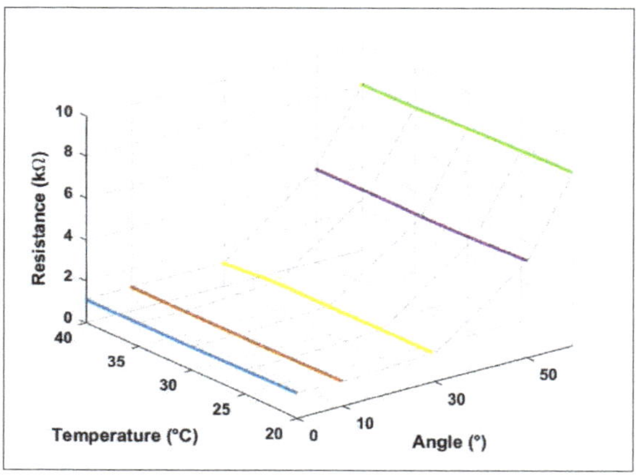

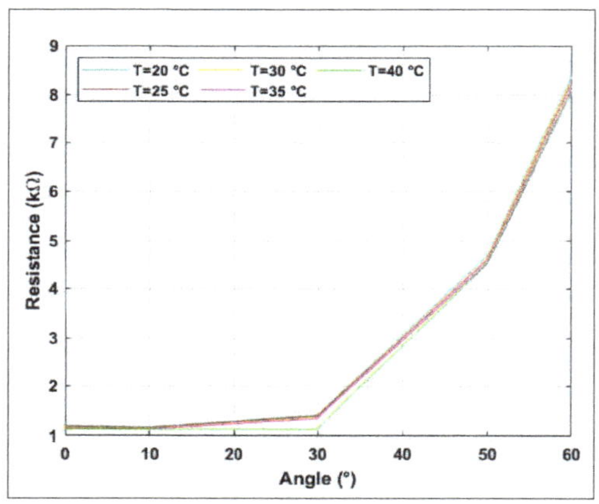

(a) (b)

Figure 16. Response of the polyester-coated bend sensor: (**a**) the average resistance between the temperature increment and decrement values for each angle through 3D measured quantities graph and (**b**) the average resistance as a function of angle of deflection.

In general, when analyzing the results presented, the bend sensor coated with polyester showed a negative linear relationship between the sensor resistance and the temperature variation; in other words, the resistance decreases when the temperature increases.

As this sensor has demonstrated a more stable behavior compared to that of polyimide, as well as a lower hysteresis value in the electrothermal characterization, in terms of applicability, it shows up as being more viable for measuring water consumption in pipes of autonomous units of buildings. For these reasons, only the polyester-coated sensor was used in the Ansys and LTspice® software simulations that follow.

3.2. Computational Simulation

3.2.1. Ansys Simulation

The sensor behavior study was carried out through a computer simulation in Ansys software to evaluate the applicability of the bend sensor for measuring water consumption in autonomous units. The aim was to obtain computationally the sensor deflection angle caused by the flow of water in the pipe and then calculate the corresponding resistances using the regression equation found in Section 3.1.1.

Thus, Table 2 presents the results of the interaction of the water flow in the pipeline with the deflection behavior of the resistive sensor with a polyester coating.

Table 2. Ansys® simulation results with the polyester-coated bend sensor.

v	Re	Pressure	Deformation	Deflection Angle	R_S
(m/s)	**(Dimensionless)**	**(Pa)**	**(mm)**	**(°)**	**(kΩ)**
0.58	21,992	287	22.92	69	9.25
0.53	19,954	240	19.17	58	4.64
0.47	17,916	189	15.09	45	2.61
0.42	15,878	151	12.06	36	1.95
0.37	13,840	117	9.34	28	1.65
0.31	11,802	81.8	6.53	20	1.49
0.26	9764	57.7	4.61	14	1.42
0.20	7726	34.1	2.72	8	1.36
0.15	5688	19.2	1.53	5	1.35

The velocity used in the flow simulation to obtain the pressure to bend the 1-inch-long sensor with polyester coating ranged from 0.15 m/s to 0.58 m/s. As a result, this sensor presented a variation of deflection between 5° and 69° and resistances (R_s) between 1.35 kΩ and 9.25 kΩ.

After completing the modeling to obtain the sensor deflection angle curved by the pressure exerted by the fluid flow inside a pipe, an evaluation more focused on the sensor's operational aspect was started. For this purpose, a simulation of the electronic circuit to be implemented was performed via LTSpice® software.

3.2.2. LTSpice Simulation

As mentioned previously, only the polyester-coated sensor was used. Thus, it is emphasized that the principal purpose of computationally simulating the electronic circuit was to obtain a relationship between the voltage from the bend sensor with the flow of water.

Therefore, considering the sensor to be one-inch-long, Equation (6)—obtained through the Solver function of the Excel® software—presents the estimated parameters to minimize the error sum of squares.

$$R(\dot{q}) = 1.376 + 0.00052\, e^{4.044\, \dot{q}}, \tag{6}$$

Thus, Figure 17 presents the resistance response obtained in the simulation via Ansys (R_S) as a function of the flow and the adjustment curve with a coefficient of determination (R^2) of 0.9972.

Given this, the bend sensor circuit was configured with Equation (6), and the generated signal (X1) passed through the voltage divider (first block of Figure 7), which transformed the resistance signal into voltage and maximized the deflection sensitivity range (R1 = 3 kΩ). After that, the signal went through an offset of 0.3145 V (V3), a gain of 16 v/v (1 + R3/R4), and a linearization process ('linearization' block in Figure 7). Finally, it again passed an offset (V4) and a gain (1 + R9/R10) of 1.897 V and 1.769 v/v, respectively.

Once the configuration above was generated with the data obtained from the simulation of the circuit and the conditioners, it was possible to obtain a relationship between voltage and flow, as shown in Figure 18.

Finally, as previously presented, this relationship between voltage and flow was used to train an artificial neural network that estimated the instantaneous flow values upon receiving the voltage values. The network training results for the one-inch-long sensor are shown in Figure 19.

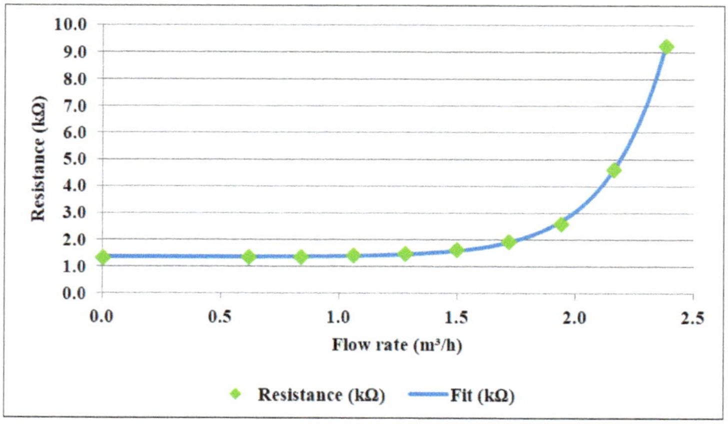

Figure 17. Resistance response as a function of flow via simulation (green) and regression (blue).

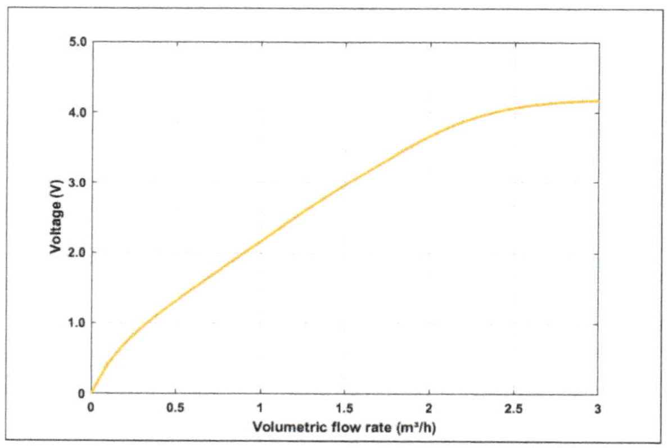

Figure 18. Polyester-coated bend sensor voltage ratio as a function of flow.

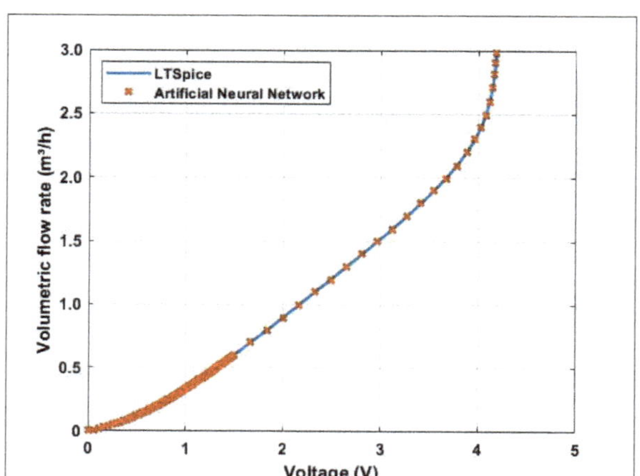

Figure 19. Result of training the network to obtain the instantaneous flow corresponding the voltage of the one-inch-long, polyester-coated bend sensor.

In this sense, as a hydrometer, conceptually, is an instrument designed for measurement and a continuous indication of the volume of water drained, the meter proposed in this work needs to indicate volume instead of flow. For this reason, in the next step, a volume estimate was performed.

3.3. Meter Operation Simulation

3.3.1. Fourth-Floor Autonomous Unit

The simulation results of the thin-film resistive sensor when the faucet is activated are shown in Figure 20. Figure 20a shows the electrical voltage signal response as a function of the faucet's operating time. In turn, Figure 20b presents a comparison between the trapezoidal pulse generated by the resistive sensor simulation via an artificial neural network (blue curve) and the faucet flow curve (red dashed curve).

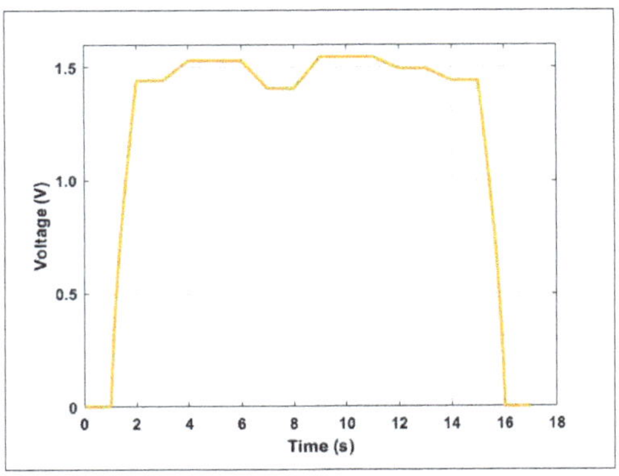

(a)

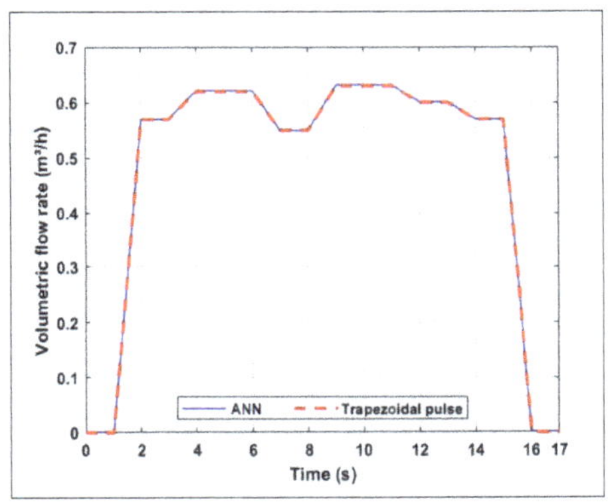

(b)

Figure 20. (**a**) Sensor response to faucet activation and (**b**) trapezoidal pulse via the bend sensor simulation on the ANN (blue curve) and faucet flow curve for the autonomous unit with 24 mH_2O (dashed red curve).

It is worth mentioning that, for the proposed hydrometer, it is necessary to estimate the water volume drained during the 15 s activation of the water tap.

Then, as explained in Section 2.5, the volume was estimated by different means from the flows found. The first method, which was the analytical calculation of the theoretical trapezoidal pulse (faucet flow curve), resulted in a volume of 0.002314 m^3 (2.314 L), and the second method, which was the numerical calculation of the trapezoidal pulse obtained via sensor simulation, resulted in 0.002319 m^3 (2.319 L) of drained volume.

For the traditional shower activation, the sensor simulation results are shown in Figure 21. In Figure 21a, the electrical voltage signal response is shown as a function of the operating time of the conventional shower. Additionally, Figure 21b presents a comparison between the trapezoidal pulse generated by the resistive sensor simulation via an artificial neural network (blue curve) and the shower flow curve (red dashed curve). By the analytical calculation, the volume of water drained during the traditional shower activation for a 15 min shower was equal to 0.3861 m^3 (386.1 L), and by the numerical calculation, it was equal to 0.3858 m^3 (385.8 L).

3.3.2. Eleventh-Floor Autonomous Unit

Concerning water tap activation, the sensor simulation results are shown Figure 22. Figure 22a shows the electrical voltage signal response as a function of the faucet operating time. In turn, Figure 22b presents a comparison between the trapezoidal pulse generated by the resistive sensor simulation via an artificial neural network (blue curve) and the faucet flow curve (red dashed curve). By analytical calculation, the volume of water drained during water tap activation for 15 s was equal to 7.472 $\times$ 10^{-4} m^3 (0.07472 L), and by numerical calculation, it was equal to 7.488 $\times$ 10^{-4} m^3 (0.07488 L).

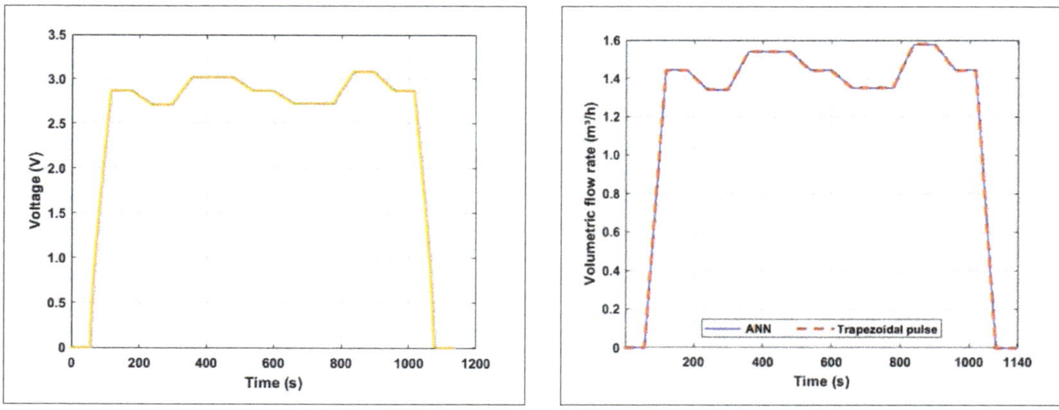

(**a**) (**b**)

Figure 21. (**a**) Sensor response to the activation of the traditional shower and (**b**) trapezoidal pulse via the bend sensor simulation on the ANN (blue curve) and traditional shower flow curve for the autonomous unit with 24 mH_2O (dashed red curve).

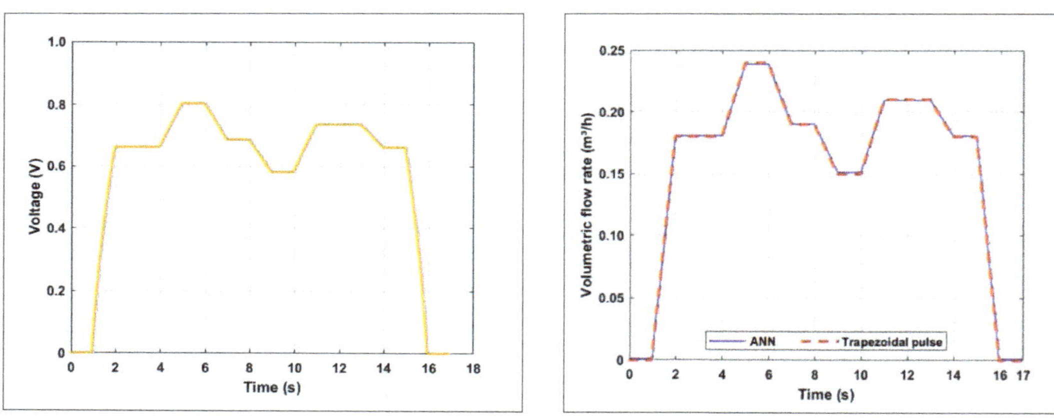

(**a**) (**b**)

Figure 22. (**a**) Sensor response to faucet activation and (**b**) trapezoidal pulse via the bend sensor simulation on the ANN (blue curve) and faucet flow curve for the autonomous unit with 3 mH_2O (dashed red curve).

In turn, the sensor simulation results before the traditional shower activation are shown in Figure 23. In Figure 23a, the electrical voltage signal response is shown as a function of the operating time of the shower. Additionally, Figure 23b shows a comparison between the trapezoidal pulse generated by the resistive sensor simulation via an artificial neural network (blue curve) and the shower flow curve (red dashed curve). By the analytical calculation, the volume of water drained during the activation of the shower for a 15 min shower was equal to 0.3427 m^3 (342.7 L), and by the numerical calculation, it was equal to 0.3423 m^3 (342.3 L).

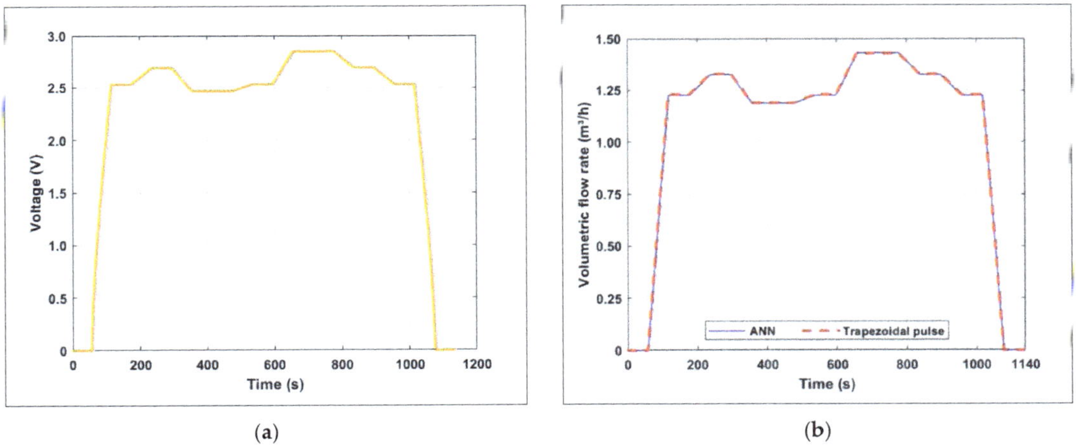

Figure 23. (**a**) Sensor response to the activation of the traditional shower and (**b**) trapezoidal pulse via simulation of the bend sensor on the ANN (blue curve) and traditional shower flow curve for the autonomous unit with 3 mH_2O (dashed red curve).

In summary, Table 3 presents the results described in this section.

Table 3. Volume calculation results.

Floor	Hydraulic Systems	Volume Estimate (L)	
		Analytical	Numerical
4th	Water tap Traditional shower	2.314 386.1	2.319 385.8
11th	Water tap Traditional shower	0.07472 342.7	0.07488 342.3

These results made it possible to compare the volume calculation in two ways (analytical and numerical). With this, it was possible to perform a simulated validation of the measurement concept.

4. Discussion and Conclusions

From an extensive exploratory study of a thin-film resistive sensor (bend sensor) characterization and simulation, the results reveal the feasibility of the proposed volumetric meter using this type of sensor for volume measurement in autonomous units.

For this, two resistive sensors that were one-inch-long and with different coatings—polyester and polyimide—were electromechanically and thermally characterized. In general, the results obtained from the behavior of the polyimide-coated bend sensor, mainly in the electrothermal characterization, can be considered unexpected because this type of polymer is classified as ultra-performance; in other words, they are super engineering plastics. Polyimide is a material that has significant and particular characteristics, such as high thermal stability, which is one of the main qualities of polyimide, allowing this material to be used for long periods at temperatures up to 200 °C and for a short term at temperatures up to 480 °C without affecting its mechanical behavior [56,57]. Additionally, study [31] corroborates this anomalous behavior of the polyimide-coated bend sensor because in this article, two two-inch-long polyimide-coated sensors were electrothermally characterized and showed the same behavior as the one-inch sensors presented in this article.

Therefore, for the proposed hydrometer, the electrothermal characterization test is more significant, as it is more similar to the reality of residential building pipes. Consequently, because of the results from this characterization, the bend sensor with polyimide coating is unfeasible for measuring the water consumption of autonomous units, as the water meters must be built with materials resistant to temperature variations between +1 °C and +40 °C according to specifications in Brazil by the National Institute of Metrology, Quality, and Technology (INMETRO) [58]. Therefore, considering the intended proposed hydrometer application and the two characterizations, the polyester-coated resistive sensor demonstrated better viability to be used as a transducer.

Concerning the computer simulation, through Ansys, it was fundamental to obtain the relation of the sensor deflection angle, which is curved by the pressure exerted by the fluid flow inside a pipe. The 1-inch-long polyester-coated sensor was simulated using a velocity range between 0.15 m/s and 0.58 m/s. This sensor presented a deflection variation between 5° and 69° and resistance (Rs) between 1.35 kΩ and 9.25 kΩ. These velocities are relevant for piping installation in buildings. In Brazil, for example, the velocity inside the pipe cannot exceed 3 m/s [55].

Regarding the proposed meter operating principle, an electronic circuit simulation was made via LTSpice® software. The simulation resulted in an electrical voltage signal with a larger intensity, a better signal-to-noise ratio, and was as linear as possible as a result of the amplification and linearization processes. In turn, an artificial neural network was trained to obtain the instantaneous volumetric flow rates from the electrical voltage signal. In addition, the results obtained showed promising behavior in the use of these sensors for the desired application.

In addition, a simulated validation of the measurement concept was made considering two autonomous units—one on the fourth floor and another on the eleventh floor—and using two hydraulic systems—a water tap with a 15 s activation and a traditional shower working for 15 min—to validate the simulations performed and estimate the volume from the flow rates of the hydraulic systems. For this, two calculations were performed by different means, the first by analytical calculations and the second by numerical calculations. These results were very similar, which allowed for validation of the simulations performed using Ansys and LTSpice® software and the trained artificial neural network. The 1-inch-long sensor presented a measurement range between 0 m^3/h and 3 m^3/h (Figure 19). In addition, according to the simulation results, this sensor proved to be capable of measuring the flows of the considered hydraulic systems, in other words, demonstrated to be adequate to work for the application proposed in this research. Furthermore, considering Portaria INMETRO n° 246, from October 17th, 2000 [58], the sensor measurement ranges are relevant for water meters, as this regulation establishes conditions for water volume meters with a nominal flow of 0.6 m^3/h at 15.0 m^3/h.

Finally, it was possible to define the meter concept by inserting the bend sensor inside the valve so that it can be adapted to work as a volumetric meter. For this, a fixed part was designed inside the valve cover, as shown in Figure 4 (in Section 2), to work as a support for the sensor, and for its fixation in this part, the use of epoxy resin was suggested, as it is highly resistant and hardens even in water, as was performed in the study by Xu et al. [28]. Furthermore, in practical terms, to standardize the proposed meter model, a wedge with a hole on the side for the passage of the support piece and an opening in the center to allow space for the sensor needs to be manufactured, as well as shown in Figure 4b. In addition, still considering the practical aspects, the electrical connection and power source necessary for the proposed hydrometer to work would be solved by a small battery-powered circuit installed next to the valve. For this, some technologies allow for increased battery life, for example, energy harvesting and polling (transmitting data only when new information is available) [59–62].

In turn, thinking in financial terms and taking into account that the proposed hydrometer will require electronic circuits to measure the voltage, store the readings, and convert them into water consumption, it was possible to estimate the costs of the work to transform the collective measurement water into individual measurements using the hydrometer from this research and a conventional one, as shown in Table 4. These estimates show that the works using the water meter proposed in this article are much cheaper when compared to those using conventional hydrometers.

Table 4. Estimate of the cost of work to the transformation from the collective measurement of water into individual measurement.

The Transformation from the Collective Measurement of Water into Individual Measurement for a Reasonably Large Building			
	With Conventional Hydrometer	**With Proposed Hydrometer**	
Number of autonomous units	160	160	48
Cost of work (USD)	180,000	32,000	9600

Therefore, the validation of the thin-film resistive sensor applicability as a transducer of the proposed meter proves to be viable to carry out measurements of the volume of water drained into pipes in autonomous units using flow control valves that are already present in the installation building hydraulics. The proposed water meter complies with Portaria INMETRO n° 246 from 17 October 2000 (Brazilian normative document) [58] under the conditions that the volume meters of cold potable water flowing in a closed pipe must satisfy.

Suggestions for future work to be developed in the continuity of this were proposed:

- The instrumentation of the gate valve through the insertion of the bend sensor inside it, as well as its adaptation as explained through the incorporation of a fixed part and a rubber wedge with a hole on the side and an opening in the center replacing existing wedges;
- To accomplish tests in a hydraulic section to experimentally corroborate the potential of the bend sensor with a polyester coating to measure water consumption in autonomous building units and evaluate the sensor's long-term behavior (useful life);
- To accomplish additional simulations implementing other scenarios, for example, using flow values lower than those used (leaks), considering air in the pipeline as in reality;
- To realize several bending cycles evaluating the variation in the behavior of the bend sensor, thereby assessing the sensor's reliability for long-term operation.

Author Contributions: Methodology, L.d.S.G., K.A.R.M. and C.R.H.B.; Software, L.d.S.G., K.A.R.M. and C.R.H.B.; Validation, L.d.S.G., K.A.R.M. and C.R.H.B.; Formal analysis, L.d.S.G., K.A.R.M. and C.R.H.B.; Investigation, L.d.S.G., K.A.R.M. and C.R.H.B.; Resources, K.A.R.M. and C.R.H.B.; Data curation, L.d.S.G. and K.A.R.M.; Writing—original draft, L.d.S.G. and K.A.R.M.; Writing—review & editing, L.d.S.G., K.A.R.M. and C.R.H.B.; Visualization, L.d.S.G.; Supervision, K.A.R.M. and C.R.H.B.; Project administration, K.A.R.M. and C.R.H.B.; Funding acquisition, C.R.H.B.. All authors have read and agreed to the published version of the manuscript.

Funding: This study was financed in part by the Coordenação de Aperfeiçoamento de Pessoal de Nível Superior-Brasil (CAPES)-Finance Code 001.

Data Availability Statement: Not applicable.

Acknowledgments: The authors thank, for the financial support provided, the Brazilian funding agencies CNPq, CAPES, FINEP, and FAPERJ. This study was financed in part by the Coordenação de Aperfeiçoamento de Pessoal de Nível Superior-Brasil (CAPES)-Finance Code 001.

Conflicts of Interest: The authors declare no conflict of interest. In addition, the funders had no role in the design of the study; in the collection, analyses, or interpretation of data; in the writing of the manuscript; or in the decision to publish the results.

References

1. Valipour, M.; Krasilnikof, J.; Yannopoulos, S.; Kumar, R.; Deng, J.; Roccaro, P.; Mays, L.; Grismer, M.E.; Angelakis, A.N. The evolution of agricultural drainage from the earliest times to the present. *Sustainability* **2020**, *12*, 416. [CrossRef]
2. Angelakis, A.N.; Zaccaria, D.; Krasilnikoff, J.; Salgot, M.; Bazza, M.; Roccaro, P.; Jimenez, B.; Kumar, A.; Yinghua, W.; Baba, A.; et al. Irrigation of world agricultural lands: Evolution through the millennia. *Water* **2020**, *12*, 1285. [CrossRef]
3. Webster, C. The sewers of Mohenjo-Daro. *J. Water Pollut. Control Fed.* **1962**, *34*, 116–123.

4. De Feo, G.; Angelakis, A.N.; Antoniou, G.P.; El-Gohary, F.; Haut, B.; Passchier, C.W.; Zheng, X.Y. Historical and technical notes on aqueducts from prehistoric to medieval times. *Water* **2013**, *5*, 1996–2025. [CrossRef]
5. Das, M.M.; Das Saikia, M.; Das, B.M. *Hydraulics and Hydraulic Machines*; PHI Learning: New Delhi, India, 2013.
6. Fagan, G.G. *Bathing in Public in the Roman World*; University of Michigan Press: Ann Arbor, MI, USA, 1999.
7. SNS. Diagnóstico Temático—Serviços de Água e Esgoto: Gestão Técnica de Água—Ano de Referência 2020. 2022. Available online: https://arquivos-snis.mdr.gov.br/DIAGNOSTICO_TEMATICO_GESTAO_TECNICA_DE_AGUA_AE_SNIS_2022.pdf (accessed on 5 September 2022).
8. de Moura, M.R.F.; Pereira, I.K.L.; da Silva, S.R.; Baptista, F.M. Measuring individual water consumption: An incentive to urban areas sustainability. *Bund. I* **2014**, *19*, 2071–2081.
9. Ali, A.; Shaikh, I.A.; Abbasi, N.A.; Firdous, N.; Ashraf, M.N. Enhancing water efficiency and wastewater treatment using sustainable technologies: A laboratory and pilot study for adhesive and leather chemicals production. *J. Water Process. Eng.* **2020**, *36*, 101308. [CrossRef]
10. Ozturk, E.; Cinperi, N.C. Water efficiency and wastewater reduction in an integrated woolen textile mill. *J. Clean. Prod.* **2018**, *201*, 686–696. [CrossRef]
11. Gabarda-Mallorquí, A.; Garcia, X.; Ribas, A. Mass tourism and water efficiency in the hotel industry: A case study. *Int. J. Hosp. Manag.* **2017**, *61*, 82–93. [CrossRef]
12. GO Associados. Perdas De Água—Desafios à Disponibilidade Hídrica e Necessidade de Avanço na Eficiência do Saneamento—Ano de Referência 2018. 2020. Available online: https://tratabrasil.org.br/wp-content/uploads/2022/09/Relatorio_Final_-_Estudo_de_Perdas_2020_-_JUNHO_2020.pdf (accessed on 6 November 2021).
13. Vesal, M.; Rahmati, M.H.; Hosseinabadi, N.T. The externality from communal metering of residential water: The case of Tehran. *Water Resour. Econ.* **2018**, *23*, 53–58. [CrossRef]
14. De Souza, C.; Kalbusch, A. Estimation of water consumption in multifamily residential buildings. *Acta Sci. Technol.* **2017**, *39*, 161–168. [CrossRef]
15. Castillo-Manzano, J.I.; Lopez-Valpuesta, L.; Marchena-Gómez, M.; Pedregal, D.J. How much does water consumption drop when each household takes charge of its own consumption? The case of the city of Seville. *Appl. Econ.* **2013**, *45*, 4465–4473. [CrossRef]
16. Orengo, G.; Sbernini, L.; Lorenzo, N.D.; Lagati, A.; Saggio, G. Curvature characterization of flex sensors for human posture recognition. *Univers. J. Biomed. Eng.* **2013**, *1*, 10–15. [CrossRef]
17. Saggio, G.; Riillo, F.; Sbernini, L.; Quitadamo, L.R. Resistive flex sensors: A survey. *Smart Mater. Struct.* **2015**, *25*, 13001. [CrossRef]
18. Saggio, G.; Orengo, G. Flex sensor characterization against shape and curvature changes. *Sens. Actuators A Phys.* **2018**, *273*, 221–231. [CrossRef]
19. Saggio, G. A novel array of flex sensors for a goniometric glove. *Sens. Actuators A Phys.* **2014**, *205*, 119–125. [CrossRef]
20. Guo, Y.R.; Zhang, X.C.; An, N. Monitoring neck posture with flex sensors. In Proceedings of the 9th International Conference on Information Science and Technology, Hulunbuir, China, 2–5 August 2019. [CrossRef]
21. Borges, L.M.; Barroca, N.; Velez, F.J.; Lebres, A.S. Smart-clothing wireless flex sensor Belt network for foetal health monitoring. In Proceedings of the 3rd International Conference on Pervasive Computing Technologies for Healthcare, London, UK, 1–3 April 2009. [CrossRef]
22. Jabin, J.; Chowdhury, A.M.; Adnan, M.E.; Islam, M.R.; Mahmud, S.S. Low cost 3D printed prosthetic for congenital amputation using flex sensor. In Proceedings of the 5th International Conference on Advances in Electrical Engineering, Dhaka, Bangladesh, 26–28 September 2019. [CrossRef]
23. Hu, Q.; Tang, X.; Tang, W. A smart chair sitting posture recognition system using flex sensors and FPGA implemented artificial neural network. *IEEE Sens. J.* **2020**, *20*, 8007–8016. [CrossRef]
24. Starck, J.R.; Murray, G.; Lawford, P.V.; Hose, D.R. An inexpensive sensor for measuring surface geometry. *Med. Eng. Phys.* **1999**, *21*, 725–729. [CrossRef]
25. Yao, C.; Hong, C.; Su, D.; Zhang, Y.; Yin, Z. Design and verification of a wireless sensing system for monitoring large-range ground movement. *Sens. Actuators A Phys.* **2020**, *303*, 111745. [CrossRef]
26. Ramesh, M.V.; Vidya, P.T.; Pradeep, P. Context aware wireless sensor system integrated with participatory sensing for real time road accident detection. In Proceedings of the 10th International Conference on Wireless and Optical Communications Networks, Bhopal, India, 26–28 July 2013. [CrossRef]
27. Fan, Y.; Dai, Z.; Xu, Z.; Zheng, S.; Dehkordy, F.M.; Bagtzoglou, A.; Zhang, W.; Gao, P.; Li, B. High resolution air flow velocity monitoring using air flow resistance-type sensor film (AFRSF). *Sens. Actuators A Phys.* **2019**, *297*, 111562. [CrossRef]
28. Xu, Z.; Fan, Y.; Wang, T.; Huang, Y.; Dehkordy, F.M.; Dai, Z.; Xia, L.; Dong, Q.; Bagtzoglou, A.; McCutcheon, J.; et al. Towards high resolution monitoring of water flow velocity using flat flexible thin mm-sized resistance-typed sensor film (MRSF). *Water Res. X* **2019**, *4*, 100028. [CrossRef]
29. Srinivasan, C.R.; Sen, S.; Kumar, A.; Saibabu, C. Measurement of flow using bend sensor. In Proceedings of the International Conference on Advances in Energy Conversion Technologies, Manipal, India, 23–25 January 2014. [CrossRef]
30. Stewart, R.L.; Fox, J.F.; Harnett, C.K. Time-average velocity and turbulence measurement using wireless bend sensors in an open channel with a rough bed. *J. Hydraul. Eng.* **2013**, *139*, 696–706. [CrossRef]

31. Gonçalves, L.d.S.; Medeiros, K.A.R.; Barbosa, C.R.H. Proposition of water meter for buildings based on a thin-film resistive sensor: Electromechanical and thermal characterizations. In Proceedings of the 11° Congresso Brasileiro de Metrologia, Rio de Janeiro, Brazil, 18–21 October 2021.
32. MatWeb. Overview of Materials for Polyester Film. Available online: http://www.matweb.com/search/DataSheet.aspx?MatGUID=40559706b4fd4aa0a43f5739799728f5&ckck=1 (accessed on 2 March 2021).
33. Mcculloch, W.S.; Pitts, W. A logical calculus of ideas immanent in nervous activity. *Bull. Math. Biophys.* **1943**, *5*, 115–133. [CrossRef]
34. Haykin, S. *Neural Networks: A Comprehensive Foundation*; Prentice Hall: Hoboken, NJ, USA, 1999.
35. Jawad, J.; Hawari, A.H.; Zaidi, S.J. Artificial neural network modeling of wastewater treatment and desalination using membrane processes: A review. *Chem. Eng. J.* **2021**, *419*, 129540. [CrossRef]
36. Taki, M.; Ajabshirchi, Y.; Ranjbar, S.F.; Rohani, A.; Matloobi, M. Heat transfer and MLP neural network models to predict inside environment variables and energy lost in a semi-solar greenhouse. *Energy Build.* **2016**, *110*, 314–329. [CrossRef]
37. Ghritlahre, H.K.; Prasad, R.K. Prediction of thermal performance of unidirectional flow porous bed solar air heater with optimal training function using artificial neural network. *Energy Procedia* **2017**, *109*, 369–376. [CrossRef]
38. Cetiner, C.; Halici, F.; Cacur, H.; Taymaz, I. Generating hot water by solar energy and application of neural network. *Appl. Therm. Eng.* **2005**, *25*, 1337–1348. [CrossRef]
39. Bala, B.K.; Ashraf, M.A.; Uddin, M.A.; Janjai, S. Experimental and neural network prediction of the performance of a solar tunnel drier for drying jackfruit bulbs and leather. *J. Food Process. Eng.* **2005**, *28*, 552–566. [CrossRef]
40. Saxena, N.; Varshney, D. Smart home security solutions using facial authentication and speaker recognition through artificial neural networks. *Int. J. Cogn. Comput. Eng.* **2021**, *2*, 154–164. [CrossRef]
41. Reddy, K.R.L.; Babu, G.R.; Kishore, L. Face Recognition based on eigen features of multi scaled face components and an artificial neural network. *Procedia Comput. Sci.* **2010**, *2*, 62–74. [CrossRef]
42. Kumar, A. Comparative analysis of RBF (Radial Basis Function) network and gaussian function in multi-layer feed-forward neural network (MLFFNN) for the case of face recognition. *Int. J. Adv. Res.* **2017**, *5*, 843–873. [CrossRef]
43. Baldo, N.; Miani, M.; Rondinella, F.; Celauro, C. Artificial neural network prediction of airport pavement moduli using interpolated surface deflection data. *IOP Conf. Ser. Mater. Sci. Eng.* **2021**, *1203*, 022112. [CrossRef]
44. Bisoi, R.; Dash, P.K. A hybrid evolutionary dynamic neural network for stock market trend analysis and prediction using unscented Kalman filter. *Appl. Soft Comput. J.* **2014**, *19*, 41–56. [CrossRef]
45. Hu, H.; Tang, L.; Zhang, S.; Wang, H. Predicting the direction of stock markets using optimized neural networks with Google Trends. *Neurocomputing* **2018**, *285*, 188–195. [CrossRef]
46. Moghar, A.; Hamiche, M. Stock market prediction using LSTM recurrent neural network. *Procedia Comput. Sci.* **2020**, *170*, 1168–1173. [CrossRef]
47. Chhajer, P.; Shah, M.; Kshirsagar, A. The applications of artificial neural networks, support vector machines, and long–short term memory for stock market prediction. *Decis. Anal. J.* **2022**, *2*, 100015. [CrossRef]
48. Wang, W.; Sun, H.; Guo, J.; Lao, L.; Wu, S.; Zhang, J. Experimental study on water pipeline leak using in-pipe acoustic signal analysis and artificial neural network Prediction. *Meas. J. Int. Meas. Confed.* **2021**, *186*, 110094. [CrossRef]
49. Pérez-Pérez, E.J.; López-Estrada, F.R.; Valencia-Palomo, G.; Torres, L.; Puig, V.; Mina-Antonio, J.D. Leak diagnosis in pipelines using a combined artificial neural network approach. *Control Eng. Pract.* **2021**, *107*, 104677. [CrossRef]
50. Ismael, M.; Mokhtar, A.; Farooq, M.; Lü, X. Assessing drinking water quality based on physical, chemical and microbial parameters in the red sea state, Sudan using a combination of water quality index and artificial neural network model. *Groundw. Sustain. Dev.* **2021**, *14*, 100612. [CrossRef]
51. Nayak, J.G.; Patil, L.G.; Patki, V.K. Artificial neural network based water quality index (WQI) for river Godavari (India). *Mater. Today Proc.* 2021, *in press*. [CrossRef]
52. Zheng, X.; Nguyen, H. A novel artificial intelligent model for predicting water treatment efficiency of various biochar systems based on artificial neural network and queuing search algorithm. *Chemosphere* **2022**, *287*, 132251. [CrossRef]
53. López, O.A.M.; López, A.M.; Crossa, J. *Multivariate Statistical Machine Learning Methods for Genomic Prediction*; Springer International Publishing: Cham, Switzerland, 2022.
54. Sutton, R.S.; Barto, A.G. *Reinforcement Learning: An Introduction*, 2nd ed.; The MIT Press: Cambridge, MA, USA, 2018.
55. *NBR 5626*; Instalação Predial de Ágia Fria. Associação Brasileira de Normas Técnicas—ABNT: Rio de Janeiro, Brazil, 1998; Volume 41.
56. Sroog, C.E. Polyimides. *Prog. Polym. Sci.* **1991**, *16*, 561–694. [CrossRef]
57. Liaw, D.J.; Wang, K.L.; Huang, Y.C.; Lee, K.R.; Lai, J.Y.; Ha, C.S. Advanced polyimide materials: Syntheses, physical properties and applications. *Prog. Polym. Sci.* **2012**, *37*, 907–974. [CrossRef]
58. Instituto Nacional De Metrologia Normalização E Qualidade—Inmetro. Portaria n° 246 de 17 de outubro de 2000. *Diário Of. da República Fed. do Bras.* 2000. Available online: http://www.inmetro.gov.br/legislacao/rtac/pdf/rtac000667.pdf (accessed on 2 July 2021).
59. Buennemeyer, T.K.; Nelson, T.M.; Gora, M.A.; Marchany, R.C.; Tront, J.G. Battery polling and trace determination for bluetooth attack detection in mobile devices. In Proceedings of the 2007 IEEE Workshop on Information Assurance, New York, NY, USA, 20–22 June 2007. [CrossRef]
60. Morris, N.M Interrupts and polling. In *Microprocessor and Microcomputer Technology*; Palgrave: London, UK, 1981; pp. 173–186.

61. Mateu, L.; Moll, F. Review of energy harvesting techniques and applications for microelectronics. In Proceedings of the SPIE 5837, VLSI Circuits and Systems II, Sevilla, Spain, 30 June 2005. [CrossRef]
62. Kim, T.Y.; Kim, S.K.; Kim, S.W. Application of ferroelectric materials for improving output power of energy harvesters. *Nano Converg.* **2018**, *5*, 30. [CrossRef] [PubMed]

water

MDPI

Article

Numerical Simulation Study on the Influence of Construction Load on the Cutoff Wall in Reservoir Engineering

Yongshuai Sun [1,*], Anping Lei [2], Ke Yang [3] and Guihe Wang [4]

1 College of Water Resources & Civil Engineering, China Agricultural University, Beijing 100083, China
2 China Highway Engineering Consultants Corporation, Beijing 100089, China
3 Land and Resources Exploration Center, Hebei Bureau of Geology and Mineral Resources Exploration, Shijiazhuang 050081, China
4 School of Engineering and Technology, China University of Geosciences, Beijing 100083, China
* Correspondence: causys666@163.com

Abstract: Relying on the Beijing-Shijiazhuang Expressway widening project near the impervious wall of a reservoir, this paper uses FLAC3D two-dimensional and three-dimensional numerical simulation methods to establish the whole process model of the impervious wall of the reservoir affected by the construction load of the high-way reconstruction section. The stress and strain state of the cut-off wall in the high-way reconstruction section and the nearby reservoir is simulated in detail, the overall deformation of the cut-off wall in the reservoir is directly reflected, and the interaction and differential deformation between the wall structures are reflected. The safety and stability of the cutoff wall of the reservoir affected by the construction load are evaluated so that various advanced mechanical behaviors of the cutoff wall can be predicted. Research results show that the horizontal displacement value of the wall gradually increases from bottom to top, and the maximum value appears at the top of the wall. The horizontal displacement value of the 1–3 walls is relatively large, with the maximum value of 22.368 mm, and the horizontal displacement value of the 4–10 walls shows little difference. This is on account of the gravity of the backfill, the strata in the whole project area having settled, and the settlement at the bottom of the cut-off wall being 2.542 mm. At the root of the rigid cut-off wall, the compressive stress concentration occurs, with the maximum value between 1.75 MPa and 2.15 MPa. Due to the size of the structure, the maximum tensile stress of 0.237 MPa appears in the local area near the guide wall of the rigid cut-off wall, which will not endanger the rigid cut-off wall because of its small value. The maximum stress in the rigid impervious wall and the plastic impervious wall are 1.90–2.15 MPa and 1.00–1.12 MPa, respectively. Apart from the small tensile stress at the connecting guide wall between the rigid cut-off wall and the plastic concrete cut-off wall, the cut-off wall is under pressure, especially the plastic cut-off wall. Combined with the analysis of the stress state of the wall, it can be determined that the anti-seepage wall (rigid cut-off wall and plastic concrete cut-off wall) is stable and safe during the construction period.

Keywords: cut-off wall; monitoring section; numerical simulation; dynamic construction stability

Citation: Sun, Y.; Lei, A.; Yang, K.; Wang, G. Numerical Simulation Study on the Influence of Construction Load on the Cutoff Wall in Reservoir Engineering. *Water* **2023**, *15*, 993. https://doi.org/10.3390/w15050993

Academic Editor: Chin H Wu

Received: 12 February 2023
Revised: 2 March 2023
Accepted: 3 March 2023
Published: 5 March 2023

1. Introduction

With the continuous construction and renovation of the surrounding projects of the reservoir [1,2], the two sides and the top surface of the cut-off wall of a reservoir will be affected by the construction disturbance of the backfilling project [3–5]. For the upper part of the upcoming large-scale construction and new load project, whether the original cut-off wall can still maintain a stable strength [6–8], whether there will be excessive deformation and damage [9–11], or whether the cut-off function is affected, reduction or loss [12–15] are important issues that concern reservoir builders and water protection workers. There are also important issues related to reservoir water storage operations and the safety and stability of cofferdams [16–18].

Due to the development of computer performance, the mathematical analysis in rock mechanics has made great progress [19–22]. Different numerical simulation methods are obtained based on different mathematical models [23–26], and the finite element method changed from linear to nonlinear. This development [27–32] has solved a great problem for rock engineering. With the rapid development of numerical simulation, the academic community also applied this method to study the impact of the construction around the cut-off wall more intuitively [33,34]. Fu Z Q et al. [35] used the linear elastic model to calculate the earth–rock dam and clay concrete cut-off wall of the hydropower station. The pebble and gravel cover layer in the calculation was used by the Duncan–Chang model and obtained the linear elastic characteristics and stress-strain of the cut-off wall. Relying on the Xiongjiagang rock dam project, Han Y [36] established the seepage model of the rock dam using Geo-studio software and analyzed the mechanism of the stability of the seepage wall of the dam. Shepherd D A [9] used the Druker–Prager model to carry out the back analysis and calculation of the measured displacement of the cut-off wall of a hydropower station. The calculation process involved the soil Poisson's ratio, density, viscous force, friction, angle, and other parameters, focusing on back-calculating the elastic modulus parameters that play a major role. Ma Xiaohua et al. [37,38] studied the influence of the change of elastic modulus on the cut-off wall by increasing the thickness parameter. Relying on the reinforcement project of the Hualiangting Reservoir. Ren Xiang et al. [39,40] simulated the operation process of the dam and analyzed the change rule of the ultra-deep concrete cut-off wall under the dam stress, water pressure, and foundation constraint conditions. Chen Xiaolian et al. [41] used the GMS simulation to study the permeability coefficient and elastic water supply of each aquifer in the research area of a hydropower station in Sichuan, evaluating the cut-off effect. Liang Yan et al. [42] selected Midas to simulate and establish the action model of the change of support conditions on the cut-off wall of the dam foundation of Longkou Station and evaluated the influence of the support stiffness coefficient on the cut-off wall. Li Mingyuan et al. [43] simulated and predicted the immersion degree and scope of the calculation area under the normal water storage level of the Xingan Navigation and Power Project and used cut-off walls and relief well-engineering measures to control the immersion of the reservoir area to evaluate the difference in the effect of immersion treatment on the reservoir area under working conditions. Yasushi Arai et al. [44–46] used three-dimensional simulation to analyze the impact of excavation construction on wall deformation and discussed the influence of two excavation sequences and three different wall thicknesses on the results. R. Schafer et al. [47,48] analyzed the influence of the construction sequence of the cut-off wall on the cut-off wall in a soft soil foundation in the numerical calculation and simulated the influence of the coupling effect of groundwater pressure and geo-stress on the wall.

The strength, stability, and deformation of the cut-off wall of the reservoir that has been built under the influence of large-scale construction and new load in the near and upper part are important factors affecting the safety and stability of the reservoir. At present, the academic community mostly focuses on the study of the cut-off performance of the cut-off wall. Yet, there is little systematic research on this aspect. Based on this, this paper, relying on the Beijing-Shijiazhuang Expressway widening project near the impervious wall of a reservoir, uses FLAC3D two-dimensional and three-dimensional numerical simulation methods to establish the whole process model of the impervious wall of the reservoir affected by the construction load of the highway reconstruction section. The stress and strain state of the cut-off wall in the highway reconstruction section and the nearby reservoir is simulated in detail. The overall deformation of the cut-off wall in the reservoir is directly reflected, and the interaction and differential deformation between the wall structures are reflected. The safety and stability of the cut-off wall of the reservoir affected by the construction load are evaluated so that various advanced mechanical behaviors of the cut-off wall can be predicted.

2. Overview of Cut-Off Wall Project

The length of the Jingshi Highway reconstruction project in a section of the reservoir is about 3094 m. The subgrade reconstruction and widening section of a reservoir area of Jingshi Highway are closely parallel to the cut-off wall of the west embankment of a storage reservoir, and the length of the intersection is about 2247 m. The large part is about 114 m long. The cut-off wall of this part is divided into upper and lower parts. The lower part is the original plastic concrete cut-off wall, and the upper part is composed of 10 reinforced concrete walls connected to each other. The joints between the walls are flexible joints, which allow slight movement between the walls.

To more clearly reflect the mechanical effect of the wall affected by the construction, the numerical simulation is based on the two typical monitoring sections selected in the expansion section, as shown in Figure 1 (located at XD0 + 744.000 (the fifth section) and XD0 + 770.000 (the seven section). Sections 1 and 2 are the main research objects, and the numerical calculation and analysis are carried out for these 3 typical sections and the project as a whole. The overall 3D numerical simulation model of the project will include all the walls of the research object, that is, the first to tenth walls from north to south and filling soil (enclosing soil) in the vicinity.

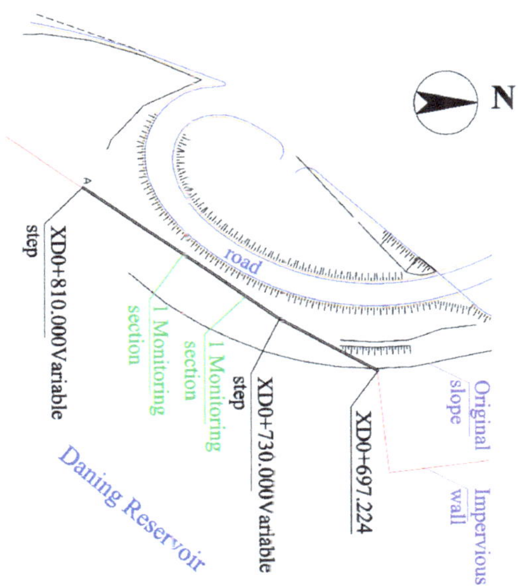

Figure 1. Location layout of monitoring sections and selection of numerical simulation sections.

The construction area is located on the platform of the west embankment of a reservoir, which belongs to the geological unit of the Xiaoqing River alluvial plain. The terrain of the site is flat, the ground elevation is about 53.41~55.42 m, slightly inclined from north to south, and the ground slope is about 1.06‰. The anti-seepage wall of the renovation and expansion section of Dujiakan is located on the west embankment platform of a reservoir area, 11 m away from the foot of the slope on the east side of the original Jingshi Expressway. The anti-seepage wall is 0.60 m thick; the wall depth is 9–12.5 m, 2.5 m into semi-cemented bedrock at the bottom of the wall.

It is constructed by two drills and one grasping method. The length of the groove section is 6.2 m and the designed compressive strength of the plastic concrete wall is 2 MPa. The height of the guide wall of the cut-off wall is 1.0 m, the width of the base is 1.0 m, and the height is 0.3 m. The thickness of the guide wall above the base is 0.4 m and it will be poured with C15 reinforced concrete. Plastic concrete anti-seepage walls have been pre-constructed. The rigid anti-seepage wall that will be heightened later is also made of C15 reinforced concrete, the wall thickness is 0.6 m, the wall height is 9–11 m, and the design compressive strength is 15 MPa. The structure between the plastic concrete cut-off wall and the rigid cut-off wall is shown in Figure 2a. The average reinforced concrete wall is 12 m, and each wall is constructed separately. Flexible joints are used between the two walls. The joint structure is shown in Figure 2b.

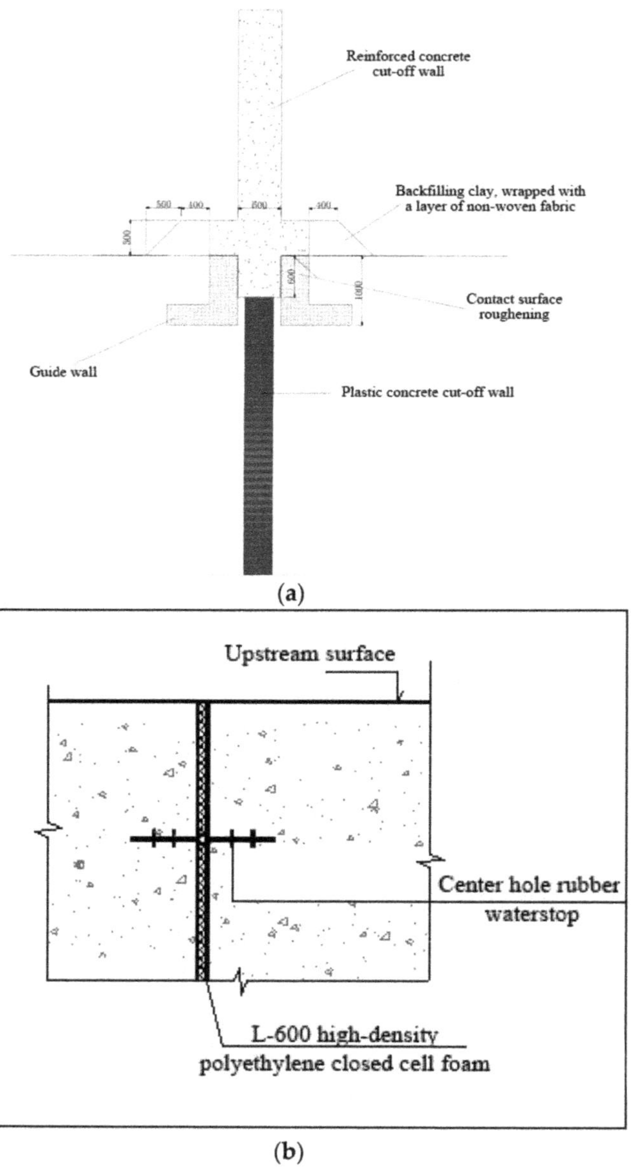

Figure 2. Schematic diagram of the cut-off wall structure. (**a**) Schematic diagram of the interface between reinforced concrete wall and plastic concrete wall. (**b**) Schematic diagram of joints between rigid cut-off walls.

3. Construction Plan and Calculation Model

3.1. Construction Plan

The simulation analysis shall consider the actual construction process of the project as much as possible. For this reason, the specific construction scheme is given.

(1) Construction scheme of the cut-off wall:

The construction plan of the cut-off wall is mainly divided into two parts. First, the original plastic concrete cut-off wall of the lower part adopts the traditional underground continuous wall construction method, that is, pouring concrete after continuous mechanical groove formation. Second, the upper reinforced concrete wall adopts a formwork pouring form. Considering the safety of the wall, the reinforced concrete wall is divided into upper and lower parts and completed step by step, as shown in Figure 3.

The detailed construction method of the cut-off wall is as follows:

1. Determine the construction position: according to the design drawings, set out the baseline of the cut-off wall;
2. Construction guide wall: The guide wall earthwork is excavated along the baseline by trench sections. The excavation depth is based on the wall elevation of each trench section. Support the formwork and pour C15 concrete. The finished guide wall should be at the same height as the surrounding ground surface to facilitate the subsequent movement of the punching and grabbing machinery;
3. Mechanical grooving: Use a punching machine to grab the wall groove section in the guide groove between the guide walls. The length of the groove section is strictly controlled within the specification requirements. Mud is used to protect the hole wall to prevent collapse;
4. Construction of plastic concrete wall: The plastic concrete wall is poured with C2 concrete. The pouring process should be coordinated with the mechanical groove to ensure the quality of the wall at the joint. It is advisable to use a conduit to facilitate the pouring;
5. Construction of reinforced concrete wall: The reinforced concrete wall is located on the surface, and the concrete label is C15. According to the design size of the wall, the wall formwork is divided into grooves, the rigid cage is bound, and the concrete is poured.

(2) Subgrade construction scheme of Dujiakan toll station:

1. Subgrade backfill both sides of cut-off wall at the same time, the thickness of each layer of fill is about 0.5 m. The thickness after compaction is about 0.3 m;
2. Both sides of the subgrade and impervious wall shall be compacted. Small equipment or manual tamping is used for narrower areas. Large-scale equipment or rolling equipment is used for relatively wide areas.

(a)

Figure 3. *Cont.*

(**b**)

Figure 3. Step-by-step construction of rigid cut-off walls. (**a**) Construction of the lower wall. (**b**) Construction of the upper wall.

3.2. Calculation Model

Since the difference in the longitudinal direction of the project is not obvious, consider each monitoring section as a plane strain problem to establish a two-dimensional calculation model. The stress-strain state of the Dujiakan reconstruction and extension section of the Beijing-Shibai Expressway and the related reservoir cut-off wall is simulated in detail. The 3D model mainly shows the overall deformation of the project, considering the joints and conduction between reinforced concrete walls. The impact of the size effect of walls and other structures reflects the interaction and differential deformation between wall structures. The fifth section (1 section) and the seventh section wall (2 section) were selected as the research objects in the analysis.

3.2.1. Monitor the Geometric Dimensions of Section Models 1 and 2

The monitoring sections 1 and 2 were located at the fifth trusses and seventh trusses of the wall, respectively. The same calculation model was used for the analysis because of the same geometrical size, backfilling process, and material parameters. The reinforced concrete wall top elevation is 59 m, with a height of 10 m. The plastic concrete wall has a depth of 11.5 m.

The horizontal distance between the slope line of the original Jingshi subgrade and the wall is 11 m, and the model is 116 m long, 2 m wide, and 49.5 m high. The construction sequence consists of pouring reinforced concrete → backfilling 0 → backfilling 1 and backfilling 2 at the same time → excavating and laying the main rainwater pipe → backfilling 3 → backfilling 4, the geometric models of sections 1 and 2 are shown in Figure 4a. Monitoring1, 2 cross-section model grid division, the positional relationship between backfill soil, and cut-off wall are shown in Figure 4b to monitor the relationship between backfill soil and the cut-off wall in sections 1 and 2.

3.2.2. Geometric Dimensions of 3D Cut-Off Wall Model

The 3D model contains all the walls of the research object, from north to south. They are the 1st to 10th wall. The model is 88.1 m long, 114 m wide, and 32 m high. The angle between the 3rd and 4th wall intersections is 174°. The elevation of the top of the reinforced concrete wall is 59 m. Walls 1–3 are 12 m high and walls 4–10 are 10 m high. The elevation of the bottom line of the plastic concrete wall is 37.5 m, of which the height of the 1st–3rd walls is 9.5 m. The height of the tree is 11.5 m. The construction sequence is initial ground stress balance → construction of plastic concrete walls → construction of reinforced concrete walls → backfilling 0→backfilling 1 and backfilling 2 simultaneously → backfilling 3→backfilling 4, as shown in Figure 5a using the grid division of the 3D model. Figure 5b shows the overall grid division of the cut-off wall in the model.

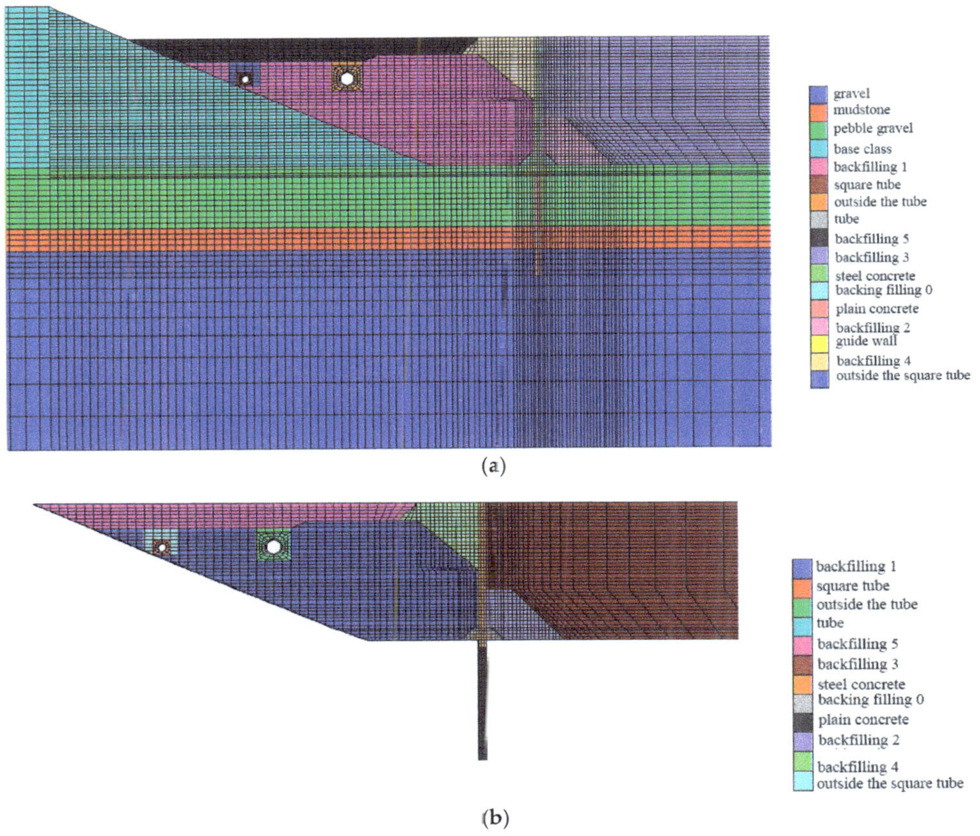

Figure 4. Monitoring 1, 2 section model. (**a**) Grid division of monitoring section model 1 and 2. (**b**) Monitoring the relationship between the backfill soil and cut-off wall at sections 1 and 2.

The main construction of this project consists of the filling and rolling process of the embankment of the Jingshi Expressway. Considering the change of the backfilling soil's physical parameters during the rolling process, the deformation characteristic parameters (bulk modulus, shear modulus) are simulated by the Duncan–Zhang Et-Kt model. As the rolling progresses, the bulk modulus and soil weight gradually increase, so the backfilled soil will have a large displacement at the initial stage of rolling, which is consistent with the actual rolling process.

Displacement constraints are used on each side of the two-dimensional model to limit the boundary horizontal displacement: the bottom surface is a fixed boundary to limit displacement; the top surface and slope are free boundaries; the contact surface element is added between the reinforced concrete wall and the soil.

The backfill soil shall be backfilled sequentially according to the actual construction conditions, and the backfill thickness of each small step shall be 0.3 m, considering the rainwater pipeline construction process during the backfill period. The four sides of the 3D model adopt displacement constraints to fix their horizontal displacement: the bottom surface boundary is fixed to limit the displacement in three directions; the top surface is a private boundary. Compared with the two-dimensional model, the construction process of the backfill is simplified, the thickness of each layer of backfill is 1 m, and the construction impact of the rainwater pipeline is ignored.

In the simulation of the rolling process, according to the actual rolling effect, the process of soil density as it gradually increases is used for simulation. In the process of carrying out the analysis, the density of the soil material is determined according to the instant depth of the fill until it reaches the maximum value (target density) and the rolling process ends.

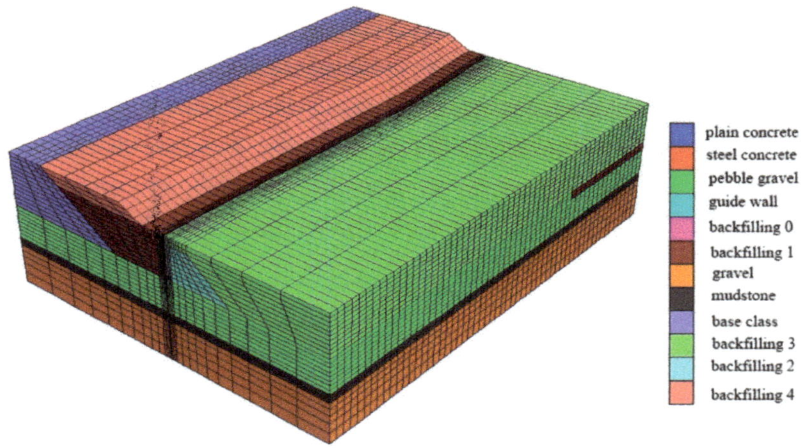

(a)

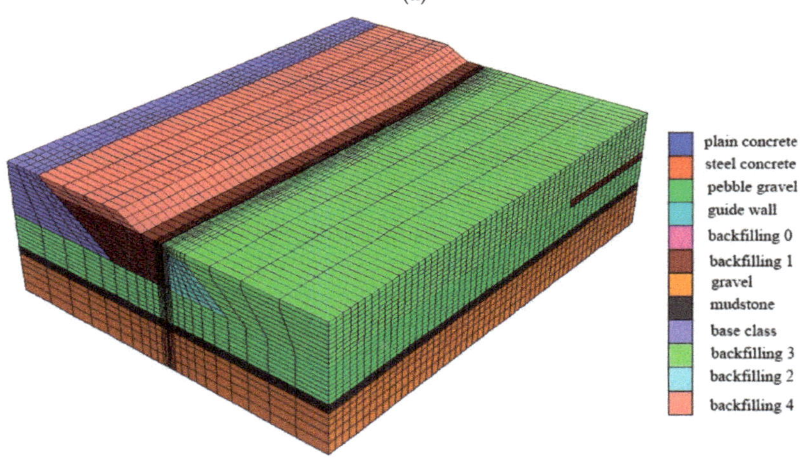

(b)

Figure 5. Three-dimensional cut-off wall model. (**a**) Meshing of 3D model. (**b**) Mesh division of cutoff wall.

4. Dynamic Construction Stability Analysis of Cut-Off Wall

4.1. Analysis of Deformation Law

The deformation analysis includes the overall deformation and the cut-off wall deformation. It mainly focuses on the study of the deformation of the cut-off wall and the surrounding soil with the construction change. The rigid cut-off wall is taken as the initial state when it is completed, that is, the deformation of the wall and the stratum in the subsequent diagrams are all caused by the backfilling construction.

Figure 6 respectively shows the cloud diagrams of the final displacement changes in the analysis domain after the construction of the cut-off wall of the reservoir. The maximum displacement of the analysis domain occurs in the backfill area, the maximum horizontal displacement is 634.171 mm, the maximum vertical displacement is 1602.385 mm, and the maximum displacement is 1675.820 mm.

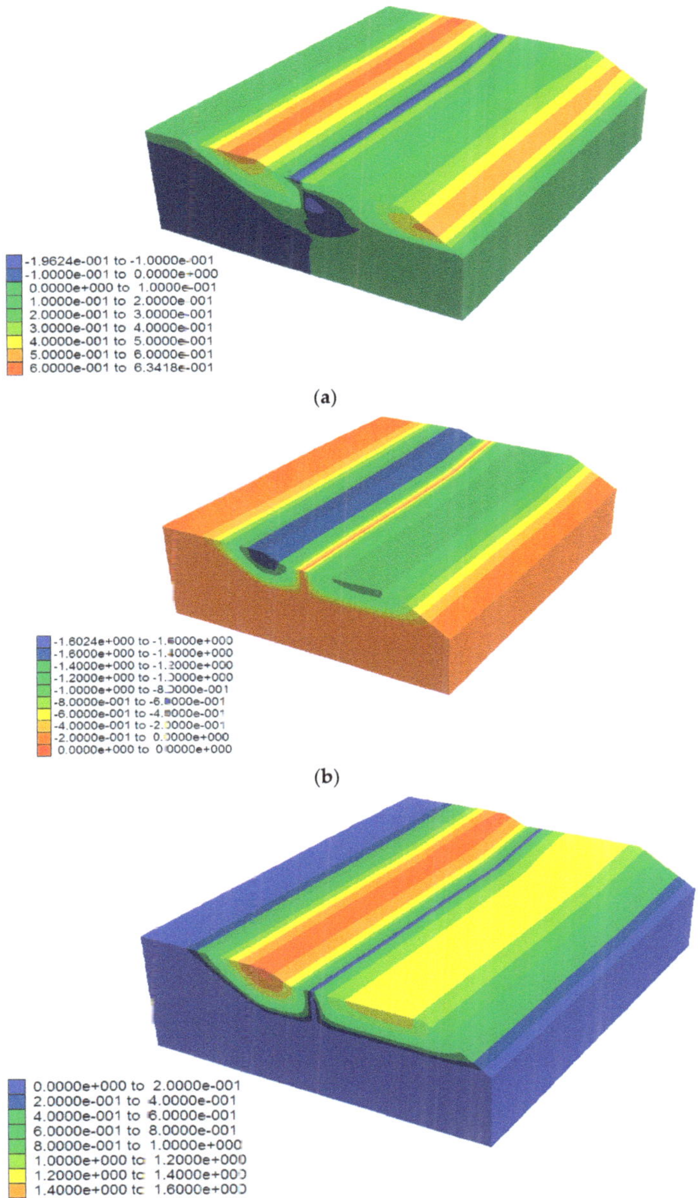

Figure 6. Cloud map of deformation law in analysis area of cut-off wall. (**a**)Analysis Domain Horizontal Displacement Contour. (**b**)Analysis domain vertical displacement cloud map. (**c**) Analyzing Domain and Displacement Contour.

Figures 7–9 show the cloud images of the deformation laws of the two monitoring sections and the cut-off walls at the monitoring sections and the comparison of the deformation laws of the two monitoring sections.

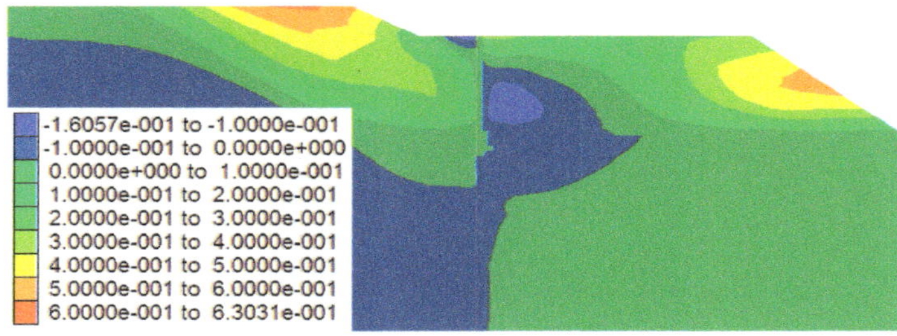

(**a**)

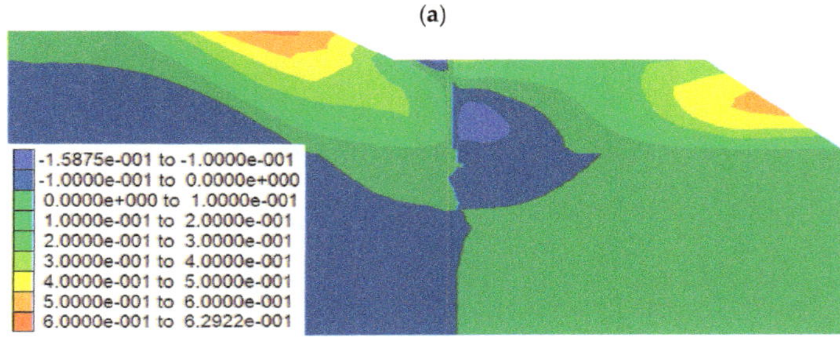

(**b**)

Figure 7. Horizontal displacement nephogram of monitoring surface. (**a**) 1. (**b**) 2.

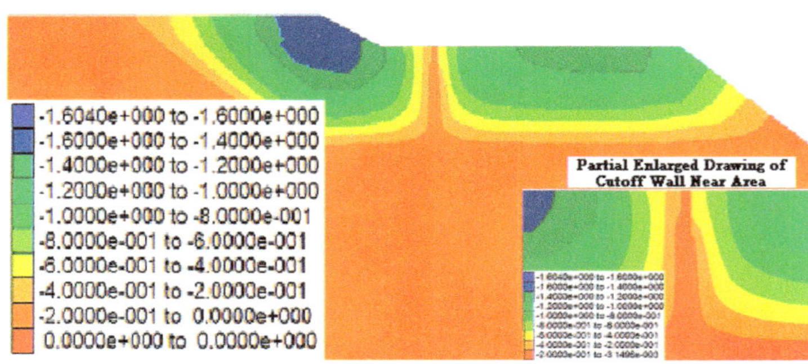

(**a**)

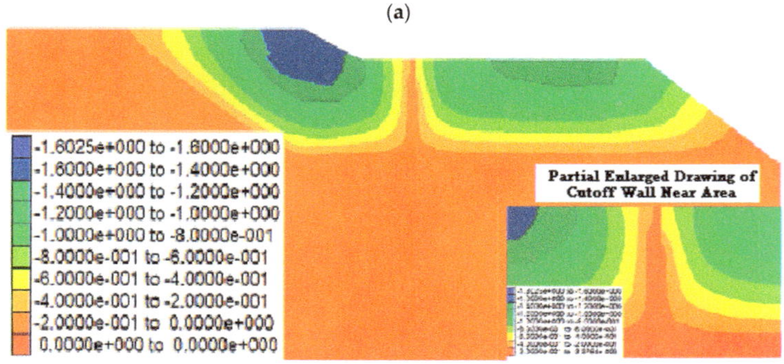

(**b**)

Figure 8. Cloud diagram of vertical displacement of monitoring surface. (**a**) 1. (**b**) 2.

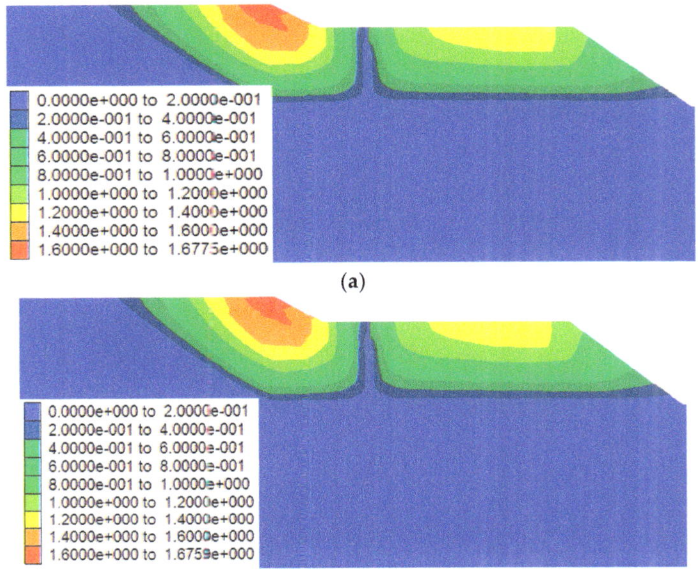

Figure 9. Monitoring surface and displacement nephogram. (**a**) 1. (**b**) 2.

It can be seen from the cloud image analysis of the deformation comparison of the 2 monitoring surface shown in Figures 8 and 9 that the deformation trends during construction are basically consistent. From the analysis of the overall displacement nephogram, it can be concluded that the maximum displacement in the horizontal direction occurs on the upper part of the backfill soil outside the wall, and the maximum vertical displacement occurs at the filling soil on the road. The combined displacement is mainly affected by the vertical displacement. Influenced by displacement, the maximum value also appears at the position where the roadside fill is close to the cut-off wall (Table 1).

Table 1. Comparison of the maximum deformation of the soil in the filling area of the monitoring surface.

Deformation Type Displacement (m)	Monitoring Surface	
	1	2
Horizontal	0.6303	0.6292
Vertical	1.6040	1.6025
Combined	1.6775	1.6759

Figure 10 shows the overall deformation contours of cut-off walls during construction. From the analysis of the overall deformation, it can be seen that horizontal deformation of the plastic cut-off wall is much smaller than that of the rigid cut-off wall, with a maximum difference of more than two orders of magnitude. The vertical displacement takes place from top to bottom. Gradually decreasing, the settlement at the top is about twice as much as that at the bottom.

The deformation trend of the wall can be divided into three stages as the construction steps proceed. The first stage: Fill on both sides of the wall, but the width of the fill on the outside of the reservoir is large. Then the wall is subject to the lateral constraint of the original reservoir embankment and the whole wall is inclined to the inside of the reservoir. The second stage: There is a large amount of fill in the inner side of the reservoir. Due to the gravity effect of the unilateral soil accumulation, the overall inclination of the wall has recovered. At this stage, the wall tends to move towards the outer side of the reservoir. The third stage: The construction stage of the Beijing-Shijiazhuang Expressway subgrade. Due to the soil piling work outside the reservoir, the wall began to tilt towards the inside of the reservoir (Table 2).

From the analysis of Figure 11, it can be seen that the horizontal plane deformation trend of the three-section cut-off wall is similar. As the construction progresses, the wall tends to displace toward the side of the reservoir as a whole. The horizontal displacement decreases from top to bottom, but it does not decrease nonlinearly or linearly. The horizontal deformation of the plastic concrete wall is small, as can be seen from the comparison curve analysis of its change in value given in Figure 11. The horizontal deformation law of the plastic wall of section 1 and section 2 is similar, with the maximum horizontal displacement occurring at the top. Figure 11 shows that the horizontal deformation of the cut-off wall is greatly affected by the construction, that is, the backfilling and rolling process of the surrounding soil. For example, whether it is a plastic cut-off wall or a rigid cut-off wall, after the completion of backfill 4, a large amount of rebound occurred.

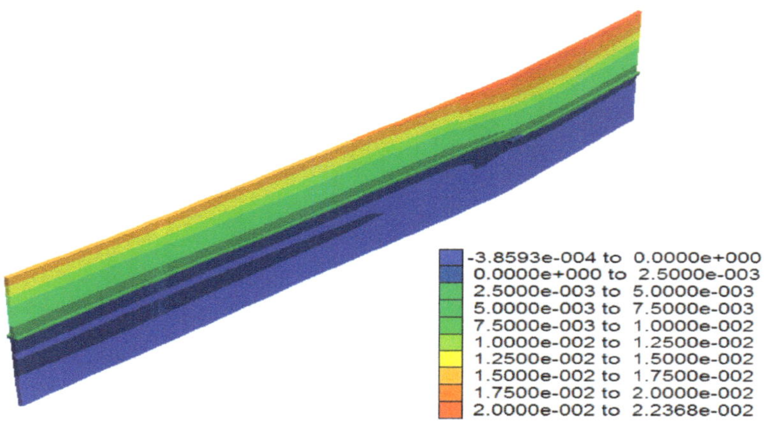

(**a**)

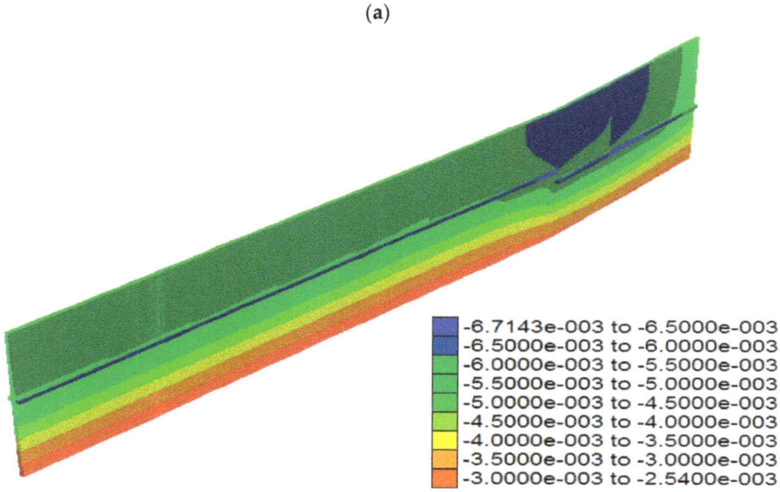

(**b**)

Figure 10. The overall deformation of cut-off wall. (**a**) horizontal displacement of cut-off wall. (**b**) vertical displacement of cut-off wall.

Table 2. Comparison of displacement and settlement of cut-off wall at different monitoring sections.

Position of Displacement	Monitoring Section	
	1	2
Horizontal displacement of the top of the rigid cut-off wall (m)	0.01801	0.01732
Horizontal displacement of the top of the plastic cut-off wall (m)	0.001704	0.001588
Top Settlement of Plastic cut-off wall (m)	0.004966	0.004932

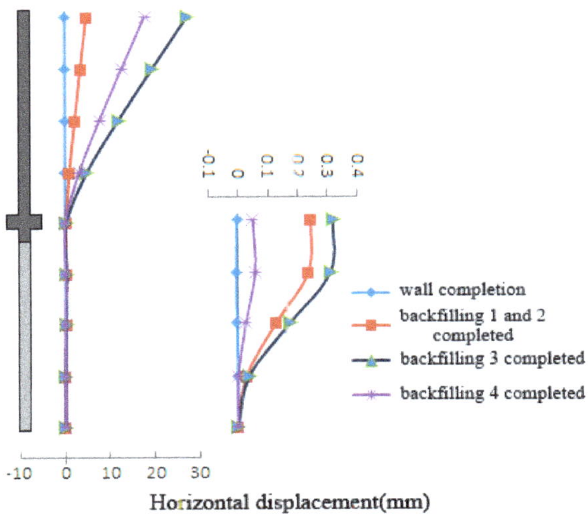

(a)

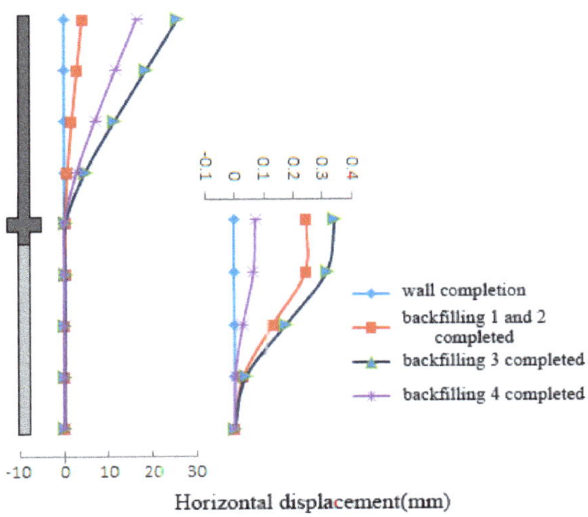

(b)

Figure 11. Horizontal displacement change of cut-off wall at monitoring section. (**a**) Monitoring 1 section cut-off wall. (**b**) Monitoring 2 section cut-off wall.

Figure 12 shows that as the construction progresses, the vertical displacement gradually increases from the lower of the wall upwards, and the vertical deformation of the upper rigid cut-off wall is slightly larger than that of the lower plastic cut-off wall. The maximum vertical deformation of the cut-off walls on the three monitoring surfaces is around 6 mm.

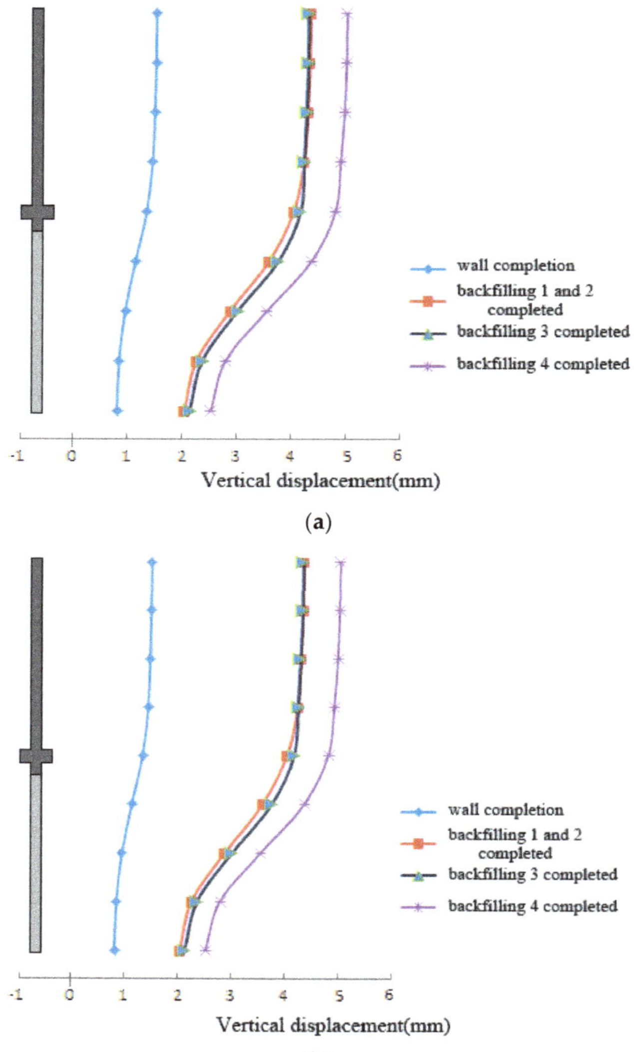

Figure 12. Vertical displacement variation of cut-off wall. (**a**) Monitoring 1 section cut-off wall. (**b**) Monitoring section 2 cut-off wall.

Figure 13 shows the displacement distribution pattern at the joints of cut-off walls. For the cut-off walls in the study section, the structural dimensions and backfilling methods of the first to third walls are the same. The fourth to tenth walls are also similar to each other. It can be seen from the displacement cloud map that the horizontal displacement of the first three walls is relatively large, so there is a displacement discontinuity at the joint between the third and fourth walls. The difference gradually decreases from the top to the bottom of the wall. The bottom gradually decreases and is almost zero at the root of the wall, and the displacement difference at the top is about 1.314 mm.

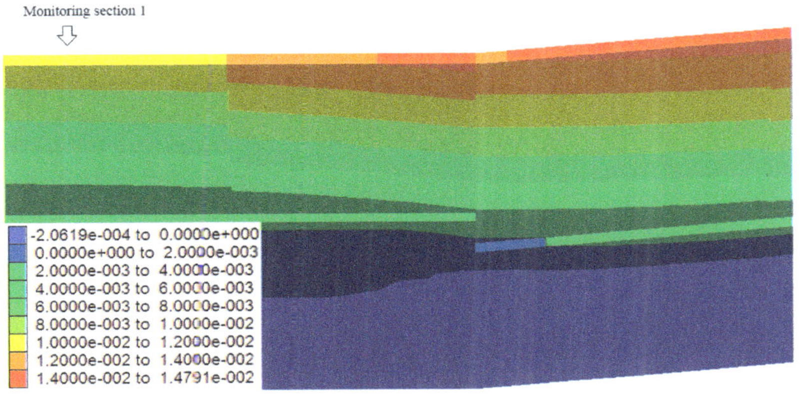

Figure 13. Horizontal displacement of wall near corner.

According to the structure of the wall joints (Figure 14), a certain friction force needs to be overcome when the walls are displaced, and then the wall displacement has a certain joint effect. There is also discontinuity of displacement between the five sections (monitoring 1 section wall), which is mainly because the length of the 4th section wall is only 10 m, which is shorter than other walls. The large deformation of the third section is caused by the connection. The friction of the seam implicates the deformation of the fourth wall. In addition, the filling soil on both sides of the third wall will also have a certain degree of influence on the deformation of the fourth wall.

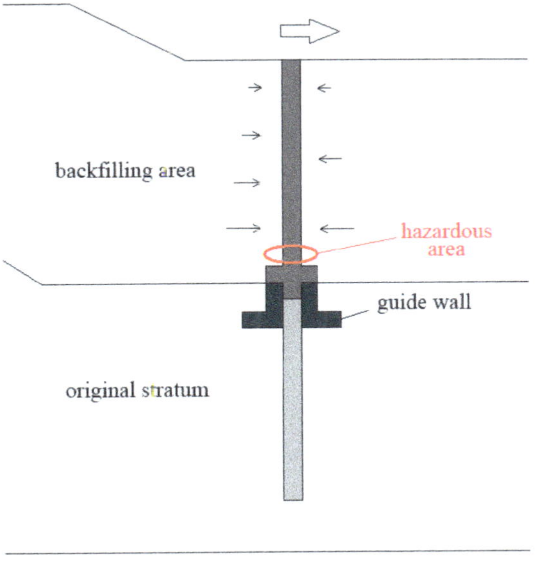

Figure 14. Schematic diagram of maximum stress area of cut-off wall.

From the above analysis, it can be seen that, as construction progresses, the deformation state of the cut-off wall shows a relatively stable change trend, and the deformation of the reinforced concrete wall is slightly larger than that of the plastic concrete. The maximum horizontal offset of the rigid wall is 2.191 mm, while that of the plastic wall is only about 0.3 mm, which will not affect the structural stability of the wall. The simulation shows that the deformation of the cut-off wall is greatly affected by the dynamic construction on both sides. In the early stage of construction, the wall tends to shift towards the reservoir side and then reaches the maximum value of about 26 mm. Until the massive filling stage at the reservoir side takes place and due to the influence of the filling pressure at the reservoir side, the existing wall is offset slightly. The final offset of the rigid cut-off wall at the end of the project remained at about 21 mm. The maximum vertical deformation of the cut-off wall is about 6 mm.

4.2. Analysis of Stress Distribution Law

Figure 15 shows the overall stress distribution in the analysis area after construction. It can be seen from the stress diagram analysis that the maximum principal stress occurs in the filling area, and the large stress value is only about 1.155 Mpa. It can be seen from Figures 16–18 of the stress distribution law of the monitoring surface that the maximum principal stress is below 1.5 Mpa, and maximum contact stress of 0.16 Mpa appears in the local area of the rigid cut-off wall close to the guide wall, but the value is small and will not endanger the rigid cut-off wall. The maximum shear stress on the ZX plane of the cut-off wall is only about 0.26 Mpa, and it is located in the lower area of the rigid cut-off wall.

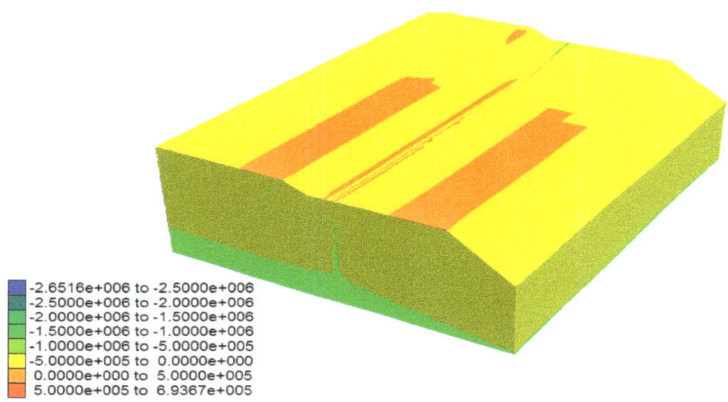

(a)

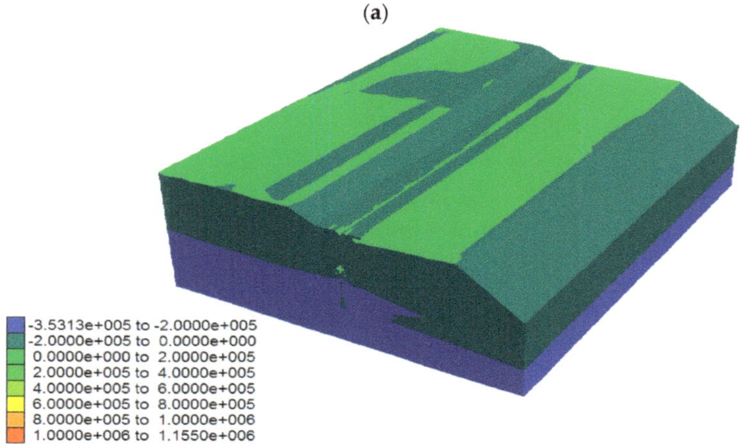

(b)

Figure 15. *Cont.*

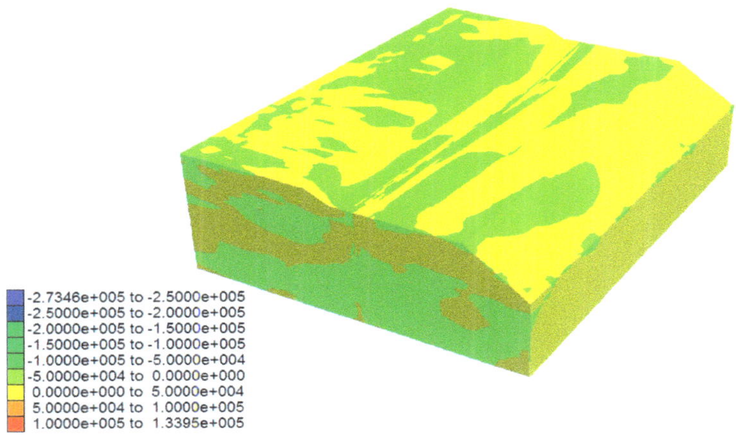

(**c**)

Figure 15. Stress distribution in the analysis area. (**a**) Nephogram of maximum principal stress. (**b**) Minimum principal stress nephogram. (**c**) YZ in-plane shear stress nephogram.

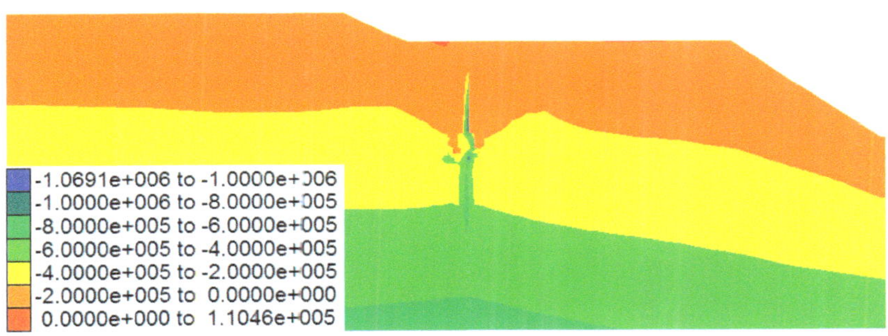

(**a**)

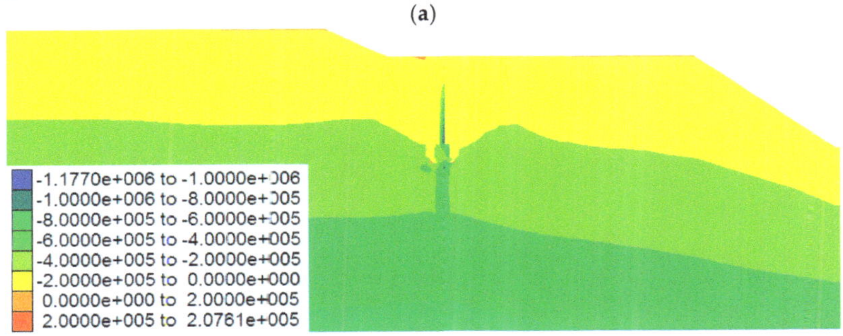

(**b**)

Figure 16. Comparison of maximum principal stress of monitoring surface. (**a**) 1. (**b**) 2.

Figure 19 shows that the maximum compressive stress mostly occurs at the lower part of the rigid cut-off wall, and the stresses in the lower part of the first to third piles are slightly greater than those of the walls at other positions. The main reason is that the filling layer in the area is thicker than other parts. The maximum stress on the rigid cut-off wall is between 1.5–2.6500 MPa, and the maximum stress on the plastic cut-off wall is between 0.5600–1.5 Mpa, both of which are within the safe range. The shear stress distribution in the ZX plane is also larger than the cut-off walls at the second and third walls, and the maximum shear stress is about 0.27 MPa.

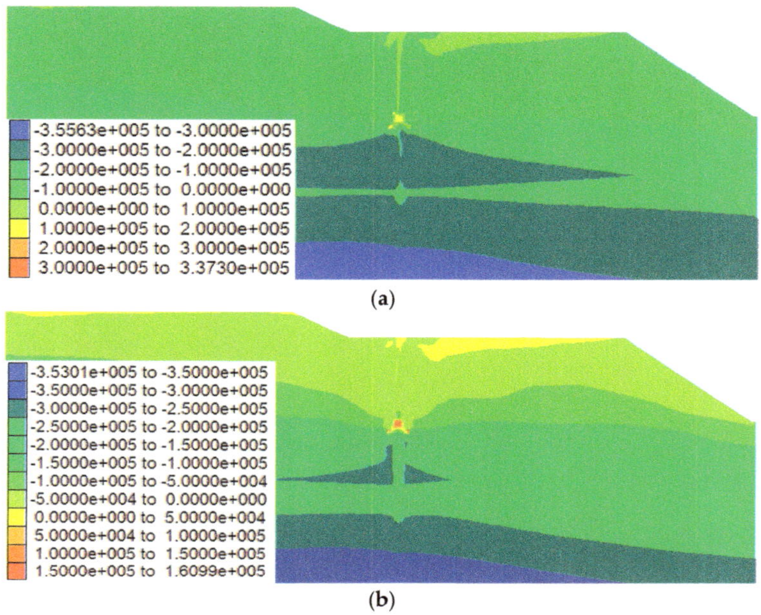

Figure 17. Comparison of minimum principal stress of monitoring surface. (**a**) 1. (**b**) 2.

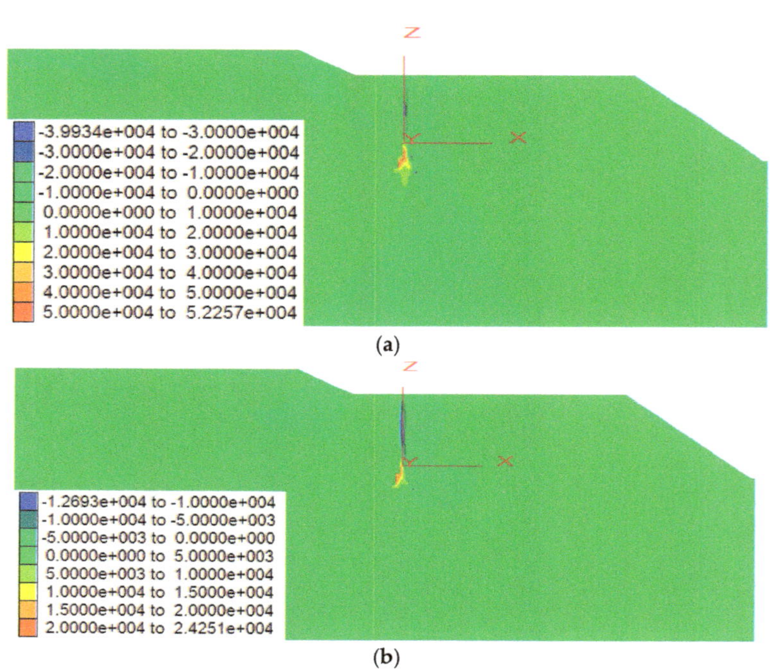

Figure 18. Comparison of XZ monitoring in-plane shear stress. (**a**) 1. (**b**) 2.

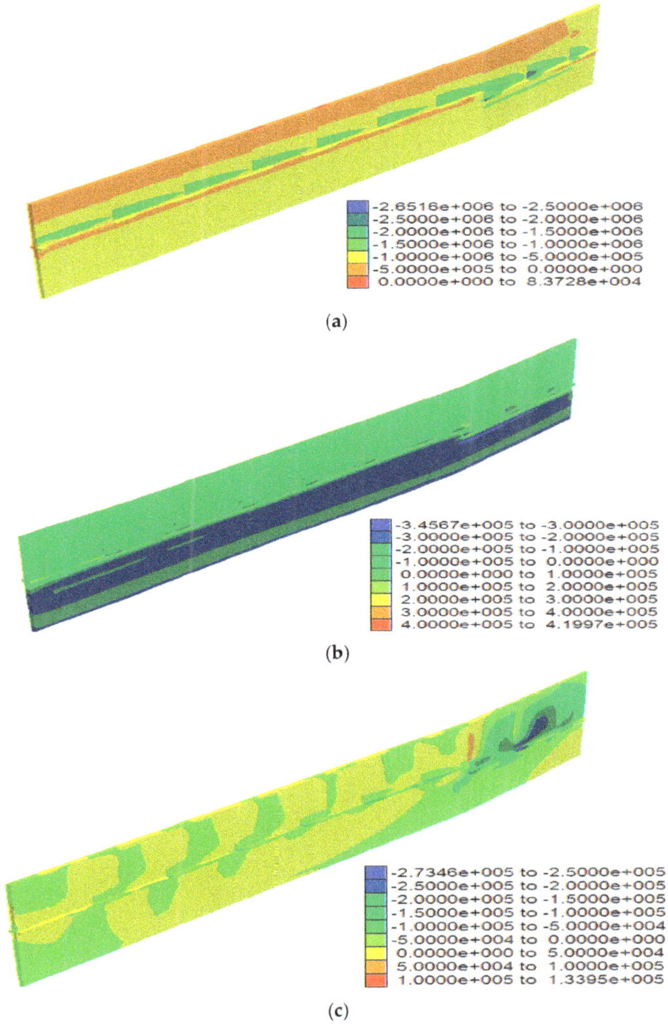

Figure 19. Schematic diagram of overall stress distribution of the cut-off wall. (**a**) Nephogram of maximum principal stress. (**b**) Minimum principal stress nephogram. (**c**) Nephogram of shear stress in YZ plane of the cut-off wall.

Figures 20–22 shows the stress distribution of the cutoff wall section at different monitoring surfaces. The maximum pressure of the rigid cut-off wall is 1.4302 MPa. The maximum pressure of the plastic cut-off wall is within 1.200 Mpa, which is smaller than that of the rigid cut-off wall. Apart from the tensile stress of less than 0.237 MPa occurring at the connecting guide wall between the rigid cut-off wall and the plastic concrete cut-off wall, the cut-off wall is basically under compression, especially the plastic cut-off wall. There is a stress concentration phenomenon in the upper part of the cross-shaped structure (guide wall) of the wall, that is, there is a large bending moment there, but there is no strong stress concentration in the lower part of the cross-shaped structure. It shows that the guide wall at the lower part of the cross-shaped structure plays a very good supporting role, offsetting some of the harmful bending moments that are unfavorable to the safety of the plastic concrete wall.

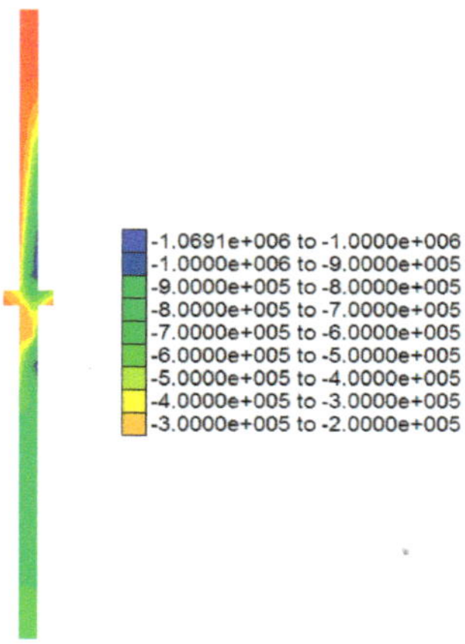

(**a**)

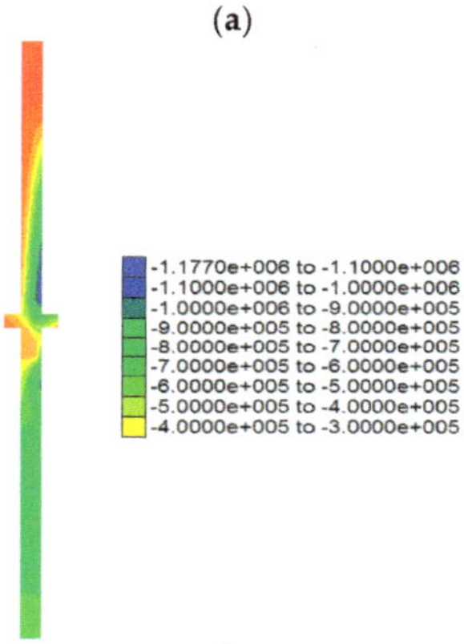

(**b**)

Figure 20. Comparison of maximum principal stress distribution of cut-off wall at monitoring surface. (**a**) 1. (**b**) 2.

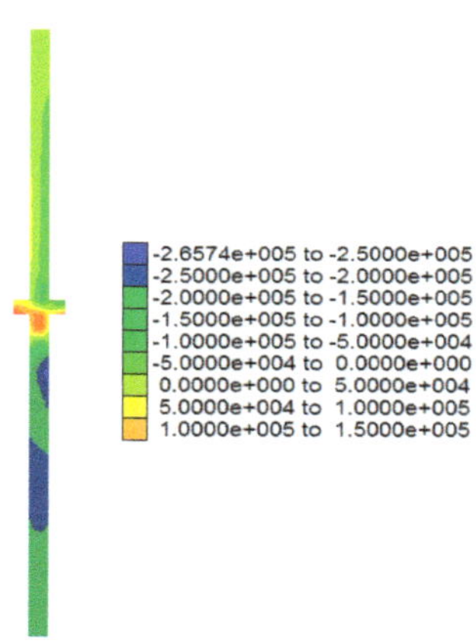

(**a**)

Figure 21. *Cont.*

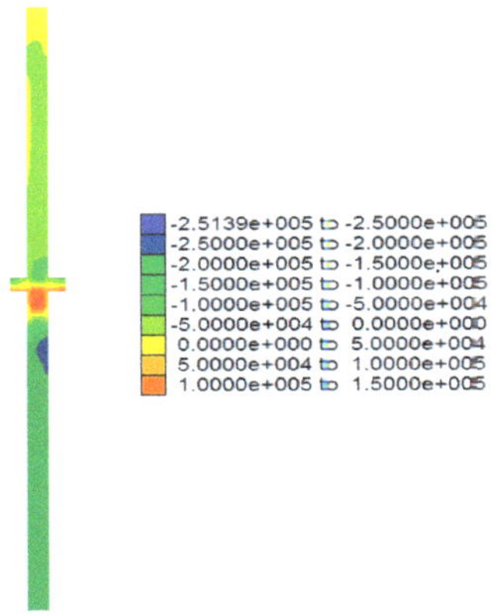

(**b**)

Figure 21. Comparison of minimum principal stress distribution of cut-off wall at monitoring surface. (**a**) 1. (**b**) 2.

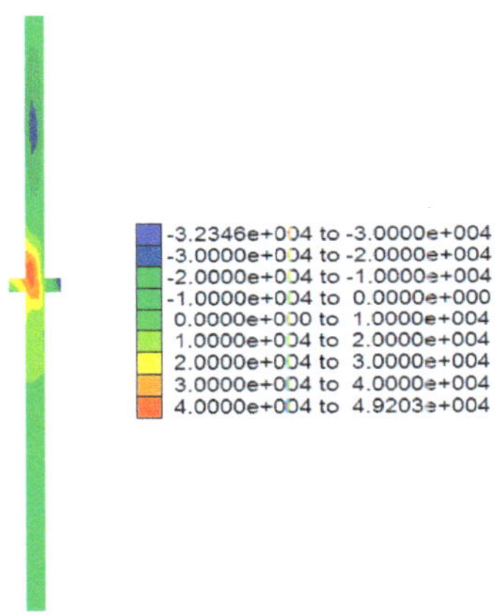

(**a**)

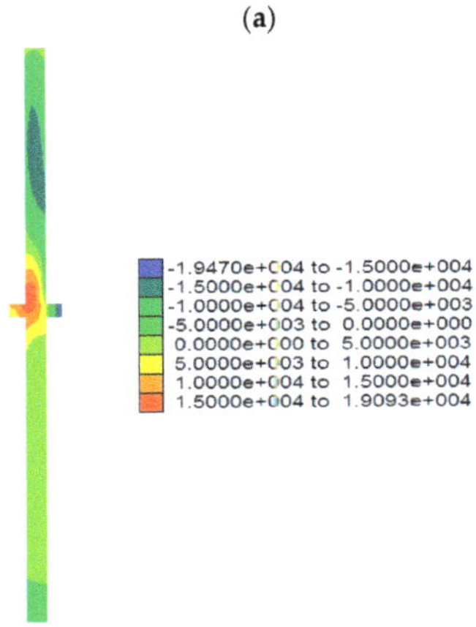

(**b**)

Figure 22. Comparison of shear stress distribution on YZ plane of cut-off l at monitoring surface. (**a**) 1. (**b**) 2.

Based on the above stress analysis, it can be seen that the overall stress on the cut-off wall gradually decreases from the bottom to the top, and the overall stress on the cut-off wall is slightly greater than that of the surrounding strata. This is not only related to the load it bears, but also to its own load. This is on account of the material properties. There is a phenomenon of compressive stress concentration at the rigid cut-off wall root, but its maximum value is only between 1.5–2.0 Mpa, which is far less than the material strength of the rigid cut-off wall of 15 Mpa. Due to the structural size, the maximum tensile stress of 0.237 Mpa appears in the local area near the guide wall of the rigid cut-off wall, but the value is small, which will not endanger the rigid cut0off wall. The shear stress in the ZX plane of the cut-off wall is slightly higher than that of the surrounding filling area, its maximum value is only about 0.3 Mpa, and it is located at the lower part of the rigid cut-off wall. The shear stress in the plastic cut-off wall is a lower order of magnitude. The maximum stress on the cut-off wall is generally below 1.5 Mpa, which is compressive stress.

The maximum stresses in the lower part of the first to third parts are slightly larger than that of the walls at other positions, and the local stress on the plastic cut-off wall in this area is also greater than that of the plastic cut-off wall in other areas. Their magnitudes are small, and the maximum stress in the rigid cut-off wall and plastic cut-off wall is between 1.5–2.65 Mpa and 0.56–1.5 Mpa, respectively. The shear stress distribution in the ZX plane is also present at the second and third walls. It is larger than other cut-off walls, and the maximum shear stress is only about 0.30 Mpa.

Apart from the small tensile stress at the connection guide wall between the rigid and the plastic concrete cut-off walls, the cut-off wall is in a state of compression. This is especially true for the plastic cut-off wall. Therefore, both the reinforced concrete wall and plastic concrete wall meet the design strength during construction. The guide wall plays an effective role in supporting the plastic cut-off wall. In conclusion, the dynamic construction effect has little influence on the strength of the cut-off wall, and the maximum stress is within the design range. Therefore, the rigid and the plastic cut-off wall are in a safe and stable state.

5. Conclusions

In this paper, through back analysis and simulation, the mechanical action of the whole construction process on a reservoir west dike cut-off wall and the mechanical effect produced by the cut-off wall have been obtained. The following main results and conclusions have been obtained:

- The deformation trend of the wall can be divided into three stages as the construction steps proceed. The first stage: Fill on both sides of the wall, but the width of the fill on the outside of the reservoir is large, and the wall is subject to the lateral constraint of the original reservoir embankment. The whole wall is inclined to the inside of the reservoir. The second stage: There is a large amount of fill in the inner side of the reservoir. Due to the gravity effect of the unilateral soil accumulation, the overall inclination of the wall has recovered. At this stage, the wall tends to move towards the outer side of the reservoir. The third stage: due to the soil piling work outside the reservoir during the construction stage of the Beijing-Shijiazhuang Expressway subgrade, the wall began to tilt towards the inside of the reservoir;
- In the process of filling construction, the whole impervious wall shows a tendency to tilt towards the inner side of the reservoir. The horizontal displacement value of the wall gradually increases from bottom to top, and the maximum value appears at the top of the wall. The horizontal displacement value of the 1–3 walls is relatively large, with the maximum value of 22.368 mm, and the horizontal displacement value of the 4–10 walls does not differ greatly. The largest difference in the horizontal displacement of the cu-off wall occurs at the joint between the third and fourth walls, and between the fourth and fifth walls, with the maximum difference of 1.314 mm;
- During the filling construction, due to the gravity of the backfill, the strata in the whole project area have settled, and the settlement at the bottom of the cut-off wall is

2.542 mm. The settlement characteristics of the rigid cut-off wall and plastic concrete cut-off wall are different. The settlement value of the rigid cut-off wall does not change along the wall height and its value is about 5.3 mm (the second) and 4.7 mm (the fifth and seventh). The settlement value of the plastic concrete wall shows a clear gradient along the elevation direction and its value changes evenly between 2.5–5.3 mm (the second) and 2.5–4.7 mm (the fifth and seventh).

- The horizontal displacement of the cut-off wall is mainly caused by the asymmetry of the fill on both sides. The fill pressure directly affects the horizontal displacement (inclination) of the reinforced concrete wall and then drives the horizontal displacement of the plastic concrete at the lower part through the structural involvement. The analysis shows that the guide wall has a great clamping effect on the deformation of the plastic cut-off wall, which makes the horizontal displacement between the rigid cut-off wall and the plastic cut-off wall interact and creates a nonlinear relationship. The vertical displacement of the cut-off wall is mainly caused by a compression deformation of the lower stratum under the action of structural gravity. On account of the elastic modulus of the plastic concrete wall material and the small surrounding stratum, the change is relatively obvious during the compression process, while the compression amount of the reinforced concrete wall itself is very small. Its settlement deformation is mainly the vertical movement of the rigid body with the settlement deformation of the plastic cut-off wall;
- The stress on the cut-off wall decreases gradually from bottom to top, and its stress value is slightly larger than that of the surrounding strata. At the root of the rigid cu-toff wall, the compressive stress concentration occurs, with the maximum value between 1.75 MPa and 2.15 MPa. Due to the size of the structure, the maximum tensile stress of 0.237 MPa appears in the local area near the guide wall of the rigid cut-off wall, which will not endanger the rigid cut-off wall because of its small value. The maximum shear stress in the ZX plane of the cut-off wall is only about 0.236 MPa, which is located in the lower part of the rigid cut-off wall. The shear stress in the plastic cut-off wall is one order of magnitude lower. The maximum stress on the plastic cut-off wall is generally below 1.5 MPa, and it is compressive stress;
- The maximum compressive stress occurs at the lower part of the rigid impervious wall, and the stress at the lower part of the first to the third is slightly greater than that of the wall at other locations. The local stress on the plastic impervious wall in this area is also greater than that of the plastic impervious wall in other areas, and their values are small. The maximum stress in the rigid impervious wall and the plastic impervious wall are 1.90–2.15 MPa and 1.00–1.12 MPa, respectively. The shear stress distribution in the ZX plane is also greater than that of the cut-off wall at other locations in the second and third walls, and the maximum shear stress is only about 0.236 MPa;
- Apart from the small tensile stress at the connecting guide wall between the rigid cut-off wall and the plastic concrete cut-off wall, the cut-off wall is basically under pressure, especially the plastic cut-off wall. During the construction process, the deformation of the backfill soil on both sides of the wall is relatively large, while the deformation of the whole wall is in a relatively small range, without obvious differential deformation. Combined with the analysis of the stress state of the wall, it can be determined that the anti-seepage wall (rigid cut-off wall and plastic concrete cut-off wall) is stable and safe during the construction period.

Author Contributions: Methodology, A.L.; Software, A.L.; Validation, G.W.; Formal analysis, K.Y.; Investigation, G.W.; Resources, K.Y.; Writing—original draft, Y.S.; Writing—review & editing, Y.S.; Visualization, A.L.; Project administration, Y.S. and G.W. All authors have read and agreed to the published version of the manuscript.

Funding: This research received no external funding.

Data Availability Statement: All data that support the findings of this study are available from the corresponding author upon reasonable request.

Conflicts of Interest: The authors declare that they have no conflicts of interest to report regarding the present study.

References

1. Shagapov, V.S.; Bashmakov, R.A.; Fokeeva, N.O. Fluid filtration in reservoirs subjected to hydraulic fracturing during transient well operation. *J. Appl. Mech. Tech. Phys.* **2022**, *63*, 474–483. [CrossRef]
2. Yan, Z.; Wang, Y.; Fan, J.; Huang, Y.; He, Y. Study on key parameters of directional long borehole layout in high-gas working face. *Shock. Vib.* **2021**, *2021*, 1–14. [CrossRef]
3. Almeida, F.; Venkatesh, P.; Gireesha, B.J. Time-Reliant Flow of Casson Nanofluid with Gyrotactic Microbes through the Contracting/Dilating Walls of the Microchannel Impelled by Chemical Reactions. *Braz. J. Phys.* **2022**, *52*, 137. [CrossRef]
4. Inazumi, S.; Shishido, K.-i.; Soralump, S. Possibility of impervious coating for the geotechnical reuse of soil and solid waste. *Environ. Geotech.* **2021**, *8*, 324–333. [CrossRef]
5. Baumstein, A.; Fendell, F. Strain-rate-free diffusion flames: Initiation, properties, and quenching. *Combust. Sci. Technol.* **1998**, *132*, 157–198. [CrossRef]
6. Yu, Z.; Chen, S.; Wong, N.H. Temporal variation in the impact of urban morphology on outdoor air temperature in the tropics: A campus case study. *Build. Environ.* **2020**, *181*, 107132. [CrossRef]
7. Sergent, P.; Prevot, G.; Mattarolo, G.; Brossard, J.; Morel, G.; Mar, F.; Benoit, M.; Ropert, F.; Kergadallan, X.; Trichet, J.-J.; et al. Adaptation of coastal structures to mean sea level rise. *Houille Blanche-Rev. Int. LEau* **2014**, *6*, 54–61. [CrossRef]
8. Guo, C.; Sun, B.; Hu, D.; Wang, F.; Shi, M.; Li, X. A Field Experimental Study on the Diffusion Behavior of Expanding Polymer Grouting Material in Soil. *Soil Mech. Found. Eng.* **2019**, *56*, 171–177. [CrossRef]
9. Shepherd, D.A.; Kotan, E.; Dehn, F. Plastic concrete for cut-off walls: A review. *Constr. Build. Mater.* **2020**, *255*, 119248. [CrossRef]
10. Wen, L.; Chai, J.; Wang, X.; Xu, Z.; Qin, Y.; Li, Y. Behaviour of concrete-face rockfill dam on sand and gravel foundation. *Proc. Inst. Civ. Eng.-Geotech. Eng.* **2015**, *168*, 439–456. [CrossRef]
11. Stahlhut, O.; Borchert, K.-M.; Voigt, R.E. Design and execution of a trough excavation pit in the Hamburg city center considering complex structural conditions. *Bautechnik* **2018**, *95*, 62–71. [CrossRef]
12. Liu, B.; Jiang, X. Detection of anti-seepage effect of building stress wall structure based on transient Rayleigh surface wave method. *Arab. J. Geosci.* **2020**, *13*, 819. [CrossRef]
13. Su, S.; Ding, J.; Zhang, T.; You, K.; Gan, Y.; Yan, M.; Feng, Z. Analysis of Factors Affecting Permeability of Cement-bentonite Mud Impervious Wall. In Proceedings of the 5th International Conference on Green Materials and Environmental Engineering (GMEE), Guangzhou, China, 27–29 December 2019; IOP Publishing: Bristol, UK, 2020.
14. Zheng, T.; Zheng, X.; Sun, Q.; Wang, L.; Walther, M. Insights of variable permeability full-section wall for enhanced control of seawater intrusion and nitrate contamination in unconfined aquifers. *J. Hydrol.* **2020**, *586*, 124831. [CrossRef]
15. Shah, Z.H.; Ullah, A.; Musa, A.; Vrinceanu, N.; Bou, S.F.; Iqbal, S.; Deebani, W.; Shah, Z.; Ullah, A.; Musa, A.; et al. Entropy optimization and thermal behavior of a porous system with considering hybrid nanofluid. *Front. Phys.* **2022**, *10*, 929463. [CrossRef]
16. Chen, K.; Guo, S.; Wang, J.; Qin, P.; He, S.; Sun, S.; Naeini, M.R. Evaluation of GloFAS-Seasonal Forecasts for Cascade Reservoir Impoundment Operation in the Upper Yangtze River. *Water* **2019**, *11*, 2539. [CrossRef]
17. He, S.; Guo, S.; Chen, K.; Deng, L.; Liao, Z.; Xiong, F.; Yin, J. Optimal impoundment operation for cascade reservoirs coupling parallel dynamic programming with importance sampling and successive approximation. *Adv. Water Resour.* **2019**, *131*, 103375. [CrossRef]
18. Arifeen, S.U.; Haq, S.; Ghafoor, A.; Ullah, A.; Kumam, P.; Chaipanya, P. Numerical solutions of higher order boundary value problems via wavelet approach. *Adv. Differ. Equ.* **2021**, *2021*, 347. [CrossRef]
19. Doghozlou, H.M.; Goodarzi, M.; Renani, H.R.; Salmi, E.F. Analysis of spalling failure in marble rock slope: A case study of Neyriz marble mine, Iran. *Environ. Earth Sci.* **2016**, *75*, 1478. [CrossRef]
20. Kumar, C.V.; Vardhan, H.; Murthy, C.S.N.; Karmakar, N.C. Estimating rock properties using sound signal dominant frequencies during diamond core drilling operations. *J. Rock Mech. Geotech. Eng.* **2019**, *11*, 850–859. [CrossRef]
21. Lawal, A.I.; Kwon, S. Application of artificial intelligence to rock mechanics: An overview. *J. Rock Mech. Geotech. Eng.* **2021**, *13*, 248–266. [CrossRef]
22. Shah, Z.; Kumam, P.; Ullah, A.; Khan, S.N.; Selim, M.M. Mesoscopic Simulation for Magnetized Nanofluid Flow Within a Permeable 3D Tank. *IEEE Access* **2021**, *9*, 135234–135244. [CrossRef]
23. Moayedi, H.; Mosallanezhad, M.; Rashid, A.S.A.; Jusoh, W.A.W.; Muazu, M.A. A systematic review and meta-analysis of artificial neural network application in geotechnical engineering: Theory and applications. *Neural Comput. Appl.* **2020**, *32*, 495–518. [CrossRef]
24. Nie, W.; Guo, W. Applications of Chapman-Richards model to geotechnical engineering. *J. Rock Mech. Geotech. Eng.* **2019**, *11*, 1286–1292. [CrossRef]
25. Vlasyuk, A.; Kuzlo, M.; Zhukovska, N.; Zhukovskyy, V.; Tarasyuk, N. Modeling of Soil Basis of Headed Hydrotechnical Structure's Deformations Under Action Of Filtration Water Flow. In Proceedings of the 10th International Conference on Advanced Computer Information Technologies (ACIT), Deggendorf, Germany, 16–18 September 2020; pp. 18–22.

26. ZeinEldin, R.A.; Ullah, A.; Khalifa, H.A.E.-W.; Ayaz, M. Analytical Study of the Energy Loss Reduction during Three-Dimensional Engine Oil-Based Hybrid Nanofluid Flow by Using Cattaneo–Christov Model. *Symmetry* **2023**, *15*, 166. [CrossRef]
27. Bhowmik, R.; Shahu, J.; Datta, M. Finite-Element Modeling of Geogrids in Trenches under Inclined Pull. *Int. J. Geomech.* **2020**, *20*, 04020129. [CrossRef]
28. Bodart, O.; Cayol, V.; Dabaghi, F.; Koko, J. An inverse problem in an elastic domain with a crack: A fictitious domain approach. *Comput. Geosci.* **2022**, *26*, 423–435. [CrossRef]
29. Shi, L.; Wang, H.; Bai, B.; Li, X. Improved finite element-based limit equilibrium method for slope stability analysis by considering nonlinear strength criteria and its application in assessing the anchoring effect. *Int. J. Numer. Anal. Methods Geomech.* **2019**, *43*, 578–598. [CrossRef]
30. Yan, G.; Hao, H. Digital Image Finite Element Analysis of Geotechnical Engineering Materials. *Integr. Ferroelectr.* **2022**, *227*, 58–73. [CrossRef]
31. Yu, P.; Hao, Q.; Wang, X.; Yu, Y.; Zhan, Z. Mixed Integration Scheme for Embedded Discontinuous Interfaces by Extended Finite Element Method. *Front. Earth Sci.* **2022**, *9*, 1404. [CrossRef]
32. Rizk, D.; Ullah, A.; Ikramullah; Elattar, S.; Alharbi, K.A.M.; Sohail, M.; Khan, R.; Khan, A.; Mlaiki, N. Impact of the KKL Correlation Model on the Activation of Thermal Energy for the Hybrid Nanofluid (GO plus ZnO plus Water) Flow through Permeable Vertically Rotating Surface. *Energies* **2022**, *15*, 2872. [CrossRef]
33. Lu, M.; Zhang, Q.; Jing, H.; Wang, Y.; Li, C. Analytical solutions for consolidation of composite ground improved by composite columns with circular and non-circular cross sections. *Eur. J. Environ. Civ. Eng.* **2022**, *26*, 2780–2796. [CrossRef]
34. Zhu, J.; Dai, G.; Xu, J.; Zhang, Z. Numerical analysis of wall deformation of impervious material based on polyvinyl alcohol modification. In Proceedings of the 1st International Conference on Environment Prevention and Pollution Control Technology (EPPCT), Tokyo, Japan, 9–11 November 2018; Tokyo University of Science: Shinjuku, Tokyo, 2018.
35. Fu, Z.Q.; Su, H.Z.; Wen, Z.P. Multi-scale numerical analysis for linear elastic behavior of clay concrete. *Int. J. Solids Struct.* **2020**, *203*, 23–45. [CrossRef]
36. Han, Y.; Ke, C.F. Study on Cut-off Effect of Xiongjiagang Earth-Rock Dam Based on Cut-off Wall Schemes. In Proceedings of the International Symposium on Multi-Field Coupling Theory of Rock and Soil Media and Its Applications, Chengdu, China, 10–11 October 2010; pp. 548–552.
37. Ma, X.; Zheng, M.; Liang, G.; Xu, C.; Mou, R. Analysis of elastic modulus impact on stress and deformation of concrete cutoff wall in dam body. *J. Hydroelectr. Eng.* **2013**, *32*, 230–236.
38. Ding, P.; Xiao, L.; Li, W.; Chen, H. Design of Dewatering Plans for Deep Excavation Pit in Thick Permeable Strata. *J. Yangtze River Sci. Res. Inst.* **2012**, *29*, 46–50.
39. Ren, X.; Gao, D.; Gao, Q.; Shen, T. Deformation and Stress Analysis of Extra-deep Concrete Cut-off Wall for Earth-rock Dam. *J. Yangtze River Sci. Res. Inst.* **2018**, *35*, 120–124.
40. Wang, L.; Li, R.; Ye, Y. Analysis of dangerous hydraulic conditions of hydraulic fill dam and impervious wall under impact of tidal action. *J. Hohai Univ. Nat. Sci.* **2012**, *40*, 426–431.
41. Chen, X.; Wen, Z.; Hu, J.; Min, Y.; Liang, X.; Sun, R.; Li, R. Application of Numerical Simulation and Analytical Methods to Estimate Hydraulic Parameters of Foundation Pit in Hydropower Stations. *Earth Sci.* **2016**, *41*, 701–710.
42. Liang, Y.; Mao, R.; Wang, Y. Numerical simulation of underground wall for retaining water at dam foundation of Longkaikou hydropower station. *Hydro-Sci. Eng.* **2018**, 57–64. Available online: http://slsy.nhri.cn/fileSLSYGCXB/journal/article/slsygcxb/2018/3/PDF/201803008.pdf (accessed on 11 February 2023).
43. Li, M.; Zhou, Z.; Yang, Y.; Xin, Y.; Liu, Y.; Guo, S.; Zhu, Y.; Jin, D. Immersion assessment and control of the right bank of Xingan Navigation and Power Junction. *J. Hohai Univ. Nat. Sci.* **2018**, *46*, 203–210.
44. Arai, Y.; Kusakabe, O.; Murata, O.; Konishi, S. A numerical study on ground displacement and stress during and after the installation of deep circular diaphragm walls and soil excavation. *Comput. Geotech.* **2008**, *35*, 791–807. [CrossRef]
45. Qiu, M.; Yang, G.; Shen, Q.; Duan, J.; Zhang, P. Deformation Characteristics of Foundation Pit Excavation under the Combined Action of Diaphragm Wall and Impervious Curtain in Water-rich Sandy Stratum. *J. Yangtze River Sci. Res. Inst.* **2020**, *37*, 81–88, 95.
46. Yanhua, X.I.A.; Shiwei, B.A.I.; Chao, Z. Seepage deformation evaluation of foundation pit excavation of power-house of a hydro-project. *Rock Soil Mech.* **2007**, *28*, 2435–2439.
47. Schaefer, R.; Triantafyllidis, T. The influence of the construction process on the deformation behaviour of diaphragm walls in soft clayey ground. *Int. J. Numer. Anal. Methods Geomech.* **2006**, *30*, 563–576. [CrossRef]
48. Lu, M.M.; Xie, K.H.; Wang, S.Y.; Li, C.X. Analytical Solution for the Consolidation of a Composite Foundation Reinforced by an Impervious Column with an Arbitrary Stress Increment. *Int. J. Geomech.* **2013**, *13*, 33–40. [CrossRef]

MDPI AG
Grosspeteranlage 5
4052 Basel
Switzerland
Tel.: +41 61 683 77 34

Water Editorial Office
E-mail: water@mdpi.com
www.mdpi.com/journal/water

www.ingramcontent.com/pod-product-compliance
Lightning Source LLC
Chambersburg PA
CBHW061805210926
51850CB00006B/621
9783725813803